STUDENT SOLUTIONS MANUAL

to accompany

Calculus
Ideas & Applications

T0235153

Alex Himonas
University of Notre Dame
Alan Howard
University of Notre Dame

WILEY

©James Randklev/The Image Bank/Getty Images

To order books or for customer service call 1-800-CALL-WILEY (225-5945).

ISBN   978-0-471-26639-6

CONTENTS

Precalculus Review

EXERCISE SET 0.2

1. $f(x) = x^3$, so that $f(2) = 2^3 = 8$, $f(-1) = (-1)^3 = -1$, $f(1/3) = (1/3)^3 = \frac{1}{27}$, $f(a) = a^3$,
   $f(-a) = (-a)^3 = -a^3$, $f(a+1) = (a+1)^3 = a^3 + 3a^2 + 3a + 1$, and

$$\frac{f(x+h) - f(x)}{h} = \frac{(x+h)^3 - x^3}{h} = \frac{x^3 + 3x^2h + 3xh^2 + h^3 - x^3}{h}$$

$$= \frac{3x^2h + 3xh^2 + h^3}{h} = \frac{h(3x^2 + 3xh + h^2)}{h}$$

$$= 3x^2 + 3xh + h^2.$$

3. $f(x) = 1/x$, so that $f(2) = \frac{1}{2}$, $f(1/2) = \frac{1}{1/2} = 2$, $f(-1) = \frac{1}{-1} = -1$, $f(a/3) = \frac{1}{a/3} = 3/a$,
   $f(3/a) = \frac{1}{3/a} = a/3$, and

$$\frac{f(a+h) - f(a)}{h} = \frac{\frac{1}{a+h} - \frac{1}{a}}{h} = \frac{\frac{a-a-h}{a(a+h)}}{h} = \frac{-h}{ha(a+h)} = \frac{-1}{a(a+h)}.$$

5. $g(x) = x^3 - 2x + 1$, so that $g(1) = 1^3 - 2 \cdot 1 + 1 = 0$, $g(-2) = (-2)^3 - 2(-2) + 1 = -8 + 4 + 1 = -3$,
   $g(-c) = (-c)^3 - 2(-c) + 1 = -c^3 + 2c + 1$,

$$g(a-1) = (a-1)^3 - 2(a-1) + 1$$
$$= a^3 - 3a^2 + 3a - 1 - 2a + 2 + 1 = a^3 - 3a^2 + a + 2, \text{ and}$$

$$g(a+h) - g(a) = (a+h)^3 - 2(a+h) + 1 - (a^3 - 2a + 1)$$
$$= a^3 + 3a^2h + 3ah^2 + h^3 - 2a - 2h + 1 - a^3 + 2a - 1$$
$$= 3a^2h + 3ah^2 + h^3 - 2h.$$

7. $h(1-a) = (1-a)^2 - (1-a) + 1 = 1^2 - 2a + a^2 - 1 + a + 1 = a^2 - a + 1 = h(a)$.

9. Since $\sqrt{1-x}$ is defined if $1 - x \geq 0$, or $1 \geq x$, the natural domain of $f(x) = \sqrt{1-x}$ is the set $(-\infty, 1]$.

11. Since $1/\sqrt{x-1}$ is defined if $x - 1 > 0$, or $x > 1$, the natural domain of $g(x) = 1/\sqrt{x-1}$ is the set $(1, \infty)$.

13. The fraction $t/(t^2 - 1)$ is defined if $t^2 - 1 \neq 0$, or $t \neq \pm 1$. So the natural domain of $f(t) = t/(t^2 - 1)$ is the set of all numbers $t \neq \pm 1$.

15. Since $u^3 - (3/u^2)$ is defined for all u with $u \neq 0$, the natural domain of $f(u) = u^3 - (3/u^2)$ is the set of all numbers $u \neq 0$.

17. $\frac{1}{1+x^2} = y \Leftrightarrow 1 + x^2 = \frac{1}{y} \Leftrightarrow x^2 = \frac{1}{y} - 1$. If $0 < y \leq 1$, then $\frac{1}{y} \geq 1$ and $\frac{1}{y} - 1 \geq 0$, so that we can take the square root of $\frac{1}{y} - 1$ and find $x = \pm\sqrt{\frac{1}{y} - 1}$. By applying the additional condition $x \geq 0$, we obtain $x = \sqrt{\frac{1}{y} - 1}$.

19. For any number x we have $x^2 \geq 0$ and $x^2 + 1 \geq 1$. Therefore, any output must satisfy $y \geq 1$. Moreover, any number $y \geq 1$ is an output of f with input $\sqrt{y-1}$. Therefore, the range of f is the set $[1, \infty)$.

21. For any $y \neq 0$, we can solve the equation $y = \frac{1}{x}$ to get $x = \frac{1}{y}$, which is the input corresponding to the output y. Therefore, the range of $f(x) = \frac{1}{x}$, $x \neq 0$, is the set of all numbers $y \neq 0$.

23. Since $x^2 \geq 0 \Rightarrow 3 + x^2 \geq 3 \Rightarrow 1/(3+x^2) \leq 1/3 \Rightarrow 2/(3+x^2) \leq 2/3$, we see that any output y must be in the interval $(0, 2/3]$. Moreover, for $0 < y \leq 2/3$ the equation $2/(3+x^2) = y$ gives $3 + x^2 = 2/y$, or $x^2 = (2/y) - 3$, or $x = \sqrt{(2/y) - 3}$ (since $2/y - 3 \geq 0$ by our assumption that $y \leq 2/3$). Thus, any number y in $(0, 2/3]$ is an output with corresponding input $\sqrt{(2/y) - 3}$, which shows that the range of $f(x) = 2/(3 + x^2)$ is the set $(0, 2/3]$.

25.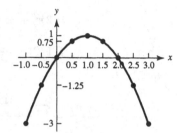

27.

x	-1	-0.8	-0.6	-0.4	-0.2	0	0.2	0.4	0.6	0.8	1
$f(x)$	1.41	1.34	1.26	1.18	1.10	1	0.89	0.77	0.63	0.44	0

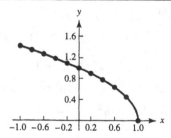

29. This curve is not the graph of a function since it fails the vertical line test. For example the y-axis meets the curve at three points.

31. This graph fails the vertical line test, since it contains a vertical line segment and so is not the graph of a function $y = f(x)$.

33. This curve is not the graph of a function $y = f(x)$ since the y-axis intersects at two points and so it fails the vertical line test.

35. This curve passes the verticle line test but fails the horizontal line test, since the x-axis meets it at two points. Therefore, it is the graph of a function that is not one-to-one.

37. This curve passes the vertical line test but fails the horizontal line test since every horizontal line above the x-axis meets the curve at two points. Therefore it is the graph of a function that is not one-to-one.

39. This curve passes both the vertical and the horizontal line tests and so it is the graph of a one-to-one function.

41. If $x > 0$ then $y = \frac{1}{x^2} \iff x = \frac{1}{\sqrt{y}}$ with $y > 0$. Therefore, $f(x) = 1/x^2$, $x > 0$, has inverse given by $g(x) = 1/\sqrt{x}$, $x > 0$. The domain of g is the set $(0, \infty)$, and its range is again the set $(0, \infty)$.

43. If $x \neq 0$ then $y = 1/x^3 \iff x = 1/\sqrt[3]{y}$ with $y \neq 0$. Therefore, $f(x) = 1/x^3$, $x \neq 0$, has inverse given by $g(x) = 1/\sqrt[3]{x}$, $x \neq 0$. Both the domain and the range of g are the set of all numbers other than zero.

45. If $x \geq 2$ then $y = \sqrt{x - 2} \iff x = y^2 + 2$ with $y \geq 0$. Therefore, $f(x) = \sqrt{x - 2}$, $x \geq 2$, has inverse given by $g(x) = x^2 + 2$, $x \geq 0$. The domain of g is $[0, \infty)$ and its range is $[2, \infty)$.

47. If $x > 0$ then $y = -1/x \iff x = -1/y$ with $y < 0$. Therefore, $f(x) = -1/x$, $x > 0$ has inverse given by $g(x) = -1/x$, $x < 0$. The domain of g is $(-\infty, 0)$ and its range is $(0, \infty)$. The graph of g is obtained by reflecting the graph of f across the diagonal $y = x$.

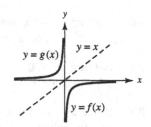

EXERCISES 0.3

1. It is increasing on the intervals $(-2, -1)$ and $(1, 2)$ and decreasing on $(-1, 1)$.

3. It is decreasing on $(-6, 0)$, increasing on $(0, 6)$.

5. It is decreasing on $(-4, 4)$.

7. $x_1 < x_2 \Rightarrow -2x_1 > -2x_2 \Rightarrow -2x_1 + 7 > -2x_2 + 7$, or $f(x_1) > f(x_2)$. So $f(x) = -2x + 7$ is decreasing on $(-\infty, \infty)$.

9. Let $-1 < x_1 < x_2$. Then $0 < x_1 + 1 < x_2 + 1 \Rightarrow (x_1 + 1)^2 < (x_2 + 1)^2$, and so $f(x) = (x + 1)^2$ is increasing on $(-1, \infty)$. On the other hand, if $x_1 < x_2 < -1$, then $x_1 + 1 < x_2 + 1 < 0$ and $(x_1 + 1)^2 > (x_2 + 1)^2$, which means that $f(x) = (x + 1)^2$ is decreasing on $(-\infty, -1)$.

11. $x_1 < x_2 < 0 \Rightarrow \frac{1}{x_1} > \frac{1}{x_2}$. Therefore, $f(x) = \frac{1}{x}$ is decreasing on $(-\infty, 0)$. Also, $0 < x_1 < x_2 \Rightarrow \frac{1}{x_1} > \frac{1}{x_2}$, so that $f(x) = \frac{1}{x}$ is also decreasing on $(0, \infty)$.

13. $x_1 < x_2 < 1 \Rightarrow x_1 - 1 < x_2 - 1 < 0 \Rightarrow (x_1 - 1)^2 > (x_2 - 1)^2 > 0 \Rightarrow \frac{1}{(x_1-1)^2} < \frac{1}{(x_2-1)^2}$. Therefore, $f(x) = \frac{1}{(x-1)^2}$ is increasing on $(-\infty, 1)$. But $1 < x_1 < x_2 \Rightarrow 0 < x_1 - 1 < x_2 - 1 \Rightarrow 0 < (x_1 - 1)^2 < (x_2 - 1)^2 \Rightarrow \frac{1}{(x_1-1)^2} > \frac{1}{(x_2-1)^2}$. Therefore, $f(x) = \frac{1}{(x-1)^2}$ is decreasing on $(1, \infty)$.

15. $f(-x) = 2(-x) = -2x = -f(x)$. So f is odd and its graph is symmetric about the origin.

17. $f(-x) = (-x)^4 - 3(-x)^2 + 1 = x^4 - 3x^2 + 1 = f(x)$. So f is even and its graph is symmetric about the y-axis.

19. $f(-x) = (-x)^3 + 5(-x) = -x^3 - 5x = -(x^3 + 5x) = -f(x)$. So f is odd and its graph is symmetric about the origin.

21. $f(-x) = \frac{(-x)^2}{(-x)^2+1} = \frac{x^2}{x^2+1} = f(x)$. Thus, f is even and its graph is symmetric about the y-axis.

23. $f(-x) = \frac{-x}{|-x|+1} = -\frac{x}{|x|+1} = -f(x)$. Thus f is odd and its graph is symmetric about the origin.

25.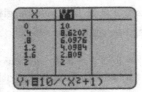

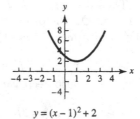

27. Setting $x = 0$, we find $y = 3 \cdot 0 + 4 = 4$, which is the y-intercept. Setting $y = 0$, we have the equation $0 = 3x + 4$. Solving it, we find that $x = -4/3$ is the only x-intercept.

29. Setting $x = 0$, we find that $y = 0$ is the y-intercept. Setting $y = 0$, we obtain the equation $0 = x^2 - x$, or $x(x - 1) = 0$. Solving it gives the two x-intercepts, $x = 0$ and $x = 1$.

31. Setting $x = 0$ gives $y = 1$, which is the y-intercept. Setting $y = 0$, we obtain $0 = x^3 + 1$, or $x^3 = -1$, or $x = -1$, which is the only x-intercept.

33. Since $x = 0$ is not in the domain of $y = \frac{1}{x}$, there is no y-intercept. And since the equation $0 = \frac{1}{x}$ has no solution, there is no x-intercept either.

35.   **(i)**   The graph of $y = (x - 2)^3$ is obtained by translating the graph of $y = x^3$ by 2 units to the right. Thus, it matches (a).

   (ii)   The graph of $y = (x + 1)^3$ is obtained by translating the graph of $y = x^3$ by 1 unit to the left. Thus, it matches (c).

   (iii)   The graph of $y = x^3 - 2$ is obtained by translating the graph of $y = x^3$ downward by 2 units. Thus, it matches (b).

37. The graph of $y = x^2 - 2$ is obtained by translating the graph of $y = x^2$ downward by 2 units.

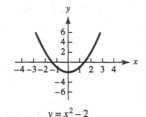

$$y = x^2 - 2$$

39. The graph of $y = (x - 1)^2 + 2$ is obtained by translating the graph of $y = x^2$ to the right by 1 unit and upward by 2 units.

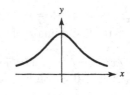

$$y = (x - 1)^2 + 2$$

41. The graph of $y = \sqrt[3]{x + 2}$ is obtained by translating the graph of $y = \sqrt[3]{x}$ to the left by 2 units.

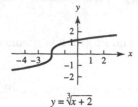

$$y = \sqrt[3]{x + 2}$$

43. The graph of $y = \sqrt[3]{x - 2} + 1$ is obtained by translating the graph of $y = \sqrt[3]{x}$ to the right by 2 units and upward by 1 unit.

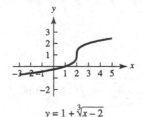

$$y = 1 + \sqrt[3]{x - 2}$$

45. $f(x) = 1 + x + x^3 \Rightarrow f(-x) = 1 - x - x^3$. So the graph of $y = 1 - x - x^3$ is a mirror image in the y-axis of the graph of $y = 1 + x + x^3$.

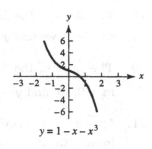

$y = 1 - x - x^3$

47. $f(x) = x^3 - 3x \Rightarrow -f(x) = 3x - x^3$. So the graph of $y = 3x - x^3$ is obtained by reflecting the graph of $y = x^3 - 3x$ in the x-axis.

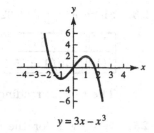

$y = 3x - x^3$

EXERCISES 0.4

1. slope $\frac{1}{2}$
y-intercept -1

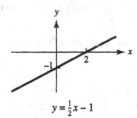

$y = \frac{1}{2}x - 1$

3. slope 0
y-intercept 4

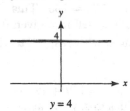

$y = 4$

5. $x - 2y = 5 \iff y = \frac{1}{2}x - \frac{5}{2}$
slope $1/2$
y-intercept $-5/2$

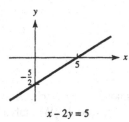

$x - 2y = 5$

7. slope 0
y-intercept -3

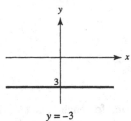

$y = -3$

9. Using the slope-intercept form, we find $y = -3x + 2$.

11. Using the point-slope form with $(x_1, y_1) = (0, 0)$ and $m = -\frac{1}{2}$, we find $y = -\frac{1}{2}x$.

13. Using the slope-intercept form with $m = 0$ and $b = -\frac{5}{4}$, we find $y = -\frac{5}{4}$.

15. Using the point-slope form with $(x_1, y_1) = (0, 3)$ and $m = -2$, we find $y - 3 = -2x$, or $y = -2x + 3$.

17. The slope of the line equals $\frac{5-1}{-1-2} = -\frac{4}{3}$. Using the point-slope form with $(x_1, y_1) = (2, 1)$ and $m = -\frac{4}{3}$, we find $y - 1 = -\frac{4}{3}(x - 2)$, or $y = -\frac{4}{3}x + \frac{11}{3}$.

19. The slope of the line equals $\frac{1-1}{-\frac{4}{7}-\frac{2}{3}} = 0$. Using the point slope form with $(x_1, y_1) = (\frac{2}{3}, 1)$ and $m = 0$, we find $y - 1 = 0(x - \frac{2}{3})$, or $y = 1$.

21. $5x - 3y = 1$ is equivalent to $y = \frac{5}{3}x - \frac{1}{3}$. The slope of the line equals $\frac{5}{3}$. Using the point-slope form with $(x_1, y_1) = (\frac{3}{2}, -\frac{1}{2})$ and $m = \frac{5}{3}$, we find $y + \frac{1}{2} = \frac{5}{3}(x - \frac{3}{2})$, or $y = \frac{5}{3}x - 3$.

23. Since $F = \frac{9}{5}C + 32$, we get $C = \frac{5}{9}F - \frac{160}{9}$

F	60	70	80	90
C	$140/9 = 15.\overline{5}$	$190/9 = 21.\overline{1}$	$240/9 = 26.\overline{6}$	$290/9 = 32.\overline{2}$

The Celsius reading changes at the rate of $\frac{5}{9}$ degrees per Fahrenheit degree.

25. Writing I for the salesman's weekly income and R for the amount of sales revenue, we have $I(r) = 0.08R + 400$, $R \geq 0$. The salesman's income changes at the rate of 0.08 dollars of income per dollar of sales revenue.

27. **(a)** Set $P(x) = 1.5x - 300 = 0$. This equation has the solution $x = 200$, which is the break-even point.

(b) Set $P(x) = 1.5x - 300 = 750$. This equation has the solution $x = 700$, which is the amount the bookstore has to sell in a given day in order to earn a profit of $750.

(c) The profit changes with respect to the number of books sold at the rate of 1.5 dollars per book sold.

29. The number of miles is $s = 1.2t + 12$, with t in minutes. Since $s(20) = 1.2 \cdot 20 + 12 = 36$, the car is 36 miles east of Chicago skyway at 20 minutes past noon. Since $s(60) = 1.2 \cdot 60 + 12 = 84$, the car is 84 miles east of the Chicago skyway at 1 P.M.

31.

Solving $-20p + 2,400 = 10p - 600$ gives $p_e = 100$, and $q_e = D(100) = 400$.

33. **(a)** Let p be the price for each ticket and let q be the number of tickets sold every week. Assume that the demand curve is given by $q = mp + b$ for some constants m and b. We solve the system of equations

$$1,000 = 30m + b$$
$$900 = 40m + b$$

to get $m = -10$ and $b = 1,300$. Thus, $q = -10p + 1,300$.

(b) We solve $-10p + 1,300 = 45p - 675$ to get $p_e = 395/11 = 35.\overline{90}$. Then $q_e = 45 \cdot (395/11) - 675 = 10,350/11 = 940.\overline{90}$. The equilibrium point is $(10,350/11, 395/11)$.

35.

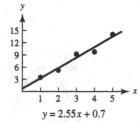

$y = 2.55x + 0.7$

37.

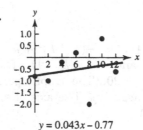

$y = 0.043x - 0.77$

39. $y = 49.95x + 674.72, \quad y(4) = (49.95) \cdot 4 + 674.72 = 874.5 \approx 875.$

41. Suppose that L_1 and L_2 are not parallel. The lines are perpendicular if and only if the lengths of the sides of the triangle determined by the points (x_0, y_0), (x_1, y_1), and (x_2, y_2) satisfy the Pythagorean formula. That is,

$$(x_1 - x_0)^2 + (y_1 - y_0)^2 + (x_2 - x_0)^2 + (y_2 - y_0)^2 = (x_2 - x_1)^2 + (y_2 - y_1)^2.$$

We rewrite this in the form

$$(y_2 - y_1)^2 - (y_1 - y_0)^2 - (y_2 - y_0)^2 = (x_1 - x_0)^2 + (x_2 - x_0)^2 - (x_2 - x_1)^2.$$

Expanding and collecting terms on each side give

$$2y_0y_1 + 2y_0y_2 - 2y_1y_2 - 2y_0^2 = 2x_0^2 - 2x_0x_1 - 2x_0x_2 + 2x_1x_2.$$

By cancelling the 2's and factoring, we get

$$(y_1 - y_0)(y_0 - y_2) - (x_1 - x_0)(x_2 - x_0).$$

On the other hand, the product of the slopes equals -1 if and only if

$$\frac{y_2 - y_0}{x_2 - x_0} \cdot \frac{y_1 - y_0}{x_1 - x_0} = -1,$$

and clearing denominators gives

$$(y_1 - y_0)(y_0 - y_2) = (x_1 - x_0)(x_2 - x_0).$$

Thus, the two conditions, *two non-vertical lines are perpendicular* and *product of the slopes is* -1, are equivalent.

Finally, if L_1 and L_2 are parallel, then they must have the same slope, say m. The product of their slopes is equal to m^2, a positive real number. Thus, the product of the slopes cannot equal -1.

43. The slope of the first line is $\frac{-1-0}{2+1} = -\frac{1}{3}$. The slope of the second line is $\frac{3-0}{0+1} = 3$. The product of their slopes equals -1 and hence the two lines are perpendicular.

45. (a) The slope of these lines equal $-\frac{3}{2}$ and $\frac{3}{2}$, respectively. They are neither parallel nor perpendicular.

(b) The slope of these lines equal $-\frac{3}{2}$ and $\frac{2}{3}$, respectively. They are perpendicular.

(c) The slope of these lines equal $-\frac{3}{2}$ and $-\frac{3}{2}$, respectively. They are parallel.

(d) The slope of these lines equal $-\frac{3}{2}$ and $-\frac{2}{3}$, respectively. They are neither perpendicular nor parallel.

EXERCISES 0.5

1. Writing $-x^2 + 6x + 11 = -(x - h)^2 + k = -x^2 + 2hx - h^2 + k$, we see that $2h = 6$ and $-h^2 + k = 11$. Therefore, $h = 3$, $k = 20$, and $y = -(x - 3)^2 + 20$. The graph opens downward, since the coefficient of x^2 is negative. The axis of symmetry is $x = 3$, and the vertex is $(3, 20)$.

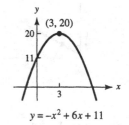

$$y = -x^2 + 6x + 11$$

3. Writing $2x^2 + 5x = 2(x - h)^2 + k = 2x^2 - 4hx + 2h^2 + k$, we see that $-4h = 5$ and $2h^2 + k = 0$. Therefore $h = -\frac{5}{4}$, $k = -\frac{25}{8}$, and $y = 2(x + \frac{5}{4})^2 - \frac{25}{8}$. The graph opens upward, since the coefficient of x^2 is positive. The axis of symmetry is $x = -\frac{5}{4}$, and the vertex is $(-\frac{5}{4}, -\frac{25}{8})$.

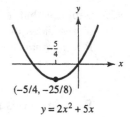

$$y = 2x^2 + 5x$$

5. Writing $-0.1x^2 - 1.2x + 3.6 = -0.1(x - h)^2 + k = -0.1x^2 + 0.2hx - 0.1h^2 + k$, we see that $0.2h = -1.2$ and $-0.1h^2 + k = 3.6$. Therefore, $h = -6$, $k = 7.2$, and $y = -0.1(x + 6)^2 + 7.2$. The graph opens downward, since the coefficient of x^2 is negative, $x = -6$ is the axis of symmetry, and $(-6, 7.2)$ is the vertex.

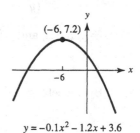

$$y = -0.1x^2 - 1.2x + 3.6$$

7. Writing $3x^2 - 2x + 1 = 3(x - h)^2 + k = 3x^2 - 6hx + 3h^2 + k$, we see that $-6h = -2$ and $3h^2 + k = 1$. Therefore, $h = \frac{1}{3}$, $k = \frac{2}{3}$, and $y = 3(x - \frac{1}{3})^2 + \frac{2}{3}$. The graph opens upward, since the coefficient of x^2 is positive, $x = \frac{1}{3}$ is the axis of symmetry of the graph and $(\frac{1}{3}, \frac{2}{3})$ is the vertex.

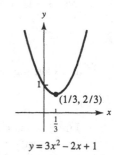

$$y = 3x^2 - 2x + 1$$

9. $f(x) = (x - 7)(x + 2)$, and has roots $x = -2$ and $x = 7$. $f(x)$ is positive is $(-\infty, -2)$ since there $x + 2 < 0$ and $x - 7 < 0$. Also, $f(x)$ is positive in $(7, \infty)$ since there $x + 2 > 0$ and $x - 7 > 0$. Finally, $f(x)$ is negative in $(-2, 7)$ since there $x + 2 > 0$ and $x - 7 < 0$.

11. $f(x) = (x - \frac{1}{2})(x - \frac{1}{3})$, and has roots $x = 1/2$ and $x = 1/3$. $f(x)$ is positive in $(-\infty, \frac{1}{3})$ and $(\frac{1}{2}, \infty)$. $f(x)$ is negative in $(\frac{1}{3}, \frac{1}{2})$.

13. $f(x) = (\frac{1}{2}x - 1)(\frac{1}{2}x - 3)$, and has roots $x = 2$ and $x = 6$. $f(x)$ is positive in $(-\infty, 2)$ and $(6, \infty)$. $f(x)$ is negative in $(2, 6)$.

15. Solving $x^2 - 2x - 1 = 0$ gives the roots $x = 1 \pm \sqrt{2}$. Therefore, $f(x) = (x - (1 - \sqrt{2}))(x - (1 + \sqrt{2}))$. $f(x)$ is positive in $(-\infty, 1 - \sqrt{2})$ and $(1 + \sqrt{2}, \infty)$. $f(x)$ is negative in $(1 - \sqrt{2}, 1 + \sqrt{2})$.

17. $f(x) = (x - \frac{2}{5})(x + \frac{1}{2})$, and has roots $x = -1/2$ and $x = 2/5$. $f(x)$ is positive in $(-\infty, -\frac{1}{2})$ and $(\frac{2}{5}, \infty)$. $f(x)$ is negative in $(-\frac{1}{2}, \frac{2}{5})$.

19. Solving $5{,}000 + 1{,}800x = -10x^2 + 2{,}400x$, or $x^2 - 60x + 500 = 0$, we obtain $x = 10$ and $x = 50$, which are the break-even points. The profit function is $P(x) = R(x) - C(x) = -10x^2 + 600x - 5{,}000 = -10(x - 10)(x - 50)$, which is positive for $10 < x < 50$. Those are the production levels at which the company makes a profit.

21. **(a)** Writing $q = mp + b$ for some constants m and b, we obtain the set of equations

$$900 = 16m + b$$
$$780 = 20m + b,$$

which has the solutions $m = -30$ and $b = 1{,}380$. Therefore $q = -30p + 1{,}380$.

(b) The revenue function is $R(p) = p(-30p + 1{,}380) = -30p^2 + 1{,}380p$.

23. **(a)** $R(p) = -20p^2 + 1{,}200p = -20(p - 30)^2 + 18{,}000$, which takes a maximum at $p = 30$. Therefore, the restaurant should charge 30 dollars on the average for each dinner to maximize the revenue.

(b) $P(p) = -20p^2 + 1{,}520p - 22{,}200 = -20(p - 38)^2 + 6{,}680$. The restaurant should charge 38 dollars on the average for each dinner to maximize the profit.

25. By completing the square, we obtain $R(p) = -10(p - 80)^2 + 64{,}000$, which has its maximum at $p = 80$. Therefore, the company should charge 80 dollars for each computer game to maximize the revenue.

27. Set $0.3(q - 20)^2 = 2q + 10$. By expanding and collecting terms, we get $0.3q^2 - 14q + 110 = 0$, which has solutions $q = 10$ and $q = 110/3$. Since $0 \leq q \leq 18$, we conclude that $q_e = 10$ and $p_e = S(10) = 2 \cdot 10 + 10 = 30$.

29. $a(x + \frac{b}{2a})^2 + \frac{4ac - b^2}{4a} = 0 \iff (x + \frac{b}{2a})^2 = \frac{b^2 - 4ac}{4a^2}$. Taking the square root of each side gives $x + \frac{b}{2a} = \pm \frac{\sqrt{b^2 - 4ac}}{2a}$. Therefore,

$$x = -\frac{b}{2a} \pm \frac{\sqrt{b^2 - 4ac}}{2a} = \frac{-b \pm \sqrt{b^2 - 4ac}}{2a}.$$

EXERCISES 0.6

1. Since $x^5 - 3x^4 + 7x^3 + 6x^2 - x - 8 \approx x^5$ when $|x|$ is very large, the graph of $f(x)$ behaves the same way as the graph of $y = x^5$, i.e., it climbs higher and higher to the right and falls lower and lower to the left.

3. Since $-0.01x^3 + x^2 + 0.1x - 10 \approx -0.01x^3$ when $|x|$ is very large, the graph of $f(x)$ behaves the same way as the graph of $y = -0.01x^3$, i.e., it falls lower and lower to the right and climbs higher and higher to the left.

5. The leading term of the polynomial is $a_n x^n$, which means that the graph of the polynomial behaves the same way as the graph of $y = a_n x^n$. Since the graph climbs to the right and falls to the left, we conclude that n is odd and $a_n > 0$.

7. n is odd and $a_n < 0$.

9. The graph has vertical asymptote $x = -1$. Since $f(x) > 0$ for $x \neq -1$, the graph climbs higher and higher as it approaches the asymptote from both sides.

11. The graph has vertical asymptote $x = -4$. Since $f(x) > 0$ for $x < -4$ and $f(x) < 0$ for $-4 < x < \frac{1}{2}$, the graph climbs higher and higher as it approaches the asymptote from the left and it falls lower and lower as it approaches the asymptote from the right.

13. The graph has vertical asymptotes $x = -2$ and $x = 2$. Since $f(x) < 0$ for $x < -2$ and $f(x) > 0$ for $-2 < x < 0$, the graph of $f(x)$ falls lower and lower as it approaches the asymptote $x = -2$ from the left and climbs higher and higher as it approaches the asymptote $x = -2$ from the right. Since $f(x) < 0$ for $0 < x < 2$ and $f(x) > 0$ for $x > 2$, the graph falls lower and lower as it approaches the asymptote $x = 2$ from the left and climbs higher and higher as it approaches the asymptote $x = 2$ from the right.

15. $f(x) = \frac{(x-2)(x-3)}{x-3} = x - 2$ for all $x \neq 3$. Therefore the graph of $f(x)$ has no vertical asymptote.

17. $f(x) = \frac{x-1}{(x-1)(x+1)} = \frac{1}{x+1}$ for all $x \neq \pm 1$. The graph has vertical asymptote $x = -1$. Since $f(x) < 0$ for $x < -1$ and $f(x) > 0$ for $-1 < x < 1$, the graph of $f(x)$ falls lower and lower as it approaches the asymptote from the left and climbs higher and higher as it approaches the asymptote from the right.

19. $f(x) = \frac{x+1}{x(x-2)(x+1)} = \frac{1}{x(x-2)}$ for all $x \neq -1, 0, 2$. The graph has vertical asymptotes $x = 0$ and $x = 2$. Since $f(x) > 0$ for $-1 < x < 0$ and $f(x) < 0$ for $0 < x < 2$, the graph climbs higher and higher as it approaches the asymptote $x = 0$ from the left and falls lower and lower as it approaches the asymptote $x = 0$ from the right. Since $f(x) < 0$ for $0 < x < 2$ and $f(x) > 0$ for $x > 2$, the graph falls lower and lower as it approaches the asymptote $x = 2$ from the left and climbs higher and higher as it approaches the asymptote $x = 2$ from the right.

21. $2^4 = 16$

23. $(0.3)^3 = 0.027$

25. $4^{-2} = \frac{1}{4^2} = \frac{1}{16}$

27. $(1/3)^{-3} = 3^3 = 27$

29. $8^{2/3} = (\sqrt[3]{8})^2 = 2^2 = 4$

31. $f(x) = (x + 7)^{1/2}$. So $f(9) = (9 + 7)^{1/2} = \sqrt{16} = 4$

33. $f(x) = 4x^{1/2}$. So $f(9) = 4 \cdot 9^{1/2} = 4\sqrt{9} = 12$

35. $f(x) = (1/x)^{3/2}$. So $f(9) = (1/9)^{3/2} = (\sqrt{1/9})^3 = (1/3)^3 = 1/27$

37. Domain: all real numbers.
No vertical asymptote.
Increasing for all x.
Symmetric about the origin.
$f(x) > 0$ for $x > 0$ and
$f(x) < 0$ for $x < 0$.

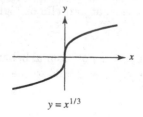

$y = x^{1/3}$

39. Domain: all real numbers
No vertical asymptote
Decreasing on $(-\infty, 0)$.
Increasing on $(0, \infty)$.
Symmetric about y-axis.
Positive for all $x \neq 0$.

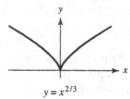

$y = x^{2/3}$

41. Domain: all real numbers $x \geq 0$
No vertical asymptote
Increasing on $(0, \infty)$
No symmetry
Positive for all $x > 0$.

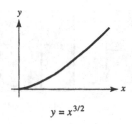

$y = x^{3/2}$

43. $P(2x) = C(2x)^{7/2} = C \cdot 2^{7/2} \cdot x^{7/2} = 2^{7/2} \cdot (Cx^{7/2}) = 2^{7/2}P(x)$. The power increases by a factor of $2^{7/2}$.

45. (a) The natural domain of $F = \frac{Gm}{r^2}$ is all the real numbers $r \neq 0$. But in applying it to gravitation, we cannot use its natural domain. Instead, we use the domain $r > 0$.

(b) $0 < r_1 < r_2 \Longrightarrow r_1^2 < r_2^2 \Longrightarrow \frac{Gm}{r_1^2} > \frac{Gm}{r_2^2}$. Therefore, F is decreasing in $(0, \infty)$. The graph of F has a vertical asymptote $r = 0$ and the graph climbs higher and higher as it approaches the asymptote.

(c) $F(1.002r) = \frac{Gm}{(1.002r)^2} = \frac{1}{(1.002)^2} \cdot \frac{GM}{r^2} = \frac{1}{(1.002)^2}F(r) \approx 0.996F(r)$. The gravitational force decreases by about 0.4%.

EXERCISES 0.7

1. (a) 2   (b) 2/9   (c) 2 ($2/3x$ is the same as $(2/3)x$.)

3. (a) $1/2 + 3 = \left(\frac{1}{2}\right) + 3 = 3.5$ (b) $1/(2 + 3) = 1/5 = 0.2$

5. (a) $3 \cdot 4 = 12$ (b) $4^2 = 16$ (c) $3 + 1^2 = 4$ (d) $(6)^2 = 36$

7.

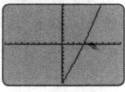

9.

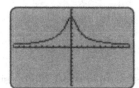

11.

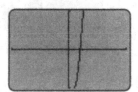

Default

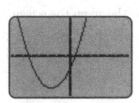

$[-40, 40] \times [-30, 30]$

13.

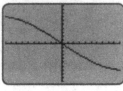

Default

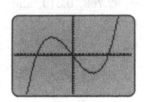

$[-30, 30] \times [-15, 15]$

15.

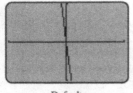 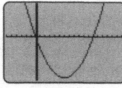

Default $[-6, 18] \times [-38, 30]$

Domain: $(-\infty, \infty)$; x-intercepts at $x = 0, 12$. By completing the square, we get $f(x) = (x-6)^2 - 36$, which shows that $f(x)$ has a minimum at $(6, -36)$

17.

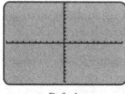 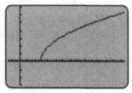

Default $[-5, 50] \times [-3, 7]$

Domain: $(10, \infty)$; Range: $(0, \infty)$; x-intercept at $x = 10$

19.

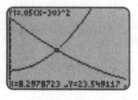 $(q_e, p_e) \approx (8.30, 23.55)$

21. The graphs of $y = x^{1/3}$ and $y = x^{1/5}$ extend to the left half of the plane and are symmetric about the origin.

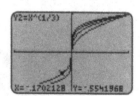

23. To 4 decimal places, solution is 0.2956 (using almost any initial guess).

25. Almost any initial guess leads to the solution shown on the screen.

27. Three solutions, which TRACE shows near -2.766, 0.319, and 2.447. Equation solver gives (to 3 decimal places) -2.778, 0.289, and 2.489.

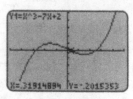

$[-5, 5] \times [-50, 50]$

29.

```
TABLE SETUP
 TblStart=0
 ΔTbl=.2
Indpnt: Auto  Ask
Depend: Auto  Ask
```

X	Y1
0	0
.2	.44721
.4	.63246
.6	.7746
.8	.89443
1	1
1.2	1.0954

Y1■√(X)

X	Y1
1.4	1.1832
1.6	1.2649
1.8	1.3416
2	1.4142
2.2	1.4832
2.4	1.5492
2.6	1.6125

Y1=1.61245154966

X	Y1
2.8	1.6733
3	1.7321
3.2	1.7889
3.4	1.8439
3.6	1.8974
3.8	1.9494
4	2

Y1=1.6733200530?

31.

```
TABLE SETUP
 TblStart=1
 ΔTbl=-.1
Indpnt: Auto  Ask
Depend: Auto  Ask
```

X	Y1
1	1
.9	1.1111
.8	1.25
.7	1.4286
.6	1.6667
.5	2
.4	2.5

Y1=1.42857142857

X	Y1
.3	3.3333
.2	5
.1	10
0	ERROR
-.1	-10
-.2	-5
-.3	-3.333

Y1=ERROR

33.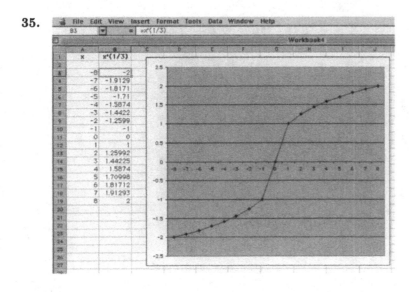

35.

37.

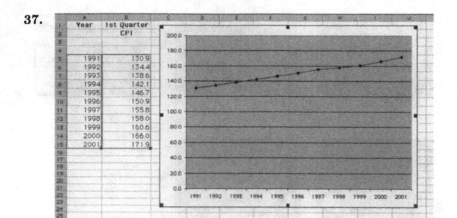

Year	1st Quarter CPI
1991	130.9
1992	134.4
1993	138.6
1994	142.1
1995	146.7
1996	150.9
1997	155.8
1998	158.0
1999	160.6
2000	166.0
2001	171.9

39.

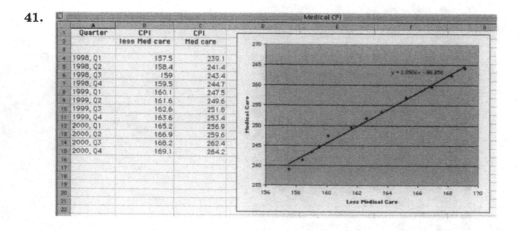

Year	1st Quarter CPI	1st Quarter Wages & Salaries
1991	130.9	106.9
1992	134.4	110.7
1993	138.6	113.5
1994	142.1	117.5
1995	146.7	120.9
1996	150.9	125.1
1997	155.8	129.0
1998	158.0	134.7
1999	160.6	138.9
2000	166.0	145.3
2001	171.9	150.9

41.

Quarter	CPI less Med care	CPI Med care
1998, Q1	157.5	239.1
1998, Q2	158.4	241.4
1998, Q3	159	243.4
1998, Q4	159.5	244.7
1999, Q1	160.1	247.5
1999, Q2	161.6	249.6
1999, Q3	162.6	251.8
1999, Q4	163.6	253.4
2000, Q1	165.2	256.9
2000, Q2	166.9	259.6
2000, Q3	168.2	262.4
2000, Q4	169.1	264.2

CHAPTER 0: REVIEW EXERCISES

1. $f(x) = x^2$. Therefore, $f(-5) = (-5)^2 = 25$, $f(-\sqrt{2}) = (-\sqrt{2})^2 = 2$, $f(-1) = (-1)^2 = 1$, $f(-\frac{3}{5}) = (-\frac{3}{5})^2 = \frac{9}{25}$, $f(0) = 0^2 = 0$, $f(3\sqrt{5}) = (3\sqrt{5})^2 = 45$, $f(t+3) - f(t-3) = (t+3)^2 - (t-3)^2 = 12t$, and
$$\frac{f(x+h) - f(x)}{h} = \frac{(x+h)^2 - x^2}{h} = \frac{h(2x+h)}{h} = 2x + h.$$

3. The natural domain of $1/(x-4)$ is the set of all real numbers with $x \neq 4$.
$f(-6) = 1/(-6-4) = -1/10$, $f(-1) = 1/(-1-4) = -1/5$,
$f(0) = 1/(0-4) = -1/4$, $f(2) = 1/(2-4) = -1/2$, $f(\frac{7}{2}) = 1/(\frac{7}{2}-4) = -2$, and
$$\frac{f(5+h) - f(5)}{h} = \frac{\frac{1}{h+1} - 1}{h} = \frac{\frac{1-h-1}{h+1}}{h} = -\frac{h}{h(h+1)} = -\frac{1}{h+1}.$$

5. (a) $f(-\frac{1}{2}) = -\frac{3}{4}$, $f(0) = \frac{5}{4}$, $f(1) = \frac{3}{2}$, $f(\frac{7}{4}) = \frac{11}{8}$

   (b) $x = \frac{1}{2}$ and $x = 2$

   (c) $x = -\frac{3}{8}$

   (d) $[-\frac{9}{16}, 2]$

   (e) $(-\frac{9}{16}, \frac{1}{2})$ and $(\frac{3}{2}, 2)$

   (f) f does not satisfy the horizontal line test.

   (g) Yes, since f satisfies the horizontal line test on $[\frac{1}{2}, \frac{3}{2}]$.

   (h) Yes, since f is one-to-one on $[-\frac{1}{2}, \frac{1}{2}]$.

7. (a)

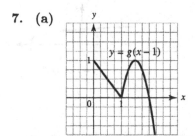

Shift 1 unit to the right

(b)

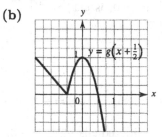

Shift 1/2 unit to the left

(c)

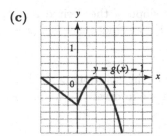

Shift 1 unit downward

(d)

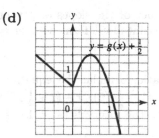

Shift 1/2 unit upward

(e)

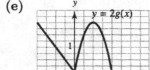
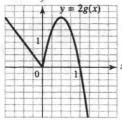

Double the height of
each point

(f)

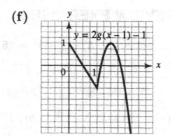

Shift 1 unit to the right, then double
the height of each point, then shift
1 unit downward

9. $f(x) = 1$ and $f(x) = x^2$ are
even functions and $f(x) = x$
and $f(x) = x^3$ are odd functions.

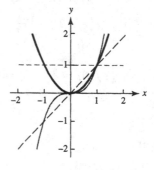

11. (a) The natural domain of $f(x) = \frac{\sqrt{x}}{x-2}$ is the set of all x satisfying both $x \geq 0$ and $x \neq 2$.

   (b) The natural domain of $f(x) = \frac{(x-9)^{1/3}}{(x-1)(x-2)}$ is the set of all x satisfying both $x \neq 1$ and $x \neq 2$.

13. $y = 1 - x^4 \iff x^4 = 1 - y \iff x = \sqrt[4]{1-y}$. The inverse of
$f(x) = 1 - x^4$, $x \geq 0$, is $g(x) = \sqrt[4]{1-x}$, whose domain is $x \leq 1$
and whose range is $y \geq 0$.

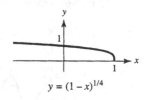

$y = (1-x)^{1/4}$

15. $2x + 3y = 5$ is equivalent to $y = -\frac{2}{3}x + \frac{5}{3}$. The slope equals $-\frac{2}{3}$.

17. The slope of the line equals $\frac{5+1}{-2-0} = -3$. Using the point-slope form with $m = -3$ and $(x_1, y_1) = (0, -1)$, we find $y + 1 = -3x$ or $y = -3x - 1$.

19. Since the product of the slopes of perpendicular lines is -1, the slope of the desired line equals $\frac{5}{2}$. Using the point-slope form with $m = \frac{5}{2}$ and $(x_1, y_1) = (4, -1)$, we find $y + 1 = \frac{5}{2}(x - 4)$ or $y = \frac{5}{2}x - 11$.

21. By the quadratic formula, the equation $-4x^2 + 8x + 1 = 0$ has the solutions $x = 1 \pm \frac{1}{2}\sqrt{5}$. $f(x) = -4(x - (1 - \frac{1}{2}\sqrt{5}))(x - (1 + \frac{1}{2}\sqrt{5}))$. Therefore, $f(x) > 0$ in the interval $(1 - \frac{1}{2}\sqrt{5}, 1 + \frac{1}{2}\sqrt{5})$ and $f(x) < 0$ in the intervals $(-\infty, 1 - \frac{1}{2}\sqrt{5})$ and $(1 + \frac{1}{2}\sqrt{5}, \infty)$.

23. Writing $-3x^2 + 10x - 8 = -3(x+h)^2 + k = -3x^2 - 6hx - 3h^2 + k$, we see that $-6h = 10$ and $-3h^2 + k = -8$. Therefore, $h = -\frac{5}{3}$, $k = \frac{1}{3}$, and $f(x) = -3(x - \frac{5}{3})^2 + \frac{1}{3}$. Since the coefficient of x^2 is negative, the graph of $f(x)$ opens downward and $f(x)$ has its maximum value $\frac{1}{3}$ at $x = \frac{5}{3}$. The y-intercept is $f(0) = -8$. To find the x-intercepts, we solve $-3x^2 + 10x - 8 = 0$, which gives $x = \frac{4}{3}$ and $x = 2$.

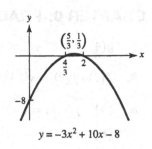

$y = -3x^2 + 10x - 8$

25. No vertical asymptote.
Even function since

$$f(-x) = \frac{-1}{(-x)^2 + 1} = \frac{-1}{x^2 + 1} = f(x).$$

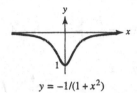

$y = -1/(1 + x^2)$

27.

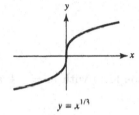

$y = x^{1/3}$

Natural domain: all real numbers

29.

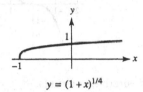

$y = (1 + x)^{1/4}$

Natural domain: $[-1, \infty)$

31. **(a)** The cost and revenue functions are:

$$C(q) = 30q + 5{,}000 \quad \text{and} \quad R(q) = 50q.$$

Therefore, the profit function is given by

$$P(q) = R(q) - C(q) = 20q - 5{,}000.$$

(b) Solving $P(q) = 20q - 5{,}000 = 0$ yields $q = 250$, the break-even point.

33. **(a)** $R(p) = pq = p(3{,}000 - 20p) = 3{,}000p - 20p^2$

(b) The cost function is given (in terms of q) by $C(q) = 20{,}000 + 40q$. Substituting $q = 3{,}000 - 20p$ converts that to $C(p) = 140{,}000 - 800p$. Therefore, $P(p) = R(p) - C(p) = -20p^2 + 3{,}800p - 140{,}000$.

(c) Completing the square gives

$$P(p) = -20p^2 + 3{,}800p - 140{,}000 = -20(p - 95)^2 + 40{,}500.$$

The maximum occurs at $p = 95$ dollars.

CHAPTER 0: PRACTICE EXAM

1. $g(4) = f(3) + 3 = 6.4 + 3 = 9.4.$

3. $f(1) \cdot f(2) = \sqrt{1} \cdot (1 + 2) = 3.$

5. **(a)** $f(-2) = (-2)^2 + 2 \cdot (-2) - 1 = -1$

 (b) $f(1-t) = (1-t)^2 + 2(1-t) - 1 = t^2 - 4t + 2$

 (c) $\dfrac{f(1+h) - f(1)}{h} = \dfrac{[(1+h)^2 + 2(1+h) - 1] - [1^2 + 2 \cdot 1 - 1]}{h}$

$$= \dfrac{h^2 + 4h}{h} = h + 4$$

7. Reflect in the line $y = x$.

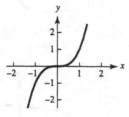

9. The slope of the desired line equals $\frac{3+1}{-7-1} = -\frac{1}{2}$. Using the point-slope form with $m = -\frac{1}{2}$ and $(x_1, y_1) = (1, -1)$, we find $y + 1 = -\frac{1}{2}(x - 1)$, or $y = -\frac{1}{2}x - \frac{1}{2}$.

11. Writing $C(x)$ and $R(x)$ for the cost and revenue of manufacturing and selling x bird feeders, respectively, we have

$$C(x) = 10x + 12{,}000$$
$$R(x) = 40x$$

Solving $10x + 12{,}000 = 40x$, we obtain $x = 400$, which is the break-even point.

13. **(a)** Write $q = mp + b$ for some constants m and b. A change in price of $\Delta p = -1$ results in a change in revenue of $\Delta q = 8$. Therefore, $m = \frac{8}{-1} = -8$. Since $q = 40$ when $p = 10$, we have $40 = 10m + b = -80 + b$. Therefore, $b = 120$ and $q = -8p + 120$.

 (b) $R(p) = pq = p(-8p + 120) = -8p^2 + 120p.$

 (c) By completing the square, we get

$$R(p) = -8p^2 + 120p = -8(p - \frac{15}{2})^2 + 450.$$

Therefore, the revenue is maximized if $p = \frac{15}{2} = 7.50$.

CHAPTER 1
Limits and Continuity

EXERCISES 1.1

1. (a) $\lim\limits_{x \to -1} f(x) = -2$ (b) $\lim\limits_{x \to 1^+} f(x) = 3$ (c) $\lim\limits_{x \to 1^-} f(x) = 2$

3. $x = 1$ and $x = 3$

5. (a) $\lim\limits_{x \to -1} g(x) = 2$ (b) $\lim\limits_{x \to 2^+} g(x) = 1$

     (c) $\lim\limits_{x \to 2^-} g(x) = 2$ (d) $g(-1) = 2$ and $g(1) = 2$

7. $x = 2$ and $x = 3$

9. (a) $\lim\limits_{q \to 10^7} A(q) = \dfrac{10,000 + 8 \times 10^7}{10^7} = 8.001$

     (b) No, since the denominator approaches zero and the numerator approaches 10,000.

11. (a) $f(t) = \begin{cases} 25 & \text{if } 0 < t \le 4 \\ 30 & \text{if } 4 < t \le 5 \\ 35 & \text{if } 5 < t \le 6 \\ 40 & \text{if } 6 < t \le 7 \end{cases}$

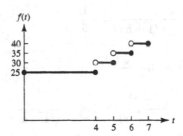

     (b) This function has no limits at $t = 4, 5, 6$.

$$\lim_{t \to 4^-} y(t) = 25, \quad \lim_{t \to 4^+} y(t) = 30;$$
$$\lim_{t \to 5^-} y(t) = 30, \quad \lim_{t \to 5^+} y(t) = 35;$$
$$\lim_{t \to 6^-} y(t) = 35, \quad \lim_{t \to 6^+} y(t) = 40.$$

13. $\lim\limits_{x \to -1} (x^3 - 2x^2 + 5x + 1) = (-1)^3 - 2(-1)^2 + 5(-1) + 1 = -7$

15. $\lim\limits_{x \to 1} \dfrac{x - 1}{x^2 - 1} = \lim\limits_{x \to 1} \dfrac{x - 1}{(x - 1)(x + 1)} = \lim\limits_{x \to 1} \dfrac{1}{x + 1} = \dfrac{1}{2}$

17. $\lim\limits_{x \to 1} \dfrac{x + 1}{x^2 + 1} = \dfrac{1 + 1}{1^2 + 1} = 1$ **19.** $\lim\limits_{x \to -1} \dfrac{x + 1}{x^2 + 1} = \dfrac{(-1) + 1}{(-1)^2 + 1} = \dfrac{0}{2} = 0$

21. $\lim\limits_{x \to 3} \dfrac{x^2 - 4x + 3}{x^2 - x - 6} = \lim\limits_{x \to 3} \dfrac{(x - 1)(x - 3)}{(x + 2)(x - 3)} = \lim\limits_{x \to 3} \dfrac{x - 1}{x + 2} = \dfrac{2}{5}$

23. $\lim\limits_{x \to -1} \dfrac{x + 1}{x^3 - 1} = \dfrac{-1 + 1}{(-1)^3 - 1} = \dfrac{0}{-2} = 0$

25. $\lim\limits_{x \to -1} \dfrac{x^3 + 1}{x + 1} = \lim\limits_{x \to -1} \dfrac{(x+1)(x^2 - x + 1)}{x + 1} = \lim\limits_{x \to -1} (x^2 - x + 1) = 3$

27. $\lim\limits_{t \to 0} \dfrac{t}{t^2 + 2t + 1} = \dfrac{0}{0^2 + 2 \cdot 0 + 1} = \dfrac{0}{1} = 0$

29. $\lim\limits_{h \to 0} \dfrac{(h+3)^2 - 9}{h} = \lim\limits_{h \to 0} \dfrac{h^2 + 6h}{h} = \lim\limits_{h \to 0} (h + 6) = 6$

31. $\lim\limits_{h \to 0} \dfrac{(h-3)^2 - 9}{(h-2)^2 - 4} = \lim\limits_{h \to 0} \dfrac{h^2 - 6h}{h^2 - 4h} = \lim\limits_{h \to 0} \dfrac{h(h-6)}{h(h-4)} = \lim\limits_{h \to 0} \dfrac{h-6}{h-4} = \dfrac{3}{2}$

33. $\lim\limits_{x \to 1} \dfrac{\sqrt{x} - 1}{x - 1} = \lim\limits_{x \to 1} \dfrac{\sqrt{x} - 1}{x - 1} \cdot \dfrac{\sqrt{x} + 1}{\sqrt{x} + 1}$

$\qquad = \lim\limits_{x \to 1} \dfrac{x - 1}{(x-1)(\sqrt{x} + 1)} = \lim\limits_{x \to 1} \dfrac{1}{(\sqrt{x} + 1)} = \dfrac{1}{2}$

35. $\lim\limits_{h \to 0} \dfrac{1 - \sqrt{1 + h}}{h^2 + 3h} = \lim\limits_{h \to 0} \dfrac{1 - \sqrt{1 + h}}{h^2 + 3h} \cdot \dfrac{1 + \sqrt{1 + h}}{1 + \sqrt{1 + h}}$

$\qquad = \lim\limits_{h \to 0} \dfrac{1 - (1 + h)}{(h^2 + 3h)(1 + \sqrt{1 + h})} = \lim\limits_{h \to 0} \dfrac{-h}{h(h + 3)(1 + \sqrt{1 + h})}$

$\qquad = \lim\limits_{h \to 0} \dfrac{-1}{(h + 3)(1 + \sqrt{1 + h})} = -\dfrac{1}{6}$

37. $\lim\limits_{x \to 0} \dfrac{\sqrt{1 + x} - \sqrt{1 - x}}{x} = \lim\limits_{x \to 0} \dfrac{\sqrt{1 + x} - \sqrt{1 - x}}{x} \cdot \dfrac{\sqrt{1 + x} + \sqrt{1 - x}}{\sqrt{1 + x} + \sqrt{1 - x}}$

$\qquad = \lim\limits_{x \to 0} \dfrac{(1 + x) - (1 - x)}{x(\sqrt{1 + x} + \sqrt{1 - x})} = \lim\limits_{x \to 0} \dfrac{2x}{x(\sqrt{1 + x} + \sqrt{1 - x})}$

$\qquad = \lim\limits_{x \to 0} \dfrac{2}{(\sqrt{1 + x} + \sqrt{1 - x})} = \dfrac{2}{\sqrt{1} + \sqrt{1}} = 1$

39. $\lim\limits_{x \to 0} \left[f(x)g(x)^2 \right] = \lim\limits_{x \to 0} f(x) \cdot \left[\lim\limits_{x \to 0} g(x) \right]^2 = 3 \cdot (-2)^2 = 12$

41. $\lim\limits_{x \to 0} (f(x)g(x) + 5) = \left(\lim\limits_{x \to 0} f(x) \right)\left(\lim\limits_{x \to 0} g(x) \right) + 5 = 3(-2) + 5 = -1$

43. $\lim\limits_{x \to 0} \dfrac{xf(x) - g(x)}{f(x) - g(x)} = \dfrac{\lim\limits_{x \to 0} x \lim\limits_{x \to 0} f(x) - \lim\limits_{x \to 0} g(x)}{\lim\limits_{x \to 0} f(x) - \lim\limits_{x \to 0} g(x)} = \dfrac{0 \cdot 3 - (-2)}{3 - (-2)} = \dfrac{2}{5}.$

45. $\lim\limits_{x \to 0^-} \dfrac{|x|}{x} = \lim\limits_{x \to 0^-} \dfrac{-x}{x} = -1.$
$\qquad\qquad$ **47.** $\lim\limits_{t \to 2^+} \dfrac{\sqrt{t} - 2}{t^2} = 0.$

49. $\lim\limits_{t \to 2^+} \dfrac{t^2 - 4}{\sqrt{t} - 2} = \lim\limits_{t \to 2^+} \dfrac{(t-2)(t+2)}{\sqrt{t} - 2} = \lim\limits_{t \to 2^+} \sqrt{t} - 2(t + 2) = 0$

51. $\lim\limits_{t \to 0^+} \dfrac{\sqrt{4 + t} - 2}{\sqrt{t}} = \lim\limits_{t \to 0^+} \dfrac{\sqrt{4 + t} - 2}{\sqrt{t}} \cdot \dfrac{\sqrt{4 + t} + 2}{\sqrt{4 + t} + 2}$

$\qquad = \lim\limits_{t \to 0^+} \dfrac{(4 + t) - 4}{\sqrt{t}(\sqrt{4 + t} + 2)} = \lim\limits_{t \to 0^+} \dfrac{t}{\sqrt{t}(\sqrt{4 + t} + 2)} = \lim\limits_{t \to 0^+} \dfrac{\sqrt{t}}{\sqrt{4 + t} + 2} = 0$

53. (a) $\lim\limits_{v\to c^-} \ell(v) = \sqrt{1 - \dfrac{c^2}{c^2}} = 0$ **(b)** $\lim\limits_{v\to 0^+} \ell(v) = \sqrt{1 - \dfrac{0}{c^2}} = 1$

55. $\lim\limits_{x\to 0^+} (f(x) - g(x)) = \lim\limits_{x\to 0^+} f(x) - \lim\limits_{x\to 0^+} g(x) = (1/2) - (-1/3) = \dfrac{5}{6}$

57. $\lim\limits_{x\to 0^+} \dfrac{|f(x)|}{f(x)} = \dfrac{\left|\frac{1}{2}\right|}{\frac{1}{2}} = 1$ **59.** $\lim\limits_{x\to 1^+} f(x-1) = \lim\limits_{y\to 0^+} f(y) = 1/2$

61. Guess: $\lim\limits_{t\to 1^-} \dfrac{1 - \sqrt{t}}{\sqrt{1-t}} = 0$

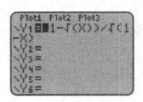

63. Guess: limit does not exist

65. Guess: $\lim\limits_{t\to 3} \dfrac{\sqrt{2t+3} - 3}{\sqrt{t+1} - 2} = 1.\overline{3}$

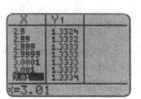

EXERCISES 1.2

1. (a) $\lim\limits_{x\to -1^-} f(x) = -1$ **(b)** $\lim\limits_{x\to -1^+} f(x) = 0$ **(c)** $\lim\limits_{x\to 0^-} f(x) = \infty$

  (d) $\lim\limits_{x\to 0^+} f(x) = -1$ **(e)** $f(-1) = 0$

3. No limit at $x = -1$, since the two one-sided limits are not equal. No limit at $x = 0$, since $\lim_{x\to 0^-} f(x) = \infty$ and $\lim_{x\to 0^+} f(x) = -1$. No limit at $x = 2$, since $\lim_{x\to 2^-} f(x) = \infty$ and $\lim_{x\to 2^+} f(x) = -\infty$.

5. (a) $\lim\limits_{x\to 1^-} g(x) = -\infty$ **(b)** $\lim\limits_{x\to 1^+} g(x) = \infty$ **(c)** $\lim\limits_{x\to 2} g(x) = 0$

  (d) $\lim\limits_{x\to \infty} g(x) = -1$ **(e)** $g(2) = 0$

7. $\lim\limits_{x\to 5^+} \dfrac{1}{(x-5)^2} = \infty$ **9.** $\lim\limits_{x\to 5^+} \dfrac{1}{x-5} = \infty$ **11.** $\lim\limits_{x\to 5^+} \dfrac{x}{x-5} = \infty$

13. $\displaystyle\lim_{x\to 2^-}\frac{x}{x^2-4}=\lim_{x\to 2^-}\frac{x}{(x-2)(x+2)}=-\infty$

15. $\displaystyle\lim_{x\to -2^-}\frac{x}{x^2-4}=\lim_{x\to -2^-}\frac{x}{(x-2)(x+2)}=-\infty$

17. $\displaystyle\lim_{t\to 2^+}\frac{\sqrt{t-2}}{t^2-4}=\lim_{t\to 2^+}\frac{\sqrt{t-2}}{(t-2)(t+2)}=\lim_{t\to 2^+}\frac{1}{\sqrt{t-2}(t+2)}=\infty$

19. $\displaystyle\lim_{x\to\infty}\frac{2x-1}{5x+1}=\lim_{x\to\infty}\frac{2x}{5x}=\lim_{x\to\infty}\frac{2}{5}=\frac{2}{5}$ 　　　**21.** $\displaystyle\lim_{x\to\infty}\frac{x^2}{x^2+1}=\lim_{x\to\infty}\frac{x^2}{x^2}=1$

23. $\displaystyle\lim_{x\to -\infty}\frac{x^3}{x^2+1}=\lim_{x\to -\infty}\frac{x^3}{x^2}=\lim_{x\to -\infty}x=-\infty$

25. $\displaystyle\lim_{x\to\infty}\frac{5x^2+1}{x^3+x+2}=\lim_{x\to\infty}\frac{5x^2}{x^3}=\lim_{x\to\infty}\frac{5}{x}=0$

27. $\displaystyle\lim_{x\to -\infty}\frac{9x+10}{x^2+1}=\lim_{x\to -\infty}\frac{9x}{x^2}=\lim_{x\to -\infty}\frac{9}{x}=0$

29. $\displaystyle\lim_{x\to -\infty}\frac{x^2+2}{2x^2+1}=\lim_{x\to -\infty}\frac{x^2}{2x^2}=\lim_{x\to -\infty}\frac{1}{2}=\frac{1}{2}$

31. **(a)** As x becomes very large, the cost gets unboundedly large, since $\lim_{x\to\infty}0.17x+100=\infty$.

　　(b) As x becomes very large, $\lim_{x\to\infty}\frac{C(x)}{x}=\lim_{x\to\infty}\left(0.17+\frac{100}{x}\right)=0.17.$ Thus, the average cost approaches 0.17 (dollars).

33. 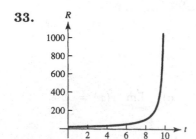　　　　Large without bound, since $\displaystyle\lim_{t\to 10^-}R(t)=\infty.$

35. **(a)** $R(q)=pq=\dfrac{2{,}400q}{q+2}$

　　(b) $\displaystyle\lim_{q\to 0^+}R(q)=0,\ \lim_{q\to 10}R(q)=\frac{2{,}400\cdot 10}{10+2}=2{,}000,\ \lim_{q\to\infty}R(q)=\lim_{q\to\infty}\frac{2{,}400q}{q+2}=2{,}400.$

　　(c) The revenue can never exceed 2,400 dollars because

$$\frac{2{,}400q}{q+2}=2{,}400\,\frac{q}{q+2}<2{,}400\cdot 1=2{,}400.$$

　　　　As demand gets larger, the revenue will be closer and closer to 2,400 dollars.

37. vertical asymptote: $x = 2$
horizontal asymptote: $y = 2$
positive on $(-\infty, -1/2)$ and $(2, \infty)$
negative on $(-1/2, 2)$

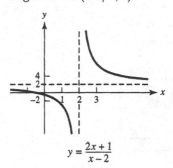

$$y = \frac{2x+1}{x-2}$$

39. vertical asymptotes: $x = 2$ and $x = -2$
horizontal asymptote: $y = 0$
positive on $(-2, 0)$ and $(2, \infty)$
negative on $(-\infty, -2)$ and $(0, 2)$

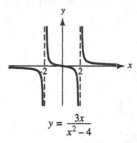

$$y = \frac{3x}{x^2 - 4}$$

41. no vertical asymptote
horizontal asymptote: $y = 1$
positive for all $x \neq 0$.

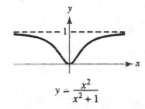

$$y = \frac{x^2}{x^2 + 1}$$

43. vertical asymptotes: $x = 1$ and $x = -1$
horizontal asymptote: $y = 1$
positive on $(-\infty, -3)$, $(-1, 1)$ and $(3, \infty)$
negative on $(-3, -1)$ and $(1, 3)$

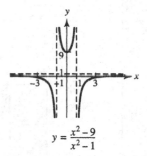

$$y = \frac{x^2 - 9}{x^2 - 1}$$

45. vertical asymptotes: $x = 3$ and $x = -1$
horizontal asymptote: $y = 0$
positive on $(-\infty, -1)$ and $(3, \infty)$
negative on $(-1, 3)$

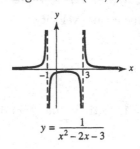

$$y = \frac{1}{x^2 - 2x - 3}$$

47. vertical asymptote: $x = 1$
horizontal asymptote: $y = 1$
positive on $(-\infty, 0)$ and $(1, \infty)$
negative on $(0, 1)$

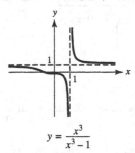

$$y = \frac{x^3}{x^3 - 1}$$

49. By dividing $3x^2 + 7$ into $x^3 - 5x^2 + 1$, with remainder, we get

$$y = \frac{1}{3}x - \frac{5}{3} + \frac{-\frac{7}{3}x + \frac{38}{3}}{3x^2 + 7}.$$

Therefore, $y = \frac{1}{3}x - \frac{5}{3}$ is a slant asymptote to the graph of $y = \frac{x^3 - 5x^2 + 1}{3x^2 + 7}$.

51. $y = 2x$ is a slant asymptote to the graph of $y = 2x + \dfrac{1}{x}$, since

$$\lim_{x \to \infty} \left[\left(2x + \frac{1}{x} \right) - 2x \right] = \lim_{x \to \infty} \frac{1}{x} = 0.$$

53. $\displaystyle \lim_{x \to \infty} \frac{\sqrt{x^2 + 9}}{x + 3} = 1$

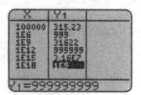

55. $\displaystyle \lim_{t \to \infty} (t - \sqrt{t}) = \infty$

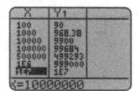

57. $\displaystyle \lim_{x \to \infty} \frac{x\sqrt{x} + 1}{x + \sqrt{x} + 1} = \infty$

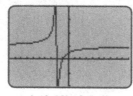

59. $\displaystyle \lim_{x \to \infty} \frac{\sqrt{x^4 + 1}}{1 - x} = -\infty$

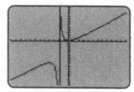

61.

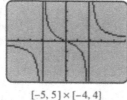

$[-10, 10] \times [-2, 4]$

63.

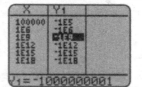

$[-5, 5] \times [-4, 4]$

65.

$[-12, 12] \times [-16, 12]$

67. If the rocket does not return to earth, then its height has no maximum, or, in other words, gets arbitrarily large. Therefore, we take $\lim_{h \to \infty} v$. Since $\lim_{h \to \infty} h/(h + R) = 1$, we have $\lim_{h \to \infty} v = \lim_{h \to \infty} \sqrt{2gR \frac{h}{h+R}} = \sqrt{2gR} \approx 11{,}182.5$ m/sec, which is the escape velocity.

EXERCISES 1.3

1. f is discontinuous at $x = -1$, since $\displaystyle \lim_{x \to -1^-} f(x) = 0$ and $\displaystyle \lim_{x \to -1^+} f(x) = 1$, so that $\displaystyle \lim_{x \to -1} f(x)$ does not exist.

f is discontinuous at $x = 1$, since $\displaystyle \lim_{x \to 1} f(x) = 2$ but $f(1) = 1$.

f is discontinuous at $x = 4$, since $\displaystyle \lim_{x \to 4} f(x)$ does not exist. In fact, $\displaystyle \lim_{x \to 4^-} f(x) = -\infty$ and $\displaystyle \lim_{x \to 4^+} f(x) = \infty$.

f is discontinuous at $x = 6$, since $\displaystyle \lim_{x \to 6} f(x) = 2$ but $f(6) = 3$.

f is discontinuous at $x = 8$, since $\displaystyle \lim_{x \to 8^-} f(x) = -1$ and $\displaystyle \lim_{x \to 8^+} f(x) = 0$, so that $\displaystyle \lim_{x \to 8} f(x)$ does not exist.

3. f is continuous in the interval $[2,3]$.

5. f has no limit at $x = 4$.

7. The limit of f is smaller than its value at $x = 6$.

9. 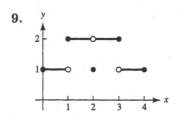   **11.**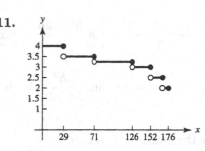

Discontinuous at $x = 29, 71, 126, 152, 168$.

13. $f(x)$ is continuous in $(-\infty, 1)$ and $(1, \infty)$.

15. $f(x)$ is continuous in $(-\infty, \infty)$.

17. Continuous at all x. In this case, $\lim\limits_{x \to -2} \dfrac{x^2 - 4}{x + 2} = -4 = f(-2)$.

19. Continuous in $(-\infty, 0)$ and $(0, \infty)$, discontinuous at $x = 0$. Same reason as in exercise 18.

21. Continuous at all x. At $x = 0$, we have $f(0) = 0$ and

$$\lim_{x \to 0^+} \frac{x^2}{|x|} = \lim_{x \to 0^+} \frac{x^2}{x} = \lim_{x \to 0^+} x = 0 \text{ and } \lim_{x \to 0^-} \frac{x^2}{|x|} = \lim_{x \to 0^-} \frac{x^2}{(-x)} = \lim_{x \to 0^-} (-x) = 0.$$

23. (a) $f(0) = 2$, $f(1) = 3$, $f(2) = \frac{1}{2}$, $f(4) = \sqrt{4} = 2$.

   (b) Discontinuous at $x = 1$ because $\lim\limits_{x \to 1^-} f(x) = \lim\limits_{x \to 1^-} (x + 2) = 3$ but $\lim\limits_{x \to 1^+} f(x) = \lim\limits_{x \to 1^+} \dfrac{1}{x} = 1$, so that that $\lim\limits_{x \to 1} f(x)$ does not exist. Also discontinuous at $x = 2$ because $\lim\limits_{x \to 2^-} f(x) = \lim\limits_{x \to 2^-} \dfrac{1}{x} = \dfrac{1}{2}$ but $\lim\limits_{x \to 2^+} f(x) = \lim\limits_{x \to 2^+} \sqrt{x} = \sqrt{2}$.

25. (a) $f(1) = 1$, $f(4) = 2$, $f(5) = 3$, $f(8) = 6$, $f(9) = 27$

   (b) Discontinuous at $x = 8$ because $\lim\limits_{x \to 8^-} f(x) = \lim\limits_{x \to 8^-} (x - 2) = 6$ but $\lim\limits_{x \to 8^+} f(x) = \lim\limits_{x \to 8^+} x^{3/2} = 16\sqrt{2}$, so that $\lim\limits_{x \to 8} f(x)$ does not exist.
   Continuous at all other x. At $x = 4$, $\lim\limits_{x \to 4^-} f(x) = \lim\limits_{x \to 4^-} \sqrt{x} = 2$ and $\lim\limits_{x \to 4^+} f(x) = \lim\limits_{x \to 4^+} (x - 2) = 2$, so that $\lim\limits_{x \to 4} f(x) = 2 = f(4)$.

27. Discontinuous at $x = 2$ because $\lim\limits_{x \to 2} \dfrac{x^2 - 3x + 2}{x - 2} = \lim\limits_{x \to 2} (x - 1) = 1$, but $f(2) = 0$. Continuous at all other x.

29. $c = \lim\limits_{x \to 3} \dfrac{x^2 - 3x}{x - 3} = \lim\limits_{x \to 3} \dfrac{x(x - 3)}{x - 3} = \lim\limits_{x \to 3} x = 3$.

31. $c = \lim\limits_{x \to 2} (x^3 - 4x^2 + 3x - 1) = -3.$

33. $c = \lim\limits_{x \to 1} \dfrac{x^4 - 1}{x^3 - 1} = \lim\limits_{x \to 1} \dfrac{(x-1)(x+1)(x^2+1)}{(x-1)(x^2+x+1)} = \lim\limits_{x \to 1} \dfrac{(x+1)(x^2+1)}{(x^2+x+1)} = \dfrac{4}{3}.$

35. $f(0) = \lim\limits_{h \to 0} \dfrac{(1-h)^2 - 1}{h} = \lim\limits_{h \to 0} \dfrac{h^2 - 2h}{h} = \lim\limits_{h \to 0} (h - 2) = -2.$

37. $f(0) = \lim\limits_{h \to 0} \dfrac{h^2}{(1+h)^2 - 1} = \lim\limits_{h \to 0} \dfrac{h^2}{h^2 + 2h} = \lim\limits_{h \to 0} \dfrac{h}{h+2} = 0.$

39. Write I for the salesman's income

$$I(x) = \begin{cases} 500 & \text{if } 0 \le x \le 5{,}000 \\ 750 & \text{if } 5{,}000 < x \le 7{,}500 \\ 1{,}000 & \text{if } 7{,}500 < x < 10{,}000 \\ 0.1x, & \text{if } x \ge 10{,}000 \end{cases}$$

$I(x)$ is discontinuous at $x = 5{,}000$ and $x = 7{,}500$.

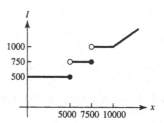

41. f takes on the value -1 twice.

43. f has two zeros in $[-3, 5]$. It crosses the x-axis between -2 and -1 and again between 1 and 2.

45. $f(x) + 2$ does not have zero in $[-3, 5]$, since $f(x)$ does not equal -2 in $[-3, 5]$.

47. The function does not have zero in $[0, 4]$, even though $g(0) < 0$ and $g(4) > 0$. The existence of zeros theorem does not apply because $g(x)$ is not continuous on $[0, 4]$.

49. $f(x)$ has at least two zeros. They lie in $(-1, 0)$ and $(2, 3)$.

51. We first compute the following table of values:

x	-3	-2	-1	0	1	2	3
$f(x)$	7	-7	-1	1	-1	17	103

From the table, we see that the equation has at least four roots. They lie in $(-3, -2)$, $(-1, 0)$, $(0, 1)$, $(1, 2)$.

53. $p = 1.98$, $q = 4.95$, obtained as shown below.

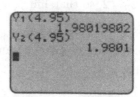

55. Root ≈ 1.703125 with $|\text{error}| < (1.71875 - 1.6875)/2 = 0.015625$.

Step	Interval	m	Sign of $f(m)$
0	$[1, 2]$	1.5	$-$
1	$[1.5, 2]$	1.75	$+$
2	$[1.5, 1.75]$	1.625	$-$
3	$[1.625, 1.75]$	1.6875	$-$
4	$[1.6875, 1.75]$	1.71875	$+$
5	$[1.6875, 1.71875]$	1.703125	

57. Root ≈ 2.671875 with $|\text{error}| < (2.71875 - 2.625)/2 = 0.046875$.

Step	Interval	m	Sign of $f(m)$
0	$[0, 3]$	1.5	$-$
1	$[1.5, 3]$	2.25	$-$
2	$[2.25, 3]$	2.625	$-$
3	$[2.625, 3]$	2.8125	$+$
4	$[2.625, 2.8125]$	2.71875	$+$
5	$[2.625, 2.71875]$	2.671875	

59.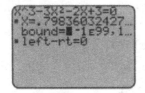

61.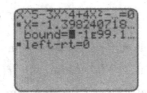

63. By checking the sign of $f(x)$ at the endpoints, we see that it changes on the intervals $[-2, 0]$, $[0, 1]$, and $[1, 3]$. Therefore, we use the equation solver with initial guesses of -1, 0.5, and 2, with the following results:

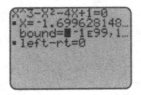 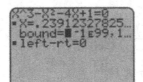 

EXERCISES 1.4

1. Given $\varepsilon > 0$, we need $\delta > 0$ such that $|3x - 6| < \varepsilon$ if $0 < |x - 2| < \delta$. Since

$$|3x - 6| < \varepsilon \iff |3(x - 2)| < \varepsilon \iff 3|x - 2| < \varepsilon \iff |x - 2| < \frac{\varepsilon}{3},$$

we can take $\delta = \varepsilon/3$.

3. Given $\varepsilon > 0$, we need $\delta > 0$ such that $\left|\frac{x+1}{2} - 1\right| < \varepsilon$ if $0 < |x - 1| < \delta$. Since

$$\left|\frac{x+1}{2} - 1\right| < \varepsilon \iff \left|\frac{x-1}{2}\right| < \varepsilon \iff \frac{1}{2}|x - 1| < \varepsilon \iff |x - 1| < 2\varepsilon,$$

we can take $\delta = 2\varepsilon$.

5. Given $\varepsilon > 0$, we need $\delta > 0$ such that $|(x^2 + 1) - 5| < \varepsilon$ if $0 < |x - 2| < \delta$. Since $|(x^2 + 1) - 5| = |x^2 - 4| = |(x - 2)(x + 2)|$, we can apply the result of the last exercise. Therefore, once again, we can let δ be any number less than or equal to both 1 and $\varepsilon/5$.

7. Given $\varepsilon > 0$, we need $\delta > 0$ such that $|x^3 - 8| < \varepsilon$ if $0 < |x - 2| < \delta$. Since $|x^3 - 8| = |(x - 2)(x^2 + 2x + 4)| = |x - 2| \cdot |x^2 + 2x + 4|$, we can use a method similar to that of exercise 4. First, take $|x - 2| < 1$, so that $x < 3$ and, therefore, $0 < |x^2 + 2x + 4| < 19$. Then $|x^3 - 8| = |x - 2| \cdot |x^2 + 2x + 4| < 19|x - 2|$. If, in addition, $|x - 2| < \varepsilon/19$, we get $|x^3 - 8| < \varepsilon$. Thus, we can let δ be any number less than or equal to both 1 and $\varepsilon/19$.

CHAPTER 1: REVIEW EXERCISES

1. (a) $\displaystyle\lim_{x \to -6} f(x) = -1$

    (b) $\displaystyle\lim_{x \to -2^+} f(x) = 1$

    (c) $\displaystyle\lim_{x \to -2^-} f(x) = -1$

    (d) $f(-2) = 4$

3. (a) $\displaystyle\lim_{x \to 1^-} f(x) = -1$

    (b) $\displaystyle\lim_{x \to 1^+} f(x) = 3$

    (c) $f(1) = 2$

5. There is no limit at $x = -2$ because $\displaystyle\lim_{x \to -2^-} f(x) = -1$ and $\displaystyle\lim_{x \to -2^+} f(x) = 1$.
There is no limit at $x = 0$ because $\displaystyle\lim_{x \to 0^-} f(x) = \infty$ and $\displaystyle\lim_{x \to 0^+} f(x) = -\infty$.
There is no limit at at $x = 1$ because $\displaystyle\lim_{x \to 1^-} f(x) = -1$ and $\displaystyle\lim_{x \to 1^+} f(x) = 3$.
There is no limit at $x = 5$ because $\displaystyle\lim_{x \to 5^-} f(x) = 1$ and $\displaystyle\lim_{x \to 5^+} f(x) = 2$.

7. At $x = -5$, $\displaystyle\lim_{x \to -5^-} f(x) = \lim_{x \to -5^+} f(x) = -2$, but $f(-5) = -3$. Therefore, the limit exists but f is not continuous.

9. $\displaystyle\lim_{x \to 3} \frac{x^2 - 9}{x - 3} = \lim_{x \to 3} \frac{(x - 3)(x + 3)}{x - 3} = \lim_{x \to 3} (x + 3) = 6.$

11. $\displaystyle\lim_{x \to -1} \frac{x + 1}{x^2 - 3x - 4} = \lim_{x \to -1} \frac{x + 1}{(x - 4)(x + 1)} = \lim_{x \to -1} \frac{1}{x - 4} = -\frac{1}{5}.$

13. $\displaystyle\lim_{h \to 0} \frac{(h - 4)^2 - 16}{h} = \lim_{h \to 0} \frac{(h^2 - 8h + 16) - 16}{h}$

$$= \lim_{h \to 0} \frac{h^2 - 8h}{h} = \lim_{h \to 0} (h - 8) = -8.$$

15. $\displaystyle\lim_{x \to 2^+} \sqrt{x - 2} = 0.$

17. $\displaystyle\lim_{x \to 7^-} \frac{x}{x - 7} = -\infty$, since $\displaystyle\lim_{x \to 7^-} x = 7$, $\displaystyle\lim_{x \to 7^-} (x - 7) = 0$, and $\dfrac{x}{x - 7} < 0$ if $0 < x < 7$.

19. $\lim\limits_{x \to 5^-} \dfrac{1}{(x-5)^3} = -\infty$, since $\lim\limits_{x \to 5^-}(x-5)^3 = 0$ and $(x-5)^3 < 0$ if $x < 5$.

21. $\lim\limits_{x \to \infty} \dfrac{8x^4 + 2x}{2x^4 + 1} = \lim\limits_{x \to \infty} \dfrac{8x^4}{2x^4} = \lim\limits_{x \to \infty} 4 = 4$.

23. $\lim\limits_{x \to \infty} \dfrac{x^6 - x^2}{x^8 + x^4 + 1} = \lim\limits_{x \to \infty} \dfrac{x^6}{x^8} = \lim\limits_{x \to \infty} \dfrac{1}{x^2} = 0$.

25. Continuous at all x. In particular, at $x = 1$, $\lim_{x \to 1^-} f(x) = \lim_{x \to 1^-}(1 - x^2) = 0$ and $\lim_{x \to 1^+} f(x) = \lim_{x \to 1^+}(x - 1) = 0$. Therefore, $\lim_{x \to 1} f(x) = 0 = f(1)$.

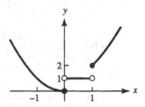

27. Discontinuous at $x = 0$ and $x = 1$. At $x = 0$, $\lim_{x \to 0^-} f(x) = \lim_{x \to 0^-} x^2 = 0$ and $\lim_{x \to 0^+} f(x) = \lim_{x \to 0^+} 1 = 1$, so that $\lim_{x \to 0} f(x)$ does not exist. At $x = 1$, $\lim_{x \to 1^-} f(x) = \lim_{x \to 1^-} 1 = 1$ and $\lim_{x \to 1^+} f(x) = \lim_{x \to 1^+}(x^2 + 1) = 2$, so that $\lim_{x \to 1} f(x)$ does not exist. The function is continuous at all other x.

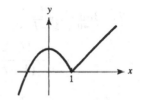

29. (d) Reasons: $f(x) = \frac{3x}{x^2-1}$ is an odd function, so its graph is symmetric about the origin. Also, the graph has vertical asymptotes $x = 1$ and $x = -1$, and horizontal asymtptote $y = 0$. Finally, $f(x) > 0$ on the intervals $(-1, 0)$ and $(1, \infty)$, and $f(x) < 0$ on $(-\infty, -1)$ and $(0, 1)$. The only graph with those properties is (d).

31. (a) Reasons: $f(x) = \frac{x^2}{x^2-1}$ is an even function, so its graph is symmetric about the y-axis. The graph has vertical asymptotes $x = 1$ and $x = -1$ and horizontal asymptote $y = 1$. Finally, $f(x) > 0$ on the intervals $(-\infty, -1)$ and $(1, \infty)$, and $f(x) < 0$ on $(-1, 1)$. The only graph with those properties is (a).

33. No vertical asymptote (the denominator is positive for all x). Horizontal asymptote is $y = 1$, since $\lim_{x \to \pm\infty} \frac{x^2-1}{x^2+1} = 1$.

35. Vertical asymptotes at $x = 10$ and $x = -10$, since $\lim_{x \to \pm 10} x^4 = 10{,}000$ and $\lim_{x \to \pm 10}(x^2 - 100) = 0$. There is no horizontal asymptote, since $\lim_{x \to \pm\infty} \frac{x^4}{x^2-100} = \lim_{x \to \pm\infty} x^2 = \infty$.

37. The graph of $f(x) = \frac{3x^4 + 15x^3 + 10x^2 + 5x + 20}{x^4+1}$ has no vertical asymptote, since the denominator is positive for all x. The horizontal asymptote is $y = 3$.

39. The graph of $f(x) = \frac{3x^2}{x^2+1}$ has no vertical asymptotes, since the denominator is always positive. In fact, $f(x) > 0$ for all x. The line $y = 3$ is a horizontal asymptote, and the graph approaches it from below as $x \to \pm\infty$. That is because $x^2 + 1 > x^2$, so that $3x^2/(x^2 + 1) < 3x^2/x^2 = 3$ for all x.

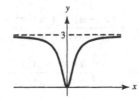

CHAPTER 1: PRACTICE EXAM

1. $\lim\limits_{x \to 1} \dfrac{x^2 - 1}{x - 1} = \lim\limits_{x \to 1} \dfrac{(x-1)(x+1)}{x - 1} = \lim\limits_{x \to 1} (x+1) = 2$

3. $\lim\limits_{x \to 0} \dfrac{x^2 + 1}{x - 1} = \dfrac{0^2 + 1}{0 - 1} = -1.$

5. Since $f(x) = \frac{x(x+1)(x+2)}{(x+3)(x-2)}$, we see that the numerator and denominator have no common factor, and, therefore, $x = -3$ and $x = 2$ are vertical asymptotes.

7. $\lim\limits_{h \to 0} \dfrac{h}{\sqrt{4+h} - 2} = \lim\limits_{h \to 0} \dfrac{h}{\sqrt{4+h} - 2} \cdot \dfrac{\sqrt{4+h} + 2}{\sqrt{4+h} + 2}$

$= \lim\limits_{h \to 0} \dfrac{h(\sqrt{4+h} + 2)}{(4+h) - 4} = \lim\limits_{h \to 0} \dfrac{h(\sqrt{4+h} + 2)}{h} = \lim\limits_{h \to 0} (\sqrt{4+h} + 2) = 4.$

9. (a) $\lim\limits_{x \to 2^-} \dfrac{2x + 3}{x - 2} = -\infty$, since $\lim\limits_{x \to 2^-} (2x + 3) = 7$, and for $-3/2 < x < 2$ we have $2x + 3 > 0$ and $x - 2 < 0$

(b) $\lim\limits_{x \to 2^+} \dfrac{2x + 3}{x - 2} = \infty$, since $\lim\limits_{x \to 2^+} (2x + 3) = 7$ and $\lim\limits_{x \to 2^+} (x - 2) = 0$, and for $x > 2$ we have $2x + 3 > 0$ and $x - 2 > 0$.

(c) $\lim\limits_{x \to \infty} \dfrac{2x + 3}{x - 2} = \lim\limits_{x \to \infty} \dfrac{2x}{x} = \lim\limits_{x \to \infty} 2 = 2.$

11. The vertical asymptotes are $x = 3$ and $x = -3$, and $y = 5$ is a horizontal asymptote.

13. The vertical asymptote is $x = -1$. There is no horizontal asymptote, since $\lim_{x \to \infty} x^2/(x+1) = \infty$ and $\lim_{x \to -\infty} x^2/(x + 1) = -\infty$

15. At $x = 0$, $\lim_{x \to 0^-} f(x) = \lim_{x \to 0^-} (x - 1) = -1$ and $\lim_{x \to 0^+} f(x) = \lim_{x \to 0^+} (x + 1) = 1$, which means that $\lim_{x \to 0} f(x)$ does not exist. Therefore, the function is discontinuous at $x = 0$

The function is continuous at all other x. In particular, at $x = 1$, we have $f(1) = 1 + 1 = 2$, and also $\lim_{x \to 1^-} f(x) = \lim_{x \to 1^-} (x + 1) = 2$, and $\lim_{x \to 1^+} f(x) = \lim_{x \to 1^+} 2x = 2$, so that $\lim_{x \to 1} f(x) = 2$. And at $x = 2$, we have $f(2) = 4$, and $\lim_{x \to 2^-} f(x) = \lim_{x \to 2^-} 2x = 4$ and $\lim_{x \to 2^+} f(x) = \lim_{x \to 2^+} x^2 = 4$, so that $\lim_{x \to 2} f(x) = 4$.

17. By the existence of zeros theorem (or the intermediate value theorem), $f(x)$ has at least three zeros in $(-2, 1)$, located in the intervals $(-2, -1.5), (-0.5, 0)$, and $(0.5, 1)$, respectively.

19. If $a = 0$, then $\lim\limits_{x \to a} \dfrac{x^4 - a^4}{x^3 - a^3} = \lim\limits_{x \to a} \dfrac{x^4}{x^3} = \lim\limits_{x \to a} x = a = 0 = \dfrac{4}{3} a.$

Next, assume $a \neq 0$. We use two identities from algebra: First,

$$x^4 - a^4 = (x^2 - a^2)(x^2 + a^2) = (x - a)(x + a)(x^2 + a^2).$$

Second, $x^3 - a^3 = (x - a)(x^2 + ax + a^2)$. Therefore,

$$\lim\limits_{x \to a} \dfrac{x^4 - a^4}{x^3 - a^3} = \lim\limits_{x \to a} \dfrac{(x - a)(x + a)(x^2 + a^2)}{(x - a)(x^2 + ax + a^2)}$$

$$= \lim\limits_{x \to a} \dfrac{(x + a)(x^2 + a^2)}{x^2 + ax + a^2} = \dfrac{2a \cdot 2a^2}{a^2 + a^2 + a^2} = \dfrac{4}{3} a.$$

CHAPTER 2
Exponentials and Logarithms

EXERCISE SET 2.1

1.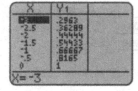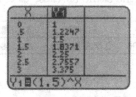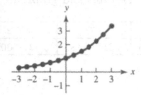

3.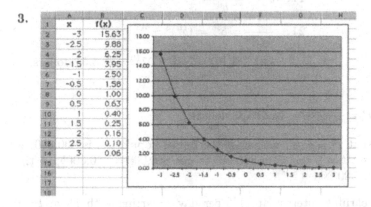

5. Apply $(\frac{2}{3})^x = (\frac{3}{2})^{-x}$ to the table of exercise 1 to get the table below (with entries rounded to 2 decimal places). The graph is obtained by reflecting the graph of exercise 1 in the y-axis.

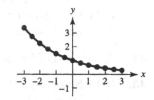

x	-3	-2.5	-2	-1.5	-1	-0.5	0	0.5	1	1.5	2	2.5	3
$f(x)$	3.38	2.76	2.25	1.84	1.5	1.22	1.0	0.82	0.67	0.54	0.44	0.36	0.30

7. Apply $(\frac{5}{2})^x = (\frac{2}{5})^{-x} = (0.4)^{-x}$ to the table of exercise 3 to get the table below (with entries rounded off). The graph is obtained by reflecting the graph of exercise 3 in the y-axis.

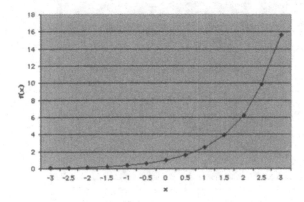

x	-3	-2.5	-2	-1.5	-1	-0.5	0	0.5	1	1.5	2	2.5	3
$f(x)$	0.06	0.10	0.16	0.25	0.40	0.63	1.0	1.58	2.50	3.95	6.25	9.88	15.63

9. Since $\left(\frac{2}{3}\right)^{-x} = \left(\frac{3}{2}\right)^{x}$, the table and graph are the same as those of exercise 1.

11. Since $\left(\frac{5}{3}\right)^{-x} = \left(\frac{3}{5}\right)^{x}$, the table and graph are the same as those of exercise 2.

13.

r	1	1.7	1.73	1.732	1.73205	1.7320508
2^r	2	3.249009585	3.317278183	3.321880096	3.321995226	3.321997068

$2^{\sqrt{3}} = 3.321997085\ldots$

15. $b^{2u} = (b^u)^2 = 3^2 = 9.$

17. $b^{v-u} = \dfrac{b^v}{b^u} = \dfrac{4}{3}.$

19. $b^0 = 1.$

21. $\dfrac{1}{b^{-3u}} = b^{3u} = (b^u)^3 = 3^3 = 27.$

23. (a) $A(t) = 500(1 + 0.07)^t$

(b) $A(10) = 500 \cdot 1.07^{10} = 983.58$ dollars.

25. $(1.06)^7 \approx 1.5,\ (1.06)^{12} \approx 2$

27. Set $50,000 = P(1 + 0.04)^{20}$. Then $P = \frac{50,000}{(1.04)^{20}} = 22,819.35$ dollars.

29. The exponential growth equation that $P = P_0 b^t$, and since the initial population is 500,000, we have $P = 500,000 b^t$. To find b we set $t = 1$ and $P = 800,000$ to get $800,000 = 500,000b$, so that $b = 8/5$. The desired formula is $P = 500,000 \cdot \left(\frac{8}{5}\right)^t$.

31. The population grows like money earning interest at 11% per day. Starting with P_0 at $t = 0$, there are $P = P_0(1 + 0.11)^t$ ants at the end of t days. Since $P = 580$ when $t = 5$, we have $580 = P_0(1.11)^5$. Solving for P_0 gives us $P_0 = 580/(1.11)^5 \approx 344$, the number at the beginning of the first day. The beginning of the second day is the end of the first day, with $P(1) = 344 \cdot 1.11 \approx 382$.

33. At the end of 5 years, $y(5) = y_0(0.88)^5 \approx 0.5277 y_0$. That is approximately 52.77% of the original y_0.

35. (a) Since y_0 is the initial amount, we have $y_0 = 50$. Next, setting $t = 10$ and $y = 30$, we get $30 = y_0 b^{10} = 50b^{10}$. Solving for b gives $b = \left(\frac{3}{5}\right)^{1/10} \approx 0.95$.

(b) The decay equation now reads $y = 50\left(\frac{3}{5}\right)^{t/10}$. Setting $t = 20$ gives $y = 50\left(\frac{3}{5}\right)^2 = 18$ mg.

(c) Replacing 50 by 80 changes the decay equation to the form $y = 80\left(\frac{3}{5}\right)^{t/10}$. Setting $t = 15$ gives $y = 80\left(\frac{3}{5}\right)^{3/2} \approx 37.18$ mg.

37.

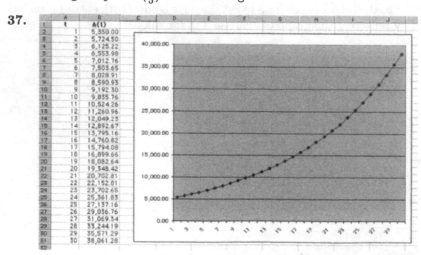

39.

| B2 | | =6.16*(1.010145)^(A2-2000) |

	A	B	C	D
1	Year	Population (in billions)		
2	2001	6.222		
3	2002	6.286		
4	2003	6.349		
5	2004	6.414		
6	2005	6.479		
7	2006	6.545		
8	2007	6.611		
9	2008	6.678		
10	2009	6.746		
11	2010	6.814		
12	2011	6.883		
13	2012	6.953		
14	2013	7.024		
15	2014	7.095		
16	2015	7.167		
17	2016	7.240		
18	2017	7.313		
19	2018	7.387		
20	2019	7.462		
21	2020	7.538		

EXERCISES 2.2

1. (a) $A(1) = 500(1 + 0.08) = 540$ dollars

(b) $A(1) = 500(1 + \frac{0.08}{12})^{12} \approx 541.50$ dollars

(c) $A(1) = 500e^{0.08} \approx 541.64$ dollars.

3. (a) $100(1 + \frac{0.07}{365})^{7,300} \approx 405.47$ dollars

(b) $100e^{0.07 \cdot 20} \approx 405.52$ dollars

5. At 6% compounded quarterly, the amount increases by a factor of $(1 + \frac{0.06}{4})^4 \approx 1.061$ after one year. At 4% compounded six times per year the factor is $(1 + \frac{0.04}{6})^6 \approx 1.041$. Interest at 6% compounded quarterly will provide a greater return after one year.

7. (a) $1,000e^{0.075/2} \approx 1,038.21$ dollars

(b) $1,000e^{2(0.075)/3} \approx 1,051.27$ dollars

(c) $1,000e^{9(0.075)/4} \approx 1,183.82$ dollars.

9. (a) $10,000e^{0.045/2} \approx 10,227.55$ dollars

(b) $10,000e^{3(0.045)/2} \approx 10,698.30$ dollars

(c) $10,000e^{3(0.045)} \approx 11,445.37$ dollars

11. With $A = 2,000$ and $t = 1/2$, we have $2,000 = Pe^{0.12/2}$. Therefore, $P = 2,000e^{-0.06} \approx 1,883.53$ dollars.

In exercises 13–21, we apply the basic formula $\lim\limits_{n \to \infty} \left(1 + \frac{r}{n}\right)^n = e^r$.

13. Setting $r = 3$, we have $\lim_{n \to \infty}(1 + \frac{3}{n})^n = e^3 \approx 20.0855$. Another way to obtain this result is to use the formula $e = \lim_{n \to \infty}(1 + \frac{1}{n})^n$, as follows. First,

$$\left(1 + \frac{3}{n}\right)^n = \left(\left(1 + \frac{1}{n/3}\right)^{n/3}\right)^3.$$

Second, if $m = n/3$, then $m \to \infty \iff n \to \infty$. Therefore,

$$\lim_{n \to \infty}\left(1 + \frac{3}{n}\right)^n = \lim_{n \to \infty}\left[\left(1 + \frac{1}{n/3}\right)^{n/3}\right]^3$$

$$= \lim_{m \to \infty}\left[\left(1 + \frac{1}{m}\right)^m\right]^3 = \left[\lim_{m \to \infty}\left(1 + \frac{1}{m}\right)^m\right]^3 = e^3.$$

15. Taking $r = -2$, we have $\displaystyle\lim_{n\to\infty}\left(1 - \frac{2}{n}\right)^n = \lim_{n\to\infty}\left(1 + \frac{-2}{n}\right)^n = e^{-2} \approx 0.1353$.

17. Taking $r = 3/4$, we have $\displaystyle\lim_{n\to\infty}\left(1 + \frac{3}{4n}\right)^n = \lim_{n\to\infty}\left(1 + \frac{3/4}{n}\right)^n = e^{3/4} \approx 2.1170$.

19. $\displaystyle\lim_{n\to\infty} 10\left(1 - \frac{1}{n}\right)^{2n} = 10\lim_{n\to\infty}\left(1 + \frac{-1}{n}\right)^{2n} = 10\lim_{n\to\infty}\left[\left(1 + \frac{-1}{n}\right)^n\right]^2$

$$= 10\left[\lim_{n\to\infty}\left(1 + \frac{-1}{n}\right)^n\right]^2 = 10e^{-2} \approx 1.3534$$

21. $\displaystyle\lim_{n\to\infty} 1{,}000\left(1 + \frac{1}{n}\right)^{n/2} = 1{,}000\lim_{n\to\infty}\left[\left(1 + \frac{1}{n}\right)^n\right]^{1/2} = 1{,}000e^{1/2} \approx 1{,}648.7213$

23. The calculator gives $1/e = 0.367879441$. The numbers in the table get close to that because $\displaystyle\lim_{n\to\infty}(1 - 1/n)^n = e^{-1}$.

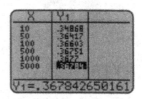

25.

$\displaystyle\lim_{n\to\infty}(1 + 3/n)^{-n/2} = e^{-3/2} \approx 0.22313016$

X	Y1
4000	.22326
5000	.22323
6000	.22321
7000	.2232
8000	.22319
9000	.22318
10000	.22318

Y1=.223180360042

27.

	File Edit View Insert Format Tools Data Window Help										9:34:19		
M2		=100*(1.003)*(L2)											

Workbook1

	A	B	C	D	E	F	G	H	I	J	K	L	M
1	Month	1	2	3	4	5	6	7	8	9	10	11	12
2	Balance	100.00	200.30	300.90	401.80	503.01	604.52	706.33	808.45	910.88	1013.61	1116.65	1220.00
3													
4	Month	13	14	15	16	17	18	19	20	21	22	23	24
5	Balance	1323.66	1427.63	1531.91	1636.51	1741.42	1846.64	1952.18	2058.04	2164.21	2270.71	2377.52	2484.65
6													
7	Month	25	26	27	28	29	30	31	32	33	34	35	36
8	Balance	2592.1	2699.88	2807.98	2916.4	3025.15	3134.23	3243.63	3353.36	3463.42	3573.81	3684.53	3795.59
9													

29.

```
e^(-X)-X=0
·X=.56714329040...
 bound={-1E99,1...
·left-rt=0
```

EXERCISES 2.3

1. $y = \log_2 8 \iff 2^y = 8 = 2^3 \iff y = 3$.
An alternate method: $\log_2 8 = \log_2 2^3 = 3\log_2 2 = 3 \cdot 1 = 3$.

3. $\log_3 1 = 0$. (Recall that $\log_b 1 = 0$ for *any* positive b except 1.)

5. $y = \log_9 27 \iff 9^y = 27 \iff y = 3/2$

7. $y = \log_{1/2} 2 \iff (\frac{1}{2})^y = 2 \iff y = -1$

9. $y = \log_{10}(0.001) \iff 10^y = 0.001 \iff y = -3$

11. $\log_7 7 = 1$ (Recall that $\log_b b = 1$ for *any* positive b except 1.)

13. $\log_2 \sqrt{8} = \log_2 2^{3/2} = \frac{3}{2}$.

15. $\log_{10} \frac{1}{\sqrt{10}} = \log_{10} 10^{-1/2} = -\frac{1}{2}$

17. (a) $\log_{10} 40 = \log_{10} 4 \cdot 10 = \log_{10} 4 + \log_{10} 10 \approx 0.602 + 1 = 1.602$

(b) $\log_{10} 0.4 = \log_{10} \frac{4}{10} = \log_{10} 4 - \log_{10} 10 \approx 0.602 - 1 = -0.398$

(c) $\log_{10} 0.25 = \log_{10} \frac{1}{4} = \log_{10} 1 - \log_{10} 4 = -\log_{10} 4 \approx -0.602$

(d) $\log_{10} 2 = \log_{10} 4^{1/2} = \frac{1}{2} \log_{10} 4 \approx \frac{1}{2} \cdot (0.602) = 0.301$

19. $\log_{10} x = 1.463 \iff x = 10^{1.463} \approx 29.04022654$

21. First, $\log_{10}(\frac{x}{100}) = \log_{10} x - \log_{10} 100 = \log_{10} x - 2$. Therefore,

$$\log_{10}\left(\frac{x}{100}\right) - 0.418 \iff \log_{10} x - 2.418 \iff x - 10^{2.418} \approx 261.8183008.$$

23. First, $100^x = (10^2)^x = 10^{2x}$. Therefore,

$$100^x = 204.5 \iff 2x = \log_{10} 204.5 \iff x = \frac{1}{2} \log_{10} 204.5 \approx 1.155346656.$$

25. $\log_2 9 = 2 \log_2 3 \approx 2 \cdot (1.585) = 3.170$

27. $\log_2 \frac{5}{9} = \log_2 5 - 2 \log_2 3 \approx 2.322 - 3.17 = -0.848$

29. $\log_2 6 = \log_2 2 + \log_2 3 \approx 1 + 1.585 = 2.585$

31. $\log_2 \sqrt{10} = \frac{1}{2} \log_2(5 \cdot 2) = \frac{1}{2}(\log_2 5 + \log_2 2) \approx \frac{1}{2}(2.322 + 1) = 1.661$

33. $\log_2 30 = \log_2(3 \cdot 5 \cdot 2) = \log_2 3 + \log_2 5 + \log_2 2 \approx 1.585 + 2.322 + 1 = 4.907$

35. $\log_2 \sqrt{15} = \frac{1}{2} \log_2(3 \cdot 5) = \frac{1}{2}(\log_2 3 + \log_2 5) \approx \frac{1}{2}(1.585 + 2.322) = 1.9535$

37. $\log_3(9A) = \log_3 9 + \log_3 A = 2 + b$ **39.** $\log_3 \sqrt{3A} = \frac{1}{2}(\log_3 3 + \log_3 A) = \frac{1}{2}(1 + b)$

41. First, $\log_9 A = v \iff 9^v = A \iff (3^2)^v = A \iff 3^{2v} = A$, so that $\log_9 A = v \iff 2v = \log_3 A = b$. Therefore, $\log_9 A = b/2$.

43. First, $\log_{\sqrt{3}} A = v \iff (\sqrt{3})^v = A \iff (3^{1/2})^v = A \iff 3^{v/2} = A$, which means that $\log_{\sqrt{3}} A = v \iff v/2 = \log_3 A = b \iff v = 2b$. Therefore, $\log_{\sqrt{3}} A = 2b$.

45. $\log_{\sqrt{A}} 3 = v \iff (\sqrt{A})^v = 3 \iff A^{v/2} = 3 \iff 3^{2/v} = A$, which means that $\log_{\sqrt{A}} 3 = v \iff 2/v = \log_3 A = b \iff v = 2/b$. Therefore, $\log_{\sqrt{A}} 3 = 2/b$.

47. Write x for the seismic amplitude of the earthquake registering 6.2 on the Richter scale. Then $\log_{10}(x/A) = 6.2$, so that

$$\log_{10}\frac{4x}{A} = \log_{10} 4 + \log_{10}\frac{x}{A} \approx 0.602 + 6.2 = 6.802,$$

which is the Richter value of an earthquake whose seismic amplitude is four times as big.

49. Write x_N for the amplitude of the sound wave of normal conversation and x_H for that of the headset. The decibel ratings tell us that

$$10\log_{10}\frac{x_H}{I} = 115 \quad\text{and}\quad 10\log_{10}\frac{x_N}{I} = 60,$$

where I is the reference amplitude. These equations are equivalent to

$$\frac{x_H}{I} = 10^{11.5} \quad\text{and}\quad \frac{x_N}{I} = 10^6.$$

Dividing the first by the second gives $x_H/x_N = 10^{11.5}/10^6 = 10^{5.5} \approx 316{,}228$, which is the factor by which the headset's sound wave exceeds that of normal conversation.

51. Let x be the intensity (i.e., sound wave amplitude) at fifty feet and $300x$ the intensity at the operator's ear. The decibel rating of 75 at fifty feet means that $10\log_{10}(x/I) = 75$. Therefore,

$$10\log_{10}\frac{300x}{I} = 10\left(\log_{10} 300 + \log_{10}\frac{x}{I}\right) \approx 10(2.477 + 7.5) = 99.77,$$

which is the decibel rating at the operator's ear.

53. Let $[H_a^+]$ be the concentration of hydrogen ions in the acid rain tested, and let $[H_n^+]$ the concentration in normal rain. We have $-\log_{10}[H_a^+] = 3.6$ and $-\log_{10}[H_n^+] = 5.6$, so that $[H_a^+] = 10^{-3.6}$ and $[H_n^+] = 10^{-5.6}$. Therefore, $[H_a^+]/[H_n^+] = 10^{-3.6}/10^{-5.6} = 10^2 = 100$. The acid rain tested was 100 times more acidic than normal rain.

EXERCISES 2.4

1. $3^7 = e^{7\ln 3}$

3. $x^5 = e^{5\ln x}$

5. $a^{-3} = e^{-3\ln a}$

7. $a^{1/3} = e^{(1/3)\ln a}$

9. $\ln\left(\frac{1}{\sqrt{e}}\right) = -\ln\sqrt{e} = -\frac{1}{2}\ln e = -\frac{1}{2}$

11. $e^{\ln 5} = 5$

13. $\ln\left(\frac{1}{e^{0.4}}\right) = -\ln e^{0.4} = -0.4$

15. $e^{2\ln 3} = e^{\ln 9} = 9$, or $e^{2\ln 3} = (e^{\ln 3})^2 = 3^2 = 9$

17. $\ln(2e) = \ln 2 + \ln e = \ln 2 + 1 \approx 1.693$

19. $\ln\frac{1}{4} = -\ln 4 = -\ln(2^2) = -2\ln 2 \approx -1.386$

21. $\ln(e^{1+\ln 4}) = 1 + \ln 4 = 1 + 2\ln 2 \approx 2.386$

23. $e^{\frac{1}{2}\ln 4} = e^{\ln\sqrt{4}} = e^{\ln 2} = 2$

25. $e^{\ln 3 + 3\ln t} = e^{\ln 3}\cdot e^{3\ln t} = 3t^3$

27. $\ln\left(\frac{1}{e^{3x}}\right) = -\ln e^{3x} = -3x$

29. $\ln(e^{1/x}) = 1/x$

31. $e^{2\ln(x-1)} = e^{\ln(x-1)^2} = (x-1)^2$

33. $e^{\ln(2x)+\ln(2/x)} = e^{\ln 2x} \cdot e^{\ln(2/x)} = 2x \cdot \dfrac{2}{x} = 4$

35. $e^{2x+1} = 7 \iff \ln e^{2x+1} = \ln 7 \iff 2x+1 = \ln 7 \iff x = \frac{1}{2}(\ln 7 - 1)$

37. $e^{x-2} = 1 \iff \ln e^{x-2} = \ln 1 \iff x - 2 = 0 \iff x = 2$

39. $\ln(x+1) = 1 \iff e^{\ln(x+1)} = e^1 \iff x+1 = e \iff x = e - 1$

41. $\ln(x^3) = 5 \iff 3\ln x = 5 \iff \ln x = 5/3 \iff x = e^{5/3}$

43. $e^{3\ln x} = 8 \iff x^3 = 8 \iff x = 2$

45. $\ln(2x-2) = 0 \iff 2x - 2 = 1 \iff x = 3/2$

47. $\ln(3x+5) = 2.47$

$3x + 5 = e^{2.47}$

$x = (e^{2.47} - 5)/3 \approx 2.27414895$

49. $e^{(1+\ln x)/2} = 8.371$

$(1 + \ln x)/2 = \ln 8.371$

$\ln x = 2 \cdot \ln 8.371 - 1$

$x = e^{2\ln 8.371 - 1} = 8.371^2 \cdot e^{-1} \approx 25.7786519$

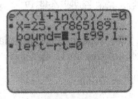

51. $e^{x+\ln x} = 3x$

$x + \ln x = \ln(3x) = \ln 3 + \ln x$

$x = \ln 3 \approx 1.09861229$

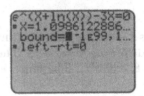

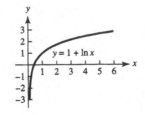

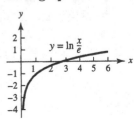

53. $\log_5 12 = \dfrac{\ln 12}{\ln 5} \approx 1.5439593$

55. $\log_2 10 = \dfrac{\ln 10}{\ln 2} \approx 3.3219281$

57. $\log_a b = \dfrac{\ln b}{\ln a}$ and $\log_b a = \dfrac{\ln a}{\ln b}$. Therefore, $\log_a b = \dfrac{1}{\log_b a}$.

59. Shift graph of $\ln x$ one unit upward.

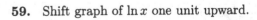

61. $\ln \dfrac{x}{e} = \ln x - 1$

Shift graph of $\ln x$ one unit downward.

63. The graph of $y = \ln(3e^{2x}) = \ln 3 + 2x$ matches graph (d). It is a line of slope 2.

65. The graph of $y = \ln(1/x^2) = -2\ln|x|$ matches graph (b). It is obtained by reflecting the graph of $y = \ln x$ in the x-axis, rescaling the height of every point by a factor of two, and adding a second branch symmetric about the y-axis.

67. The amount after t years is given by $y = Pe^{0.075t}$, where P is the initial amount. Then, $3P = Pe^{0.075t} \iff e^{0.075t} = 3 \iff t = \frac{1}{0.075}\ln 3 \approx 14.65$ years.

69. (a) Substituting each of the data pairs into the model gives $\begin{cases} 210 = Ae^{-k} \\ 86 = Ae^{-2k} \end{cases}$. By dividing the first equation by the second, we get $e^k = 210/86 = 105/43$, so that $k = \ln\frac{105}{43}$. Next, substituting that value in the first equation yields $210 = Ae^{-\ln(105/43)} = 43A/105$. Therefore, $A = 22{,}050/43$.

(b) Setting $A = 22{,}050/43$ and $k = \ln\frac{105}{43}$ in the basic equation gives

$$T = \frac{22{,}050}{43}e^{-t\ln(105/43)} = \frac{22{,}050}{43}\left(\frac{43}{105}\right)^t.$$

With $t = 9$, we get $\frac{22{,}050}{43}\left(\frac{43}{105}\right)^9 \approx 0.166$ second.

71. (a) First, $A = 4{,}000$ (the initial population). Next, setting $t = 5$ and $y = 6{,}000$ gives $6{,}000 = 4{,}000e^{5k}$, and solving for k gives $k = \frac{1}{5}\ln\frac{3}{2} \approx 0.0811$. Therefore, $y = 4{,}000e^{0.0811t}$.

(b) $4{,}000e^{(0.0811)\cdot 8} \approx 7653$

(c) Setting $y = 12{,}000$ gives $12{,}000 = 4{,}000e^{0.0811t}$, and solving for t yields $t = (\ln 3)/0.0811 \approx 13.55$ days.

73. First, $p(t) \to 8$ as $t \to \infty$, so that $8 = \lim_{t\to\infty} p(t) = \lim_{t\to\infty}(a - be^{-rt}) = a$. Next, setting $t = 0$ and $p(0) = 5$ yields $5 = a - b = 8 - b$. Therefore, $b = 3$. Finally, setting $t = 10$ and $p(10) = 6$ gives $6 = a - be^{-10r} = 8 - 3e^{-10r}$. Therefore, $e^{-10r} = 2/3$, and $r = (\ln\frac{3}{2})/10 \approx 0.04$.

75.

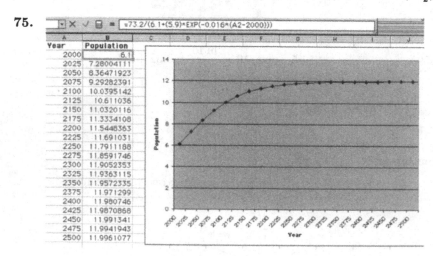

77.

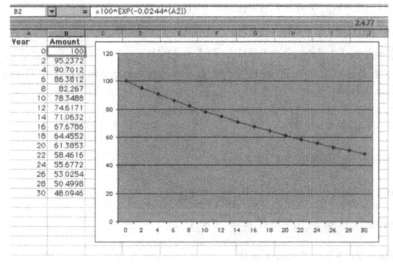

79. $67 = 100e^{-k5} \iff k = \frac{1}{5}\ln\frac{100}{67}$. Therefore, the decay equation is $y = 100e^{-(t/5)\ln(100/67)}$. With $t = 7$, we get $100e^{-(7/5)\ln(100/67)} = 100(\frac{67}{100})^{7/5} \approx 57.08$ mg.

CHAPTER 2: REVIEW EXERCISES

1. (a) $(\frac{1}{9})^{1/2} = \frac{1}{3}$ (b) $(\frac{1}{9})^{-2} = 81$ (c) $(0.1)^3 = 0.001$ (d) $(0.1)^{-3} = 1,000$

3. (a) The population is given by a formula of the form $P(t) = 600b^t$, with t in days. To determine b, we substitute $t = 3$ and $P(3) = 1,000$, which gives $1,000 = 600b^3$. Solving for b, we get $b = (5/3)^{1/3}$. Thus, $P(t) = 600(5/3)^{t/3}$.

   (b) $P(6) = 600(5/3)^2 \approx 1,667$

5. (a) $5,000(1 + \frac{0.06}{12})^{12} \approx 5,308.39$ dollars   (b) $5,000(1 + \frac{0.06}{4})^{12} \approx 5,978.09$ dollars

7. $100(1 + \frac{0.06}{n})^n$ dollars

9. (a) We solve $1,000 = P(1 + 0.05)^2$ to get $P = 1,000(1.05)^{-2} \approx 907.03$ dollars.

   (b) We solve $1,000 = P(1 + \frac{0.05}{6})^{12}$, to get $P = 1,000(1 + \frac{0.05}{6})^{-12} \approx 905.21$ dollars.

   (c) We solve $1,000 = Pe^{0.05 \cdot 2}$ to get $P = 1,000e^{-0.1} \approx 904.84$ dollars.

11. $\lim\limits_{n\to\infty}\left(1 - \frac{1}{n}\right)^{2n} = \lim\limits_{n\to\infty}\left(1 + \frac{-1}{n}\right)^{2n} = \left[\lim\limits_{n\to\infty}\left(1 + \frac{-1}{n}\right)^n\right]^2 = (e^{-1})^2 = e^{-2}$

13. Let $n = 1/h$. Then $h \to 0 \iff n \to \infty$ and (see exercise 11)

$$\lim_{h\to 0}(1 - h)^{2/h} = \lim_{n\to\infty}(1 - 1/n)^{2n} = e^{-2}.$$

15. **(a)** $\log_3(1/4) = -\log_3(2^2) = -2\log_3 2 \approx -2 \cdot 0.631 = -1.262$

(b) $\log_3 \sqrt{14} = \frac{1}{2}(\log_3 2 + \log_3 7) \approx \frac{1}{2}(0.631 + 1.771) = 1.201$

(c) $\log_3 18 = \log_3(2 \cdot 3^2) = \log_3 2 + 2\log_3 3 \approx 0.631 + 2 \cdot 1 = 2.631$

(d) $y = \log_9 2 \iff 9^y = 2 \iff (3^2)^y = 2 \iff 3^{2y} = 2 \iff 2y = \log_3 2 \approx 0.631$. Therefore, $\log_9 2 = y \approx 0.631/2 = 0.3155$.

(e) $y = \log_{1/3} 7 \iff (1/3)^y = 7 \iff 3^{-y} = 7 \iff -y = \log_3 7$
Therefore, $\log_{1/3} 7 = -\log_3 7 \approx -1.771$.

17. **(b)** The graph is positive and decreasing for all x.

19. **(b)** The graph is the same as that of $y = (1/2)^x$ in exercise 17.

21. $\ln(1 + \ln x) = 0 \iff 1 + \ln x = 1 \iff \ln x = 0 \iff x = 1$

23. $e^{\ln(2x+3)} = 2x + 3$. Therefore, $2x + 3 = 7$ and $x = 2$.

25. $\ln(e^x) = x$. Therefore, $x = 2x - 3$ and $x = 3$.

27. $\ln \sqrt{e} = \frac{1}{2}\ln e = 1/2$ **29.** $\sqrt{e^{\ln 4}} = \sqrt{4} = 2$

31. With t in hours past noon, we have $A = 10{,}000$ (the initial amount). Substituting $t = 2$ and $y = 40{,}000$, we get $40{,}000 = 10{,}000e^{2k}$, which means that $k = \frac{1}{2}\ln 4 = \ln 2$. Then $y = 10{,}000e^{t\ln 2}$, and $y(3) = 10{,}000e^{3 \cdot \ln 2} = 10{,}000 \cdot 2^3 = 80{,}000$.

33. The amount after t years is $y = 250e^{0.08t}$. Setting $y = 1{,}000$ and solving for t yields $t = (\ln 4)/0.08 \approx 17.33$ years.

35. **(a)** $\lim_{x \to \infty} e^{0.1x} = \infty$ **(b)** $\lim_{x \to \infty} e^{-0.1x} = 0$ **(c)** $\lim_{x \to -\infty} e^{0.1x} = 0$

(d) $\lim_{x \to -\infty} e^{-0.1x} = \infty$ **(e)** $\lim_{x \to \infty} \ln(2x) = \infty$ **(f)** $\lim_{x \to 0^+} \ln(2x) = -\infty$

37. Taking the natural logarithm of both sides gives $(x + 3)\ln 2 = \ln 31$. Therefore, $x = \frac{\ln 31}{\ln 2} - 3 \approx 1.9542$.

39. $0.1y \ln \frac{8}{y} - 0.01y = 0.1y(\ln \frac{8}{y} - 0.1) = 0$. If $y \neq 0$ then $\ln \frac{8}{y} - 0.1 = 0$ and $y = 8/e^{0.1}$. The equation has two solutions, $y = 0$ and $y = 8e^{-0.1}$.

41. We set $25{,}000 = 10{,}000e^{0.04t}$ and solve for t to get $t = \frac{1}{0.04}\ln 2.5 \approx 23.9$ years.

43. In this case, $A = 100$ (the initial amount). Then $25 = 100e^{-10k}$, so that $k = (\ln 4)/10 = \ln 2/5$. Therefore, at $t = 15$ we have $y = 100e^{-(15\ln 2)/5} = 100e^{-3\ln 2} = 100 \cdot 2^{-3} = 12.5$ grams.

45. Letting r denote the interest rate, we have $3P = Pe^{14r}$, so that $e^{14r} = 3$ and $r = (\ln 3)/14 \approx 0.0785$.

CHAPTER 2: EXAM

1. $1,000(1 + \frac{0.04}{12})^{36} \approx 1,127.27$ dollars.

3. $1,000e^{20 \cdot (0.065)} \approx 3,669.30$ dollars.

5. (a) $\displaystyle\lim_{n \to \infty} \left(1 + \frac{1}{4n}\right)^n = \lim_{n \to \infty} \left(1 + \frac{1/4}{n}\right)^n = e^{1/4}$

    (b) $\displaystyle\lim_{n \to \infty} \left(\frac{n-3}{n}\right)^{2n} = \left[\lim_{n \to \infty} \left(1 - \frac{3}{n}\right)^n\right]^2 = (e^{-3})^2 = e^{-6}.$

7. (a) $\log_3 \frac{1}{25} = -\log_3 25 = -2\log_3 5 \approx -2 \cdot 1.465 = -2.930$

    (b) $\log_3(45) = \log_3 9 + \log_3 5 = 2\log_3 3 + \log_3 5 \approx 2 + 1.465 = 3.465$

9. (a) $e^{3\ln(2x+1)} = e^{[\ln(2x+1)]^3} = (2x+1)^3$

    (b) $\ln(\frac{1}{2}e^{3x^2+4}) = \ln\frac{1}{2} + \ln(e^{3x^2+4}) = 3x^2 + 4 - \ln 2$

11. (a)   (b)   (c)

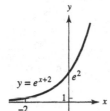

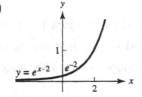

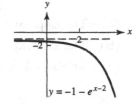

13. Rewriting the equation as $(e^x)^2 - 2e^x - 8 = 0$, we see that it is quadratic in e^x. By factoring, we obtain $(e^x - 4)(e^x + 2) = 0$ with solution $e^x = 4$, so that $x = \ln 4 = 2\ln 2$. Note that the second factor cannot be zero because $e^x > 0$.

15. (a) No vertical asymptote. Since $e^{-x} > 0$ for all x, we have $1 + 2e^{-x} > 0$. Thus, the denominator is positive for all x, which means the quotient is defined throughout the x-axis.
         Horizontal asymptote $y = 2$ (as $x \to \infty$). Since $\displaystyle\lim_{x \to \infty} e^{-x} = 0$, we have $\displaystyle\lim_{x \to \infty} f(x) = 2$.
         Horizontal asymptote $y = 0$ (as $x \to -\infty$). Since $\displaystyle\lim_{x \to -\infty} e^{-x} = \infty$, we have $\displaystyle\lim_{x \to -\infty} f(x) = 0$.

     (b) Vertical asymptote $x = \ln 2$. For $2e^{-x} - 1 = 0 \iff e^{-x} = 1/2 \iff e^x = 2 \iff x = \ln 2$.
         Horizontal asymptote $y = -2$ (as $x \to \infty$). Since $\displaystyle\lim_{x \to \infty} e^{-x} = 0$, we have $\displaystyle\lim_{x \to \infty} g(x) = -2$.
         Horizontal asymptote $y = 0$ (as $x \to -\infty$). Since $\displaystyle\lim_{x \to -\infty} e^{-x} = \infty$, we have $\displaystyle\lim_{x \to -\infty} g(x) = 0$.

17. Letting P be the initial amount, we have $4P = Pe^{0.07t}$. Thus, $e^{0.07t} = 4$ and $t = \frac{1}{0.07}\ln 4 \approx 19.8$ years.

19. Write r for the growth constant. The population at the end of t days is given by $y = 4,000e^{rt}$. In particular, $9,000 = 4,000e^{2r}$, and solving for r gives $r = \frac{1}{2}\ln\frac{9}{4} = \ln\sqrt{\frac{9}{4}} = \ln\frac{3}{2}$. Therefore, $y = 4,000e^{t\ln(3/2)}$, and setting $y = 64,000$ yields $16 = e^{t\ln(3/2)}$. Taking natural logarithms, we obtain $\ln 16 = t\ln\frac{3}{2}$, so that $t = (\ln 16)/(\ln\frac{3}{2}) \approx 6.838$ days.

CHAPTER 3
The Derivative

EXERCISES 3.1

1. The tangent line passes through the points $(0.5, 0)$ and $(1.5, 2)$, so the slope is $\frac{2-0}{1.5-0.5} = 2$.

3. C and D

5. The slope at B is 0 because the tangent is horizontal. The slope at D is 0 for the same reason. The slope at C is $\frac{1-1.75}{1.25-0.25} = -0.75$ because the tangent passes through $(1.25, 1)$ and $(0.25, 1.75)$

7. C, B, A

In exercise 9 we use the fact that the slope of the graph of $y = x^2$ at (x, y) is $2x$.

9. **(a)** The slope at $(\frac{1}{2}, \frac{1}{4})$ is 1, and the tangent has equation $y = x - \frac{1}{4}$.
   (b) The slope at $(0, 0)$ is 0, and the tangent has equation $y = 0$.
   (c) At $x = -3$, $y = 9$ and the slope is -6. The tangent has equation $y = -6x - 9$.

In exercises 11–17 we use the fact that the slope of the graph of $y = x^3$ at (x, y) equals $3x^2$.

11. The slope at $(-2, -8)$ is $3 \cdot (-2)^2 = 12$, and the equation of the tangent line is $y = 12x + 16$.

13. If $x = \frac{1}{2}$, then the slope is $\frac{3}{4}$ and $y = \frac{1}{8}$. Therefore, the equation of the tangent line is $y = \frac{3}{4}x - \frac{1}{4}$.

15. Set $3x^2 = 3$. Then $x = \pm 1$, and the points are $(1, 1)$ and $(-1, -1)$.

17. Set $3x^2 = 0$. Then $x = 0$ and the point is $(0, 0)$.

In exercises 19–23 we use the fact that the slope of the graph of $y = \sqrt{x}$ at (x, y) equals $1/(2\sqrt{x})$.

19. The slope at $(9, 3)$ is $1/(2\sqrt{9}) = 1/6$ and the equation of the tangent line is $y = \frac{1}{6}x + \frac{3}{2}$.

21. Set $1/(2\sqrt{x}) = 2$. Then $x = \frac{1}{16}$, and the point is $(\frac{1}{16}, \frac{1}{4})$.

23. Since $1/(2\sqrt{x}) = 0$ has no solution, there is no point on the graph where the tangent line is horizontal.

25. **(a)**
$$\frac{f(1+h) - f(1)}{h} = \frac{[(1+h)^2 + (1+h)] - [1^2 + 1]}{h}$$
$$= \frac{(h^2 + 3h + 2) - 2}{h} = \frac{h(h+3)}{h} = h + 3.$$

   (b) $\displaystyle \lim_{h \to 0} \frac{f(1+h) - f(1)}{h} = \lim_{h \to 0} (h + 3) = 3.$

(c) $\lim\limits_{h \to 0} \dfrac{f(x+h) - f(x)}{h} = \lim\limits_{h \to 0} \dfrac{[(x+h)^2 + (x+h)] - (x^2 + x)}{h}$

$\qquad\qquad\qquad = \lim\limits_{h \to 0} \dfrac{(x^2 + 2xh + h^2 + x + h) - (x^2 + x)}{h}$

$\qquad\qquad\qquad = \lim\limits_{h \to 0} \dfrac{h(2x + h + 1)}{h} = \lim\limits_{h \to 0} (2x + h + 1) = 2x + 1.$

(d) **(i)** The slope at $(1,2)$ is 3, and the equation of the tangent line is $y = 3x - 1$.

   (ii) The slope at $(-2,2)$ is -3, and the equation of the tangent line is $y = -3x - 4$.

   (iii) The slope at $(-1,0)$ is -1, and the equation of the tangent line is $y = -x - 1$.

27. (a) $\lim\limits_{h \to 0} \dfrac{f(x+h) - f(x)}{h} = \lim\limits_{h \to 0} \dfrac{[(x+h)^2 + 3(x+h) - 1] - [x^2 + 3x - 1]}{h}$

$\qquad\qquad\qquad = \lim\limits_{h \to 0} \dfrac{(x^2 + 2xh + h^2 + 3x + 3h - 1) - (x^2 + 3x - 1)}{h}$

$\qquad\qquad\qquad = \lim\limits_{h \to 0} \dfrac{h(2x + h + 3)}{h} = \lim\limits_{h \to 0} (2x + h + 3) = 2x + 3.$

(b) The slope at $(1,3)$ is 5.

(c) The slope at $x = -2$ is -1, and $y = -3$. Therefore, the equation of the tangent line is $y = -x - 5$.

29. (a) $\lim\limits_{h \to 0} \dfrac{f(x+h) - f(x)}{h} = \lim\limits_{h \to 0} \dfrac{\sqrt{x+h+4} - \sqrt{x+4}}{h}$

$\qquad\qquad\qquad = \lim\limits_{h \to 0} \dfrac{\sqrt{x+h+4} - \sqrt{x+4}}{h} \cdot \dfrac{\sqrt{x+h+4} + \sqrt{x+4}}{\sqrt{x+h+4} + \sqrt{x+4}}$

$\qquad\qquad\qquad = \lim\limits_{h \to 0} \dfrac{(x+h+4) - (x+4)}{h(\sqrt{x+h+4} + \sqrt{x+4})}$

$\qquad\qquad\qquad = \lim\limits_{h \to 0} \dfrac{h}{h\left(\sqrt{x+h+4} + \sqrt{x+4}\right)}$

$\qquad\qquad\qquad = \lim\limits_{h \to 0} \dfrac{1}{\left(\sqrt{x+h+4} + \sqrt{x+4}\right)} = \dfrac{1}{2\sqrt{x+4}}$

(b) **(i)** The slope at $(0,2)$ is $\frac{1}{2\sqrt{0+4}} = \frac{1}{4}$, and the equation of the tangent line is $y = \frac{1}{4}x + 2$.

   (ii) The slope at $(-3,1)$ is $\frac{1}{2}$, and the equation of the tangent line is $y = \frac{1}{2}x + \frac{5}{2}$.

   (iii) The slope at $(5,3)$ is $\frac{1}{6}$, and the equation of the tangent line is $y = \frac{1}{6}x + \frac{13}{6}$.

31. (a) $\lim\limits_{h \to 0} \dfrac{f(x+h) - f(x)}{h} = \lim\limits_{h \to 0} \dfrac{[(x+h)^3 - 2(x+h) + 1] - [x^3 - 2x + 1]}{h}$

$\qquad\qquad\qquad = \lim\limits_{h \to 0} \dfrac{(x^3 + 3x^2 h + 3xh^2 + h^3 - 2x - 2h + 1) - (x^3 - 2x + 1)}{h}$

$\qquad\qquad\qquad = \lim\limits_{h \to 0} \dfrac{h(3x^2 + 3xh + h^2 - 2)}{h} = \lim\limits_{h \to 0} (3x^2 + 3xh + h^2 - 2) = 3x^2 - 2.$

(b) The slope at $(0,1)$ is -2, and the equation of the tangent line is $y = -2x + 1$.

33. (a) $\displaystyle\lim_{h\to 0}\frac{L(t+h)-L(t)}{h}=\lim_{h\to 0}\frac{0.01(t+h)^2-0.01t^2}{h}$

$$=\lim_{h\to 0}\frac{0.01(t^2+2th+h^2)-0.01t^2}{h}=\lim_{h\to 0}\frac{0.02th+0.01h^2}{h}$$

$$=\lim_{h\to 0}\frac{h(0.02t+0.01h)}{h}=\lim_{h\to 0}(0.02t+0.01h)=0.02t$$

(b) Setting $0.02t=0.08$, we get $t=4$. The astronaut is losing bone matter at the rate of 0.08% per month at the end of 4 months.

35. (a) $\displaystyle\lim_{h\to 0}\frac{f(t+h)-f(t)}{h}=\lim_{h\to 0}\frac{12(t+h)^3-12t^3}{h}$

$$=\lim_{h\to 0}\frac{12(t^3+3t^2h+3th^2+h^3)-12t^3}{h}=\lim_{h\to 0}\frac{36t^2h+36th^2+12h^3}{h}$$

$$=\lim_{h\to 0}(36t^2+36th+12h^2)=36t^2.$$

(b) $36\cdot 5^2=900$

(c) The rate of grape production is equal to $36t^2=12(3t^2)=12\cdot(\text{ the rate of wine production})$.

37. (a) The slope of AB is $1/1=1$, which is the average rate of change of sales per year (in millions per year per year) over $[2,3]$. The slope of AC is $2/3$, which is the average rate of sales per year (in millions per year) over $[2,5]$. The slope of the graph at A is approximately $\frac{6-1.5}{3.5-0}=\frac{9}{7}$, which is the instantaneous rate of change (in millions per year per year) at $t=2$.

(b) The slope of ED is $1.5/(-1)=-1.5$, which is the average rate of change of sales per year (in millions per year per year) over $[6,7]$. The slope of EF is $(-1)/1=-1$, which is the average rate of change of sales per year (in millions per year per year) over $[7,8]$. The slope of the graph at E is approximately $\frac{1-4}{9-7}=-1.5$, which is the instantaneous rate of change (in millions per year per year) at $t=7$.

(c) The sales per year are increasing fastest at $t=2$ and decreasing fastest at $t=7$. The sales change from increasing to decreasing at $t=5$, and the rate of change is zero.

39. (a) $R(i)=\begin{cases}0.05i & 0\le i\le 6,000,\\ 300 & i>6,000\end{cases}$

(b) The slope at $i=4,000$ is 0.05, and the slope at $i=8,000$ is 0.

(c) The possible values for the slope are 0.05 and 0. The positive slope of 0.05 means that R is increasing as i gets large but the slope of 0 means that R is equal to a constant, 300 dollars.

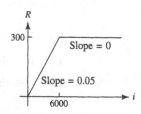

41. (a)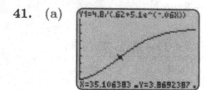

(b)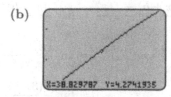

(c) $\lim_{t\to\infty}p=\frac{4.8}{0.62}\approx 7.742$. The population asymptotically approaches an upper bound of about 7,742. The slope approaches zero.

EXERCISES 3.2

1. $f'(x) = \lim\limits_{h \to 0} \dfrac{\frac{1}{3(x+h)} - \frac{1}{3x}}{h} = \lim\limits_{h \to 0} \dfrac{-1}{3x(x+h)} = -\dfrac{1}{3x^2}$

3. $f'(x) = \lim\limits_{h \to 0} \dfrac{\frac{1}{2-(x+h)} - \frac{1}{2-x}}{h} = \lim\limits_{h \to 0} \dfrac{1}{(2-x)[2-(x+h)]} = \dfrac{1}{(2-x)^2}$

5. $f'(x) = \dfrac{5}{3}x^{2/3}$ **7.** $f'(x) = 1.4x^{0.4}$

9. $f'(x) = \dfrac{d}{dx}x^{-1/2} = -\dfrac{1}{2}x^{-3/2} = -\dfrac{1}{2x^{3/2}}$ **11.** $f'(x) = -4x^{-5}$

13. $f'(x) = -\dfrac{1}{4}x^{-5/4} = -\dfrac{1}{4x^{5/4}}$

15. Since the derivative is the slope of the tangent line, $f'(a) = -1$.

17. Using the central difference formula,

$$f'(0.7) \approx \frac{1.92 - 2}{0.8 - 0.6} = -0.4 \quad \text{and} \quad f'(0.9) \approx \frac{1.7 - 1.92}{1 - 0.8} = -1.1.$$

Using the backward difference formula, $f'(1.1) \approx \frac{1.56 - 1.7}{1.1 - 1} = -1.4$.

19. Since $y' = -3x^{-4} = -3/x^4$, the slope of the graph of $y = x^{-3}$ at $x = \frac{1}{2}$ is $-3/(\frac{1}{2})^4 = -48$.

21. Since $y' = 3x^2$, the slope of the graph of $y = x^3$ at $(2, 8)$ is 12. The equation of the tangent line is $y = 12x - 16$.

23. Solving $f'(x) = \frac{2}{3}x^{-1/3} = \frac{1}{6}$ yields $x = 64$. Therefore, $(64, 16)$ is the only point at which the graph has slope $1/6$.

25. **(a)** $m'(t) = \dfrac{d}{dt}0.92t^{1/2} = 0.46t^{-1/2} = \dfrac{0.46}{\sqrt{t}}$

(b) $m'(5) = \dfrac{0.46}{\sqrt{5}} \approx 0.2057$ and $m'(50) = 0.06505$

The rate at which the student can grasp new concepts decreases over time.

27. All three limits are equivalent ways of defining $f'(2)$ if $f(2) = -3$. Since the tangent line at $(2, -3)$ has slope -1, we have $f'(2) = -1$, and each of the three limits equals -1.

29. All three limits are equivalent ways of defining $f'(-1)$ if $f(-1) = 3$. Since the tangent line at $x = -1$ has slope -2, we have $f'(-1) = -2$, and each of the three limits equals -2.

31. $g'(t) = 15t^4 + 6t^2 - 5$ **33.** $g'(x) = \dfrac{3}{2}(x^2 - x^{-4})$ **35.** $f'(s) = 2s + 6s^{-3} + \dfrac{5}{2}s^{-1/2}$

37. $\dfrac{d}{du}\left(\dfrac{\sqrt{u}}{2} + \dfrac{2}{\sqrt{u}}\right) = \dfrac{d}{du}\left(\dfrac{1}{2}u^{1/2} + 2u^{-1/2}\right) = \dfrac{1}{4}u^{-1/2} - u^{-3/2} = \dfrac{1}{4\sqrt{u}} - \dfrac{1}{u^{3/2}}$

39. $\dfrac{d}{dz}\left(\dfrac{z^3}{3} - \dfrac{z^2}{4} + \dfrac{z}{2} - \dfrac{3}{2}\right) = z^2 - \dfrac{1}{2}z + \dfrac{1}{2}$

41. Since $y' = 4x^3 + 6x^2 + 10x + 4$, the slope of the graph at $x = -1$ is -4. And since $y(-1) = -2$, the equation of the tangent line is $y = -4x - 6$.

43. Since $y' = 6x - 5$, the slope of the graph at $(2, 3)$ is 7. The tangent line has equation $y = 7x - 11$.

45. Solving $y' = 6x - 4 = 8$, we find $x = 2$. The corresponding y value is $y(2) = -1$. Therefore, the graph has slope 8 at $(2, -1)$.

47. The given line has slope $7/4$ and the tangent line has slope $y' = 3x^2 + 1$ at any point (x, y) on the graph. Solving $3x^2 + 1 = \frac{7}{4}$ yields $x = \pm\frac{1}{2}$ with corresponding y values $y(\pm 1/2) = \pm 5/8$. Thus, the tangent line to the graph at the points $(\frac{1}{2}, \frac{5}{8})$ and $(-\frac{1}{2}, -\frac{5}{8})$ is parallel to the line $7x - 4y = 2$.

49. The rate of change of depth with respect to time (in months) is given by the derivative $D'(m) = 12.5 - 10.16m + 2.136m^2$. In particular,

$$D'(1) = 4.476, \quad D'(3) = 1.244, \quad \text{and} \quad D'(7) = 46.044.$$

The depth is increasing fastest at $m = 7$ and slowest at $m = 3$.

51. $q'(k) = 5.4k^{-0.7}$ is the rate at which the output per worker changes per unit change of capital stock. This rate varies with k, and $q'(8) \approx 1.26$ is the rate when $k = 8$. It is also the approximate change in output per worker resulting from a one-unit change in capital stock.

53. Using the central difference formula, we obtain $y'(1) \approx \dfrac{7.54 - 7.26}{0.02} = 14$.

55.

t	2	4	6	8	10	12	14	16
$M'(t)$	-0.27	-0.125	-0.08	-0.06	-0.055	-0.045	-0.014	-0.0005

The multiplier $M(t)$ is always decreasing, but more and more slowly as the girl gets older.

57. 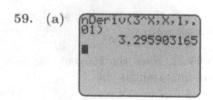 $f'(x) = \dfrac{d}{dx} x^{1/3} = \dfrac{1}{3} x^{-2/3}$

$f'(8) = \dfrac{1}{3 \cdot 8^{2/3}} = \dfrac{1}{12} = 0.08\overline{3}$

59. (a)    (b) $y = 3.296x - 0.296$ (c)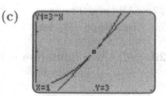

61. 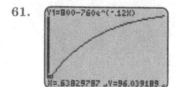   (a) At $t = 0$ (b) It decreases.   (c) [nDeriv(Y1,X,20,.001) 8.27347736]

63.

C3	▼	= =(D2-B2)/(D1-B1)															
								3.2.63									
x	1.92	1.93	1.94	1.95	1.96	1.97	1.98	1.99	2	2.01	2.02	2.03	2.04	2.05	2.06	2.07	2.08
f(x)	3.337	3.3722	3.4075	3.4482	3.4792	3.5156	3.5523	3.5894	3.6269	3.6647	3.7028	3.7414	3.7803	3.8196	3.85926	3.89982	3.9398
f'(x)		3.5151	3.55	3.585	3.62	3.655	3.69	3.73	3.765	3.795	3.835	3.875	3.91	3.948	3.986	4.0255	

EXERCISES 3.3

1. (a) The derivative is $y' = -32t + 128$. At the end of 5 seconds
   - the height of the rocket is $y(5) = -16 \cdot 5^2 + 128 \cdot 5 = 240$ feet,
   - the velocity is $y'(5) = -32 \cdot 5 + 128 = -32$ ft/sec, and
   - the speed is $|y'(5)| = 32$ ft/sec.

   (b) We solve $y'(t) = -32t + 128 = 112$ to get $t = 1/2$ sec.

   (c) We solve $y'(t) = -32t + 128 = -128$ to get $t = 8$ sec.

3. (a)
$$\frac{y(1.1) - y(1)}{0.1} = \frac{(-16(1.1)^2 + 64) - 48}{0.1} = -33.6 \text{ ft/sec};$$

$$\frac{y(1.01) - y(1)}{0.01} = \frac{(-16(1.01)^2 + 64) - 48}{0.01} = -32.16 \text{ ft/sec};$$

$$\frac{y(1.001) - y(1)}{0.001} = \frac{(-16(1.001)^2 + 64) - 48}{0.001} = -32.016 \text{ ft/sec};$$

   (b) $v = y' = -32t$, so that $v(1) = -32$ ft/sec;

5. The velocity is $v(t) = s'(t) = -1.66t$.

   (a) $s'(2) = -3.32$ m/s
   (b) We solve $100 - 0.83t^2 = 0$ to get $t = \sqrt{100/0.83} \approx 11$ sec
   (c) Velocity is $s'(10.98) = -18.2268$ m/s; speed is $|s'(10.98)| = 18.2268$ m/s

7. The (instantaneous) velocity is $H'(t) = -4t^3 - 4.08t^2 - 1.2$.

   (a) $H'(1) = -9.28$ ft/sec, $H'(3) = -145.92$ ft/sec.
   (b) Since there are 5,280 feet in a mile and 3,600 seconds in an hour, 145.92 ft/sec $= (145.92) \cdot 3,600/5,280 = 99.49$ mi/hr. The model is within the guideline given.

9. (a) $A(30) = B(30) + 10$ (b) $B'(10) = A'(10) + 5$
   (c) $A'(20) = 2B'(20)$ (d) $A''(1) = B''(1)$

11. Since $y'' = \frac{d}{dt}(-32t + 50) = -32$, both answers are -32 ft/sec^2.

13. The velocity is $y'(t) = -9.8t + 16.4$ and the acceleration is $y''(t) = -9.8$. At $t = 1$ and 2, we have

t	1	2
$y'(t)$	6.6	-3.2
$y''(t)$	-9.8	-9.8

   (a) Velocity and its speed both decreasing at $t = 1$. (Acceleration negative $\Rightarrow$ velocity decreasing, and, since the velocity is positive, its absolute value is also decreasing.)
   (b) Its velocity is decreasing and its speed is increasing at $t = 2$. (Acceleration negative $\Rightarrow$ velocity decreasing, and, since the velocity is negative, its absolute value is increasing.)

15. (a) $dx/dt > 0$ and $d^2x/dt^2 > 0$ (b) $dx/dt > 0$ and $d^2x/dt^2 < 0$

   (c) $dx/dt < 0$ and $d^2x/dt^2 < 0$ (d) $dx/dt < 0$ and $d^2x/dt^2 > 0$

17. (a)   **19.** (b)   **21.**

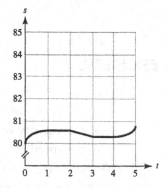

23. Since $f'(t) = 2t - 17$, we have $f''(t) = 2$.

25. Since $f'(t) = -1/t^2$, we have $f''(t) = 2/t^3$.

27. Since $f'(t) = 3$, we have $f''(t) = 0$.

29. Since $f'(x) = 5x^4 + 12x^2 - 7$, we have $f''(x) = 20x^3 + 24x$ and $f''(1) = 44$.

31. Since $\frac{d}{dx} x^{1/3} = \frac{1}{3} x^{-2/3}$, we have $\frac{d^2}{dx^2} x^{1/3} = -\frac{2}{9} x^{-5/3}$.

33. $f'(x) = 5x^4 - 12x^3 + 6x^2 - 14x + 1 \Rightarrow f''(x) = 20x^3 - 36x^2 + 12x - 14$

      $\Rightarrow f'''(x) = 60x^2 - 72x + 12$

35. $f'(u) = -\frac{1}{2} u^{-3/2} \Rightarrow f''(u) = \frac{3}{4} u^{-5/2} \Rightarrow f'''(u) = -\frac{15}{8} u^{-7/2}$

37. Cost: $C(x) = 1,000 + 4.50x$,   Marginal cost: $C'(x) = 4.50$,
    Revenue: $R(x) = 10x - 0.01x^2$,   Marginal revenue: $R'(x) = 10 - 0.02x$,
    Marginal profit: $P'(x) = R'(x) - C'(x) = 5.5 - 0.02x$, $0 < x < 800$.

39. (a) According to the given data, q changes by $\Delta q = -100$ when p changes by $\Delta p = 50$. Therefore, the slope is $\Delta q / \Delta p = -100/50 = -2$. Since the point $(p, q) = (900, 3{,}400)$ is on the line, we get the equation
$$q - 3{,}400 = -2(p - 900) \quad \text{or} \quad q = -2p + 5{,}200.$$

   (b) $R(p) = pq = -2p^2 + 5{,}200p$, $MR(p) = R'(p) = -4p + 5{,}200$

   (c) $C(p) = 400q = 400(-2p + 5{,}200) = -800p + 2{,}080{,}000$, $MC(p) = C'(p) = -800$

   (d) $MP(p) = MR(p) - MC(p) = -4p + 6{,}000$

   (e) $MP(1{,}050) = 1{,}800 > 0$. The company should increase its price. The profit is changing at a positive rate at $p = 1{,}050$, which means a small increase in p causes an increase in P.

   (f) $MP(1{,}550) = -200 < 0$. The company should decrease its price. The profit is changing at a negative rate at $p = 1{,}550$, which means a small increase in p causes a decrease in P.

41. Since $A'(t) = -0.06 + 0.06t - 0.03t^2$ and $A'(2) = -0.06$, the calcium is being eliminated at a rate of 0.06 mg/hr.

43. (a) $V_A'(1) = -2.4$ (b) $V_A'(2) = 3$

   (c) $V_B'(3) = V_A'(3) + 1.1$ (d) $V_B'(4) = 2V_A'(4)$

45. (a) $C(2) = 50$ **(b)** $C'(2) = -3.5$
(c) $C'(t) = -kC(t)$, where $k > 0$ is proportionality constant

47.

C5		=(D2-B2)/0.5									
						3.347					
A	B	C	D	E	F	G	H	I	J	K	L
t	0.00	0.25	0.50	0.75	1.00	1.25	1.50	1.75	2.00	2.25	2.50
v	0.00	0.23	0.66	1.21	1.86	2.60	3.42	4.31	5.26	6.28	7.35
		1.32	1.96	2.40	2.78	3.12	3.42	3.68	3.94	4.18	4.40
t	2.75	3.00	3.25	3.50	3.75	4.00	4.25	4.50	4.75	5.00	
v	8.48	9.66	10.90	12.18	13.51	14.88	16.30	17.76	19.26	20.80	
	4.52	4.84	5.04	5.22	5.4	5.58	5.76	5.92	6.08		

49.

C3		=(D2-B2)/(D1-B1)															
								Workbook2									
A	B	C	D	E	F	G	H	I	J	K	L	M	N	O	P	Q	R
x	-1.6	-1.4	-1.2	-1	-0.8	-0.6	-0.4	-0.2	0	0.2	0.4	0.6	0.8	1	1.2	1.4	1.6
y	0.1109	0.1497	0.1942	0.2420	0.2897	0.3332	0.3603	0.3910	0.3909	0.3910	0.3603	0.3332	0.2897	0.2420	0.1942	0.1497	0.1109
		0.2083	0.2306	0.2388	0.226	0.1965	0.1445	0.0765	0	-0.076	-0.145	-0.197	-0.226	-0.239	-0.231	-0.208	

51.

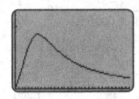 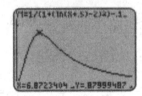   **(a)**

(b) Sales are increasing at the end of 3 months (positive derivative) and decreasing at the end of 17 months (negative slope) as the market gets saturated.

(c) Using the TRACE: At $t \approx 6.87$

53. For an object falling from rest, its height (in feet) is given by $h(t) = -16t^2 + h_0$, where h_0 is the initial height. (The initial velocity satisfies $v_0 = 0$.) During the nth second, the distance traveled is given by

$$h(n) - h(n-1) = -16n^2 + h_0 - [-16(n^2 - 2n + 1) + h_0] = -16(2n - 1).$$

Similarly, during the $(n+1)$th second, the distance traveled is given by

$$h(n+1) - h(n) = -16(n^2 + 2n + 1) + h_0 - [-16n^2 + h_0] = -16(2n + 1).$$

The ratio of these distances is

$$\frac{h(n+1) - h(n)}{h(n) - h(n-1)} = \frac{2n+1}{2n-1},$$

which is exactly the ratio of the $(n+1)$st to the nth odd integer.

EXERCISES 3.4

1. Here we take $f(x) = \sqrt{x}$. Since $f'(x) = \frac{1}{2\sqrt{x}}$, linear approximation gives $\sqrt{x} \approx \sqrt{a} + \frac{1}{2\sqrt{a}}(x - a)$. Taking $a = 36$, we obtain

(a) $\sqrt{38} \approx 6 + \frac{2}{12} = \frac{37}{6}$.   **(b)** $\sqrt{35} \approx 6 - \frac{1}{12} = \frac{71}{12}$.

3. As in exercise 1, $\sqrt{x} \approx \sqrt{a} + \frac{1}{2\sqrt{a}}(x-a)$.

   (a) With $a = 400$, we have $\sqrt{405} \approx 20 + \frac{1}{40} \cdot 5 = \frac{161}{8}$.

   (b) With $a = 4$, we have $\sqrt{3.95} \approx 2 + \frac{1}{4} \cdot (-0.05) = 1.9875$.

   (c) With $a = 25$, we have $\sqrt{25.03} \approx 5 + \frac{1}{10} \cdot (0.03) = 5.003$.

5. Since $f'(x) = 2x$, we have $x^2 \approx a^2 + 2a(x-a)$.

   (a) With $a = 5$, we have $(5.0031468)^2 \approx 25 + 10 \cdot (0.0031468) = 25.031468$.

   (b) With $a = \frac{16}{32} = \frac{1}{2}$, we have $(\frac{17}{32})^2 \approx \frac{1}{4} + \frac{1}{32} = \frac{9}{32}$.

   (c) With $a = 1$, we have $(0.999973)^2 \approx 1 - 2 \cdot (0.000027) = 0.999946$.

7. Take $f(x) = \sqrt[3]{x} = x^{1/3}$ and $a = 1{,}000$. Then $f'(x) = \frac{1}{3}x^{-2/3} = \frac{1}{3x^{2/3}}$ and

$$\sqrt[3]{997} \approx \sqrt[3]{1{,}000} + \frac{(-3)}{3 \cdot 1{,}000^{2/3}} = 10 - \frac{1}{100} = 9.99.$$

9. Take $f(x) = 1/\sqrt[3]{x}$ and $a = 8$. Then $f'(x) = -1/(3x^{4/3})$ and

$$\frac{1}{\sqrt[3]{7.24}} \approx \frac{1}{2} + \frac{1}{48} \cdot (0.76) = \frac{619}{1{,}200}.$$

11. Take $f(x) = 1/x$ and $a = 100$. Then $f'(x) = -1/x^2$ and

$$\frac{1}{102} \approx \frac{1}{100} - \frac{2}{100^2} = 0.0098.$$

13. Take $f(x) = \sqrt[4]{x}$ and $a = 81$. Then $f'(x) = 1/(4x^{3/4})$ and

$$\sqrt[4]{80} \approx \sqrt[4]{81} - \frac{1}{4 \cdot (81)^{3/4}} = 3 - \frac{1}{108} = \frac{323}{108}.$$

15. Take $f(x) = x^{0.3}$ and $a = 1$. Then $f'(x) = 0.3x^{-0.7}$ and

$$(0.99)^{0.3} \approx 1 - (0.3) \cdot (0.01) = 0.997.$$

17. $f(3) = 2.5$ and $f'(3) \approx \frac{4-2.5}{4-3} = 1.5$. Therefore $f(2.9) \approx 2.5 + 1.5 \cdot (-0.1) = 2.35$, and $f(2.99) \approx 2.5 + 1.5 \cdot (-0.01) = 2.485$. The second approximation should be more accurate because x is closer to a in that case. We would guess the approximations are smaller than the true values because the tangent line lies below the graph.

19. $f(5) = 6$ and $f'(5) \approx \frac{6-4}{5-2} = \frac{2}{3}$. Therefore $f(4.9) = 6 + \frac{2}{3} \cdot (-0.1) \approx 5.933$, and $f(4.95) = 6 + \frac{2}{3} \cdot (-0.05) \approx 5.967$. The second approximation should be more accurate because x is closer to a in that case. We would guess the approximations are greater than the true values because the tangent line lies above the graph.

21. $R(9{,}800) \approx R(10{,}000) + R'(10{,}000) \cdot (-200) = 8{,}750 - 200 \cdot (0.64) = 8{,}622$

23. (a) $C'(x) = 500\left(1 - \frac{1}{2\sqrt{x}}\right)$, so that $C'(16) = 500 \cdot \left(\frac{7}{8}\right) = 437.50$

   (b) $C(17) - C(16) \approx C'(16) \cdot 1 = MC(16) = 437.50$

   (c) Calculator gives $C(17) - C(16) \approx 438.45$ dollars.

25. $v(5.2) \approx v(5) + v'(5)(5.2 - 5) = v(5) + a(5) \cdot (0.2) = 40 - 3 \cdot (0.2) = 39.4$ ft/sec

27. Let v be the volume of a cube with each edge having length s. Then $s = v^{1/3}$ and $s' = \frac{1}{3}v^{-2/3}$. Therefore, $s(7.88) \approx 8^{1/3} + \frac{1}{3}8^{-2/3}(7.88 - 8)$, and $\Delta s = s(7.88) - 8^{1/3} \approx \frac{1}{3}8^{-2/3}(7.88 - 8) = -\frac{1}{12} \cdot 0.12 = -0.01$. Thus, the length of each edge decreases by about 0.01 inches.

29. $5.9 + (0.02) \cdot 5.9 \cdot 10 = 7.8$ billion

31. (a) $\dfrac{M(700) - M(650)}{700 - 650} = \dfrac{0.0575 - 0.0525}{50} = 0.0001$

   (b) $\dfrac{M(750) - M(700)}{750 - 700} = \dfrac{0.0624 - 0.0575}{50} = 0.000098$

   (c) $M'(700) = \dfrac{1}{2}(0.0001 + 0.000098) = 0.000099.$

   (d) $M(710) \approx M(700) + M'(700) \cdot 10 = 0.0575 + (0.000099) \cdot 10 = 0.05849$

33. The error equals $(0.04)^2 = 0.0016.$

35.

t	v	s
0.00	0.00	0
0.25	0.23	0
0.50	0.66	0.0575
0.75	1.21	0.2225
1.00	1.86	0.525
1.25	2.60	0.99
1.50	3.42	1.64
1.75	4.31	2.495
2.00	5.26	3.5725
2.25	6.26	4.6875
2.50	7.35	6.4575
2.75	8.48	8.295
3.00	9.66	10.415
3.25	10.90	12.83
3.50	12.18	15.555
3.75	13.51	18.6
4.00	14.88	21.9775
4.25	16.30	25.6975
4.50	17.76	29.7725
4.75	19.26	34.2125
5.00	20.80	39.0275

37. (a) Since $y' = 1/(3x^{2/3})$, we have $y(8) = 2$ and $y'(8) = 1/12$. Therefore, $y - 2 = \frac{1}{12}(x - 8)$, or $y = \frac{1}{12}x + \frac{4}{3}$ is the equation of the tangent line at $x = 8$. The error is less than or equal to 0.1 if and only if $x^{1/3} - 0.1 \le \frac{1}{12}x + \frac{4}{3} \le x^{1/3} + 0.1$. Using a $[0, 18] \times [1, 3]$ window, we get

   (b) Lower bound:

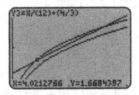

With TRACE   With repeated ZOOM

   Upper bound:

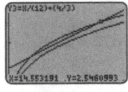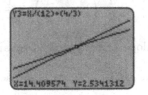

With TRACE   With repeated ZOOM

Conclude: $|\text{error}| \le 0.1$ for approximately $3.6 < x < 14.4$.

39. $f(x)$ is not differentiable at $x = 0$ since

$$\lim_{x \to 0^-} \frac{f(x) - f(0)}{x - 0} = \lim_{x \to 0^-} \frac{x^2 - x}{x - 0} = -1 \quad \text{and} \quad \lim_{x \to 0^+} \frac{f(x) - f(0)}{x - 0} = \lim_{x \to 0^+} \frac{x^2 + x}{x - 0} = 1.$$

Therefore, $\lim_{x \to 0} \frac{f(x) - f(0)}{x - 0}$ does not exist. However, $f(x)$ is continuous at $x = 0$ since $f(0) = 0$ and

$$\lim_{x \to 0^+} f(x) = \lim_{x \to 0^+} (x^2 + x) = 0 \quad \text{and} \quad \lim_{x \to 0^-} f(x) = \lim_{x \to 0^-} (x^2 - x) = 0.$$

41. $f(x)$ is not continuous at $x = 0$ since

$$\lim_{x \to 0^-} f(x) = \lim_{x \to 0^-} (1 - x^2) = 1 \quad \text{and} \quad \lim_{x \to 0^+} f(x) = \lim_{x \to 0^+} x^2 = 0.$$

Therefore, according to theorem 3.4.2, $f(x)$ cannot be differentiable at $x = 0$.

43. (a) The graph is discontinuous at $t = 1, 2, 3, 4, 7, 8, 9$.

(b) The slope is zero at all other points in the domain.

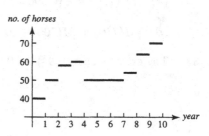

45. (a) $f(x) = \begin{cases} 0.07x & \text{if } 0 \le x \le 100{,}000 \\ 3{,}000 + 0.04x & \text{if } 100{,}000 < x \end{cases}$

(b) $f(x)$ is continuous for all $x > 0$.

(c) $f(x)$ differentiable at all $x > 0$ except $x = 100{,}000$.

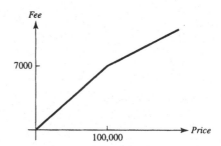

47.

x	-2	-1	0	1	2	3	4	5	6	7	8
f continuous ?	yes	yes	yes	yes	no	yes	yes	yes	no	no	yes
f differentiable ?	yes	yes	no	no	no	no	no	no	no	no	yes

EXERCISES 3.5

1. $\dfrac{d}{dx}(x^2 + 4x + \ln x) = 2x + 4 + \dfrac{1}{x}$

3. $\dfrac{d}{dx}(10 - \ln x) = -\dfrac{1}{x}$

5. $\dfrac{d}{dx}\left(\dfrac{x \ln x + 1}{x}\right) = \dfrac{d}{dx}\left(\ln x + \dfrac{1}{x}\right) = \dfrac{1}{x} - \dfrac{1}{x^2}$

7. $y' = 1/x$. The slope of the graph is equal to $y'(1/2) = 2$.

9. $y' = \frac{d}{dx}(x^2 + \ln x) = 2x + \frac{1}{x}$, so that the slope is $y'(1) = 3$. The equation of the tangent line is $y = 3x - 2$.

11. Solving $y' = 1/x = 2$ yields $x = 1/2$. The point is $\left(\frac{1}{2}, -\ln 2\right)$.

13. Solving $y' = 2x - \frac{1}{x} = 0$ yields $x = \frac{\sqrt{2}}{2}$, and the point is $\left(\frac{\sqrt{2}}{2}, \frac{1}{2} + \frac{1}{2}\ln 2\right)$.

15. $\dfrac{d}{dx}\left(\ln\dfrac{5}{x}\right) = \dfrac{d}{dx}(\ln 5 - \ln x) = -\dfrac{1}{x}$

17. $\dfrac{d}{dx}(\ln\sqrt{x}) = \dfrac{d}{dx}\left(\dfrac{1}{2}\ln x\right) = \dfrac{1}{2x}$

19. $\dfrac{d}{dt}\left(\ln\dfrac{2}{t^3}\right) = \dfrac{d}{dt}(\ln 2 - 3\ln t) = -\dfrac{3}{t}$

21. (a) Tangent line: $y = x - 1$

(b)

x	1.1	1.2	0.9	0.8
lin. approx.	0.1	0.2	−0.1	−0.2
calc.	0.0953	0.18232	−0.10536	−0.22314

(c) The linear approximation is the y-coordinate on the tangent line, which is always above the graph.

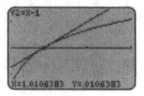

23. $\ln 2 \approx 0.69315$

(a) $\ln(2.01) \approx 0.69315 + \dfrac{1}{2}\cdot(0.01) = 0.69815$

$\ln(1.9) \approx 0.69315 - \dfrac{1}{2}\cdot(0.1) = 0.64315$

(b) $y = \dfrac{1}{2}x + \ln 2 - 1 \approx \dfrac{1}{2}x - 0.30685$

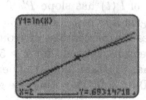

(c) Greater, because the tangent line is above the graph.

25. $\dfrac{d}{dx}\left(2e^x - \dfrac{1}{\sqrt{x}}\right) = 2e^x + \dfrac{1}{2x^{3/2}}$

27. $\dfrac{d}{dt}\left(\dfrac{e^t + 1}{2}\right) = \dfrac{d}{dt}\left(\dfrac{e^t}{2} + \dfrac{1}{2}\right) = \dfrac{e^t}{2}$

29. $\dfrac{d}{dx}(e^{\ln x} + e^x + \ln e^x) = \dfrac{d}{dx}(x + e^x + x) = 2 + e^x$

31. Since $y' = e^x$, the slope at $x = 1$ equals $y'(1) = e$. Also $y(1) = 0$. The equation of the tangent line is $y = e(x - 1) = ex - e$.

33. Solving $y' = e^x = 3$ yields $x = \ln 3$, and the point is $(\ln 3, 3)$.

35. Solving $y' = 2e^x - 1 = 0$ yields $e^x = 1/2$, or $x = -\ln 2$. The y-coordinate is $y = 2e^{-\ln 2} + \ln 2 = 1 + \ln 2$. Therefore, the point where the tangent line is horizontal is $(-\ln 2, 1 + \ln 2)$.

37.

```
nDeriv(e^(X),X,0
..000000001)
            1
■
```

39. $\dfrac{d}{dx}e^{x/3} = \dfrac{1}{3}e^{x/3}$

41. $\dfrac{d}{dx} e^{3x+1} = \dfrac{d}{dx}(e^{3x} \cdot e) = e \cdot \dfrac{d}{dx} e^{3x} = e \cdot 3e^{3x} = 3e^{3x+1}$

43. $\dfrac{d}{dt}\left(\dfrac{1}{e^t}\right) = \dfrac{d}{dt} e^{-t} = -e^{-t}$

45. If $y = e^{-x}$, then $y' = -e^{-x}$, and the slope of the graph at $x = -a$ is $-e^{-(-a)} = -e^{a}$. On the other hand, the slope of the graph of $y = e^x$ at $x = a$ is e^a.

47. Set $y' = 2e^{2x} - e^{-x} = 0$. Then $2e^{2x} = e^{-x}$, or $e^{3x} = 1/2$. Taking the natural logarithm of both sides gives $3x = \ln\frac{1}{2} = -\ln 2$. Therefore, $x = -\frac{1}{3}\ln 2$.

49. $\dfrac{d}{dx}\left(\dfrac{2^x}{3} + \dfrac{2}{3x}\right) = \dfrac{2^x}{3}\ln 2 - \dfrac{2}{3x^2}$ **51.** $\dfrac{d}{dt}\left(\dfrac{1}{2^t}\right) = \dfrac{d}{dt}\left(\dfrac{1}{2}\right)^t = \left(\dfrac{1}{2}\right)^t \cdot \ln\dfrac{1}{2} = -\dfrac{\ln 2}{2^t}$

53. $\dfrac{d}{dt}(2^{1+t}) = \dfrac{d}{dt}(2 \cdot 2^t) = 2\dfrac{d}{dt} 2^t = 2 \cdot 2^t \ln 2 = 2^{1+t}\ln 2$

55. Let $P(t) = 1{,}000(1.023)^t$ be the population. Then $P'(t) = 1{,}000(1.023)^t \ln 1.023$. At the end of 10 days, the size of the population will be $P(10) = 1{,}000(1.023)^{10} \approx 1{,}255$, and it will be increasing at the rate of $P'(10) = 1{,}000(1.023)^{10}\ln(1.023) \approx 29$ insects per day.

57. **(a)** $P'(t) = -20e^{-0.2t}$

    (b) At $t = 5$ the graph of $P(t)$ has slope $P'(5) = -20e^{-1}$, and $P(5) = 100e^{-1}$. Therefore, the equation of the tangent line is $y - 100e^{-1} = -20e^{-1}(t - 5)$, or $y = -20e^{-1}t + 200e^{-1}$. That linear function gives the percentage of trout remaining after the fifth day, assuming that the removal rate remains consant for $t \ge 5$. (That is, constant removal rate means the population graph has constant slope.) Therefore, to find when the population is zero, we solve $-20e^{-1}t + 200e^{-1} = 0$. Factoring, we get $20e^{-1}(-t + 10) = 0$, or $t = 10$ days.

59. **(a)** The derivative of H is positive, because the function H is increasing as time goes by. (The longer the subject has the habit, the harder it is to break.)

    (b) $H'(t) = (1 - e^{-0.1t})' = 0.1e^{-0.1t} > 0$, as predicted.

    (c) To make the difficulty index increase more slowly, we would replace 0.1 by a smaller constant. To make the difficulty index increase faster, we should replace 0.1 by a larger constant.

61. **(a)** The temperature is given by a formula of the form $H(t) = 40 + Ae^{-kt}$, where t is the time (in minutes) and k and A are constants. To find A, we set $t = 0$ and $H(0) = 65$ (the initial temperature) to get $65 = 40 + A \cdot e^0 = 40 + A$. Therefore, $A = 25$. To find k, we set $t = 20$ and $H(20) = 55$ to get $55 = 40 + 25e^{-20k}$, or $e^{-20k} = 3/5$. Taking the natural logarithm gives $k = (\ln\frac{5}{3})/20$. We have now found a formula for the temperature of the water :

$$H(t) = 40 + 25e^{-t(\ln 5/3)/20}.$$

    (b) At $t = 30$, the temperature is $H(30) = 40 + 25e^{-3(\ln 5/3)/2} \approx 51.62$. To find the rate, we take the derivative: $H'(t) = -\frac{25}{20}(\ln\frac{5}{3})e^{-t(\ln 5/3)/20}$. Then $H'(30) = -\frac{5}{4}(\ln\frac{5}{3})e^{-3(\ln 5/3)/2} = -\frac{5}{4}(\ln\frac{5}{3})(\frac{5}{3})^{-3/2} \approx -0.30$ degrees per minute.

    (c) Set $45 = 40 + 25e^{-t(\ln 5/3)/20}$. Then $e^{-t(\ln 5/3)/20} = 5/25 = 1/5$. Solving for t gives $t = (20\ln 5)/(\ln\frac{5}{3}) \approx 63$ minutes.

63. (a) Newton's law says that: $dH/dt = k(400 - H) = -k(H - 400)$, where k is a constant. Comparing this differential equation to equation (41) suggests that a solution is $H = 400 + Ae^{-kt}$. To check that is, we take the derivative, $H'(t) = -kAe^{-kt}$, which is indeed equal to $-k(H - 400)$. To find the constants A and k, we use the given data $H(20) = 320$ and $H'(20) = 6$, which leads to the equations $Ae^{-20k} = -80$ and $kAe^{-20k} = -6$. Dividing the first into the second gives $k = 3/40$ and $A = -80e^{3/2}$. Therefore,

$$H(t) = 400 - 80e^{3/2}e^{-3t/40}.$$

(b) $H(0) = 400 - 80e^{3/2} \approx 41.465$, and $H'(0) = -kA = 6e^{3/2} \approx 26.89$.

(c) The rate at which the temperature increases is $H'(t) = 6e^{3/2}e^{-3t/40}$, which is a decreasing function. That makes intuitive sense; it says that as the potato's temperature gets close to that of the oven, it rises more slowly.

65. (a) Setting $T_a = 72$, we get $T = 72 + 26.6(0.6)^h$, and $T' = 26.6(\ln 0.6)(0.6)^h$. Then $T'(4.5) = 26.6(\ln 0.6)(0.6)^{4.5} \approx -1.364$ degrees per hour.

(b) With $T_a = 98$, $T = 98 + 0.6(0.6)^h$ and $T' = 0.6(\ln 0.6)(0.6)^h$, so that $T'(4.5) = 0.6(\ln 0.6)(0.6)^{4.5} \approx -0.03$ degrees per hour.

(c) $T' = (98.6 - T_a)(\ln 0.6)(0.6)^h = (\ln 0.6)(T - T_a) = k(T - T_a)$, where $k = \ln 0.6$.

67. As in the previous exercise, we have $R = y'(t)/y'(0) = e^{-t(\ln 2)/5,568}$. In this case, we are given $t = 4,600$. Therefore, $R = e^{-4,600(\ln 2)/5,568} \approx 0.5640$.

69. (a) $16/2,000 = 0.008$. In other words, 0.8%.

(b) $dP/dt = -0.008P$, with t in years after 1999. The initial condition is $P(0) - 2,000$.

(c) $P = 2,000e^{-0.008t}$

EXERCISES 3.6

1. $\dfrac{d}{dx}(xe^{-x}) = x\dfrac{d}{dx}e^{-x} + e^{-x}\dfrac{d}{dx}x = -xe^{-x} + e^{-x} = e^{-x}(1 - x)$

3. $\dfrac{d}{dx}\left(\dfrac{1}{x}\ln x\right) = \dfrac{1}{x}\dfrac{d}{dx}\ln x + \ln x\dfrac{d}{dx}\dfrac{1}{x} = \dfrac{1}{x^2} - \dfrac{1}{x^2}\ln x = \dfrac{1 - \ln x}{x^2}$

5. $\dfrac{d}{dx}(e^{-3x}\ln x) = e^{-3x}\dfrac{d}{dx}\ln x + \ln x\dfrac{d}{dx}e^{-3x} = e^{-3x}\left(\dfrac{1}{x} - 3\ln x\right)$

7. $\dfrac{d}{dx}\left(\dfrac{x^2 - 2}{x^2 + 4}\right) = \dfrac{(x^2 + 4)\frac{d}{dx}(x^2 - 2) - (x^2 - 2)\frac{d}{dx}(x^2 + 4)}{(x^2 + 4)^2}$

$= \dfrac{2x(x^2 + 4) - 2x(x^2 - 2)}{(x^2 + 4)^2} = \dfrac{12x}{(x^2 + 4)^2}$

9. $\dfrac{d}{dx}\left(\dfrac{x}{e^x + 1}\right) = \dfrac{(e^x + 1)\frac{d}{dx}x - x\frac{d}{dx}(e^x + 1)}{(e^x + 1)^2} = \dfrac{e^x + 1 - xe^x}{(e^x + 1)^2}$

11. $\dfrac{d}{dx}\left(\dfrac{e^x - 1}{e^x + 1}\right) = \dfrac{(e^x + 1)\frac{d}{dx}(e^x - 1) - (e^x - 1)\frac{d}{dx}(e^x + 1)}{(e^x + 1)^2}$

$= \dfrac{(e^x + 1)e^x - (e^x - 1)e^x}{(e^x + 1)^2} = \dfrac{2e^x}{(e^x + 1)^2}$

13. $\frac{d}{dx}(x^2 \ln x) = x^2 \frac{d}{dx} \ln x + \ln x \frac{d}{dx}(x^2) = x + 2x \ln x$

15. $\frac{d}{dx}\left(\frac{x \ln x}{x+1}\right) = \frac{(x+1)\frac{d}{dx}(x \ln x) - x \ln x \frac{d}{dx}(x+1)}{(x+1)^2}$

$$= \frac{(x+1)\left(x \cdot \frac{1}{x} + \ln x\right) - x \ln x}{(x+1)^2} = \frac{x + \ln x + 1}{(x+1)^2}$$

17. $\frac{f(2)g'(2) - g(2)f'(2)}{[f(2)]^2} = \frac{3(\frac{2}{3}) - \frac{1}{2} \cdot (-1)}{3^2} = \frac{5}{18}$

19. **(a)** $y' = x\frac{d}{dx}(\sqrt{2x+1}) + \sqrt{2x+1}\frac{d}{dx}x = x(2x+1)^{-1/2} + \sqrt{2x+1}$

Therefore, the slope of the graph at $x = 12$ equals $y'(12) = \frac{12}{\sqrt{25}} + \sqrt{25} = \frac{37}{5}$.

(b) $y' = \frac{\sqrt{2x+1} \cdot \frac{d}{dx}x - x\frac{d}{dx}\sqrt{2x+1}}{(\sqrt{2x+1})^2} = \frac{\sqrt{2x+1} - x(2x+1)^{-1/2}}{2x+1} = \frac{2x+1-x}{(2x+1)^{3/2}} = \frac{x+1}{(2x+1)^{3/2}}$

Therefore, the slope of the graph at $x = 4$ equals $5/27$.

21. $y' = (x^2 + 2x + 3)\frac{d}{dx}e^x + e^x\frac{d}{dx}(x^2 + 2x + 3) = (x^2 + 2x + 3)e^x + e^x(2x + 2) = (x^2 + 4x + 5)e^x$

Therefore, the slope is $y'(0) = 5e^0 = 5$. Also, $y(0) = 3e^0 = 3$, and the equation of the tangent line is $y - 3 = 5x$, or $y = 5x + 3$.

23. $y' = \frac{(1 + \ln x)\frac{d}{dx}(x \ln x) - x \ln x \frac{d}{dx}(1 + \ln x)}{(1 + \ln x)^2}$

$$= \frac{(1 + \ln x)\left(x \cdot \frac{1}{x} + \ln x\right) - (x \ln x) \cdot \frac{1}{x}}{(1 + \ln x)^2} = \frac{(\ln x)^2 + (\ln x) + 1}{(1 + \ln x)2}$$

Therefore, the slope is $y'(e) = 3/4$. Also, $y(e) = e/2$, and the equation of the tangent line is $y - \frac{e}{2} = \frac{3}{4}(x - e)$, or $y = \frac{3}{4}x - \frac{e}{4}$.

25. $y' = x^2 e^x + 2xe^x = xe^x(x + 2)$. Therefore, the slope at $x = 1$ is $y'(1) = 3e$. Also, $y(1) = e$, and the equation of the tangent line is $y - e = 3e(x - 1)$, or $y = 3ex - 2e$.

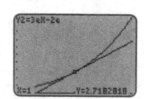

27. Set $y' = xe^x + e^x = e^x(x + 1) = 0$. The only solution is $x = -1$.

29. **(a)** $R = pq = 5,000pe^{-0.01p}$

   (b) $MR(p) = R'(p) = 5,000e^{-0.01p} - 50pe^{-0.01p} = 50e^{-0.01p}(100 - p)$

   (c) $R(200) = 1,000,000e^{-2}$ and $MR(200) = R'(200) = -5,000e^{-2}$

   (d) The company ought to decrease the price to raise the revenue.

31. $E'(w) = \frac{(9 + 0.3w) \cdot w - 0.3(0.5w^2)}{(9 + 0.3w)^2} = \frac{9w + 0.15w^2}{(9 + 0.3w)^2}$

33. **(a)** $P'(t) = 0.47(2te^{0.0035t} + 0.0035t^2 e^{0.0035t})$

   (b) $P'(30) = 0.47(60e^{0.105} + 3.15e^{0.105}) = 32.97 > 20$. An eruption is highly likely to occur at the end of 30 months.

EXERCISES 3.7

1. $\dfrac{d}{dx}\,e^{1/x} = e^{1/x}\,\dfrac{d}{dx}\,\dfrac{1}{x} = -\dfrac{1}{x^2}\,e^{1/x}$

3. $\dfrac{d}{dx}\,(e^x + e^{-x})^2 = 2(e^x + e^{-x})\,\dfrac{d}{dx}\,(e^x + e^{-x}) = 2(e^x + e^{-x})(e^x - e^{-x})$

5. $\dfrac{d}{dx}\,\ln(1 + e^x) = \dfrac{1}{1 + e^x}\,\dfrac{d}{dx}\,(1 + e^x) = \dfrac{e^x}{1 + e^x}$

7. $y = (x^2 + x + 1)^{10}$, and
$$\frac{dy}{dx} = \frac{dy}{du}\cdot\frac{du}{dx} = 10u^9\cdot(2x + 1) = 10(x^2 + x + 1)^9(2x + 1)$$

9. $y = \ln(x^4 + 1)$, and $\dfrac{dy}{dx} = \dfrac{dy}{du}\cdot\dfrac{du}{dx} = \dfrac{1}{u}\cdot 4x^3 = \dfrac{4x^3}{x^4 + 1}$

11. $\dfrac{d}{dx}\left[x(1 + x^4)^{1/2}\right] = x\cdot\dfrac{d}{dx}\,(1 + x^4)^{1/2} + \sqrt{1 + x^4}$

$$= x\cdot\frac{1}{2\sqrt{1 + x^4}}\cdot\frac{d}{dx}\,(1 + x^4) + \sqrt{1 + x^4}$$

$$= x\cdot\frac{1}{2\sqrt{1 + x^4}}\cdot 4x^3 + \sqrt{1 + x^4} = \frac{2x^4}{\sqrt{1 + x^4}} + \sqrt{1 + x^4}$$

13. $\dfrac{d}{dx}\,e^{-\sqrt{x}} = e^{-\sqrt{x}}\cdot\dfrac{d}{dx}\,(-\sqrt{x}) = -\dfrac{1}{2\sqrt{x}}e^{-\sqrt{x}}$

15. $\dfrac{d}{dx}\,[x\ln(2x + 1)] = x\cdot\dfrac{d}{dx}\,\ln(2x + 1) + \ln(2x + 1)$

$$= x\cdot\frac{1}{2x + 1}\cdot\frac{d}{dx}\,(2x + 1) + \ln(2x + 1) = \frac{2x}{2x + 1} + \ln(2x + 1)$$

17. $\dfrac{d}{dx}\,(e^{2x} + e^x + 1)^{-1} = -\dfrac{1}{(e^{2x} + e^x + 1)^2}\cdot\dfrac{d}{dx}\,(e^{2x} + e^x + 1) = -\dfrac{2e^{2x} + e^x}{(e^{2x} + e^x + 1)^2}$

19. $\dfrac{d}{dx}\,\dfrac{e^{2x} - 1}{e^{2x} + 1} = \dfrac{(e^{2x} + 1)\frac{d}{dx}(e^{2x} - 1) - (e^{2x} - 1)\frac{d}{dx}(e^{2x} + 1)}{(e^{2x} + 1)^2}$

$$= \frac{(e^{2x} + 1)\cdot 2e^{2x} - (e^{2x} - 1)\cdot 2e^{2x}}{(e^{2x} + 1)^2} = \frac{4e^{2x}}{(e^{2x} + 1)^2}$$

21. $\dfrac{d}{dx}\,\ln(x^2 + 3x + 5) = \dfrac{1}{x^2 + 3x + 5}\cdot\dfrac{d}{dx}(x^2 + 3x + 5) = \dfrac{2x + 3}{x^2 + 3x + 5}$

23. $\dfrac{d}{dx}\,\ln(1 + \sqrt{x}) = \dfrac{1}{1 + \sqrt{x}}\cdot\dfrac{d}{dx}(1 + \sqrt{x}) = \dfrac{1}{1 + \sqrt{x}}\cdot\dfrac{1}{2\sqrt{x}} = \dfrac{1}{2(x + \sqrt{x})}$

25. $x^2\,\dfrac{d}{dx}\,(\sqrt{2x + 1}) + 2x\sqrt{2x + 1} = x^2\cdot\dfrac{1}{2\sqrt{2x + 1}}\cdot 2 + 2x\sqrt{2x + 1}$

$$= \frac{x^2}{\sqrt{2x + 1}} + 2x\sqrt{2x + 1} = \frac{5x^2 + 2x}{\sqrt{2x + 1}}$$

27. $y' = \dfrac{1}{2\sqrt{5+x^2}} \dfrac{d}{dx}(5+x^2) = \dfrac{1}{2\sqrt{5+x^2}} \cdot 2x = \dfrac{x}{\sqrt{5+x^2}}$

Therefore, the slope at $x = 2$ is $y'(2) = 2/3$. Also, $y(2) = 3$, and the equation of the tangent line is $y - 3 = \frac{2}{3}(x - 2)$, or $y = \frac{2}{3}x + \frac{5}{3}$.

29. $y' = xe^{2x-1}\dfrac{d}{dx}(2x-1) + e^{2x-1} = xe^{2x-1} \cdot 2 + e^{2x-1} = e^{2x-1}(2x+1)$

Therefore, the slope at $x = 1/2$ is $y'(1/2) = 2$. Also, $y(1/2) = 1/2$, and the equation of the tangent line is $y - \frac{1}{2} = 2\left(x - \frac{1}{2}\right)$, or $y = 2x - \frac{1}{2}$.

31. $y' = \dfrac{1}{2\sqrt{4x+3}} \cdot \dfrac{d}{dx}(4x+3) = \dfrac{2}{\sqrt{4x+3}}$

Therefore, the slope at $x = 3$ is $y'(3) = 2/\sqrt{15}$. Also, $y(3) = \sqrt{15}$, and the equation of the tangent line is $y = \frac{2}{\sqrt{15}}(x - 3) + \sqrt{15}$.

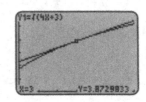

33. $y' = x\dfrac{d}{dx}e^{-x^2/2} + e^{-x^2/2} = -x^2 e^{-x^2/2} + e^{-x^2/2} = e^{-x^2/2}(1 - x^2)$

Therefore $y' = 0 \iff 1 - x^2 = 0 \iff x = \pm 1$.

35. $\dfrac{d}{dx}f(1+x^3) = f'(1+x^3) \cdot 3x^2$

Therefore, the slope at $x = 1$ is $f'(2) \cdot 3 = (-1) \cdot 3 = -3$.

37. $\dfrac{d}{dx}[2 + f(x)]^{1/2} = \dfrac{f'(x)}{2\sqrt{2+f(x)}}$.

Therefore, the slope at $x = 0$ is $\dfrac{f'(0)}{2\sqrt{2+f(0)}} = -\dfrac{3}{2\sqrt{2+2}} = -\dfrac{3}{4}$. Also, when $x = 0$, $y = \sqrt{2 + f(0)}$ $= \sqrt{2 + 2} = 2$, and the equation of the tangent line is $y - 2 = -\frac{3}{4}x$, or $y = -\frac{3}{4}x + 2$.

39. $\dfrac{d}{dx}f(e^{2x} - 1) = f'(e^{2x} - 1) \cdot 2e^{2x}$.

Therefore the slope at $x = 0$ is $f'(0) \cdot 2 = (-3) \cdot 2 = -6$

41. $f'(g(0)) \cdot g'(0) = f'(1) \cdot 3 = \dfrac{1}{2} \cdot 3 = \dfrac{3}{2}$

43. $2 \cdot [f(3) \cdot g(3)] \cdot (f(3)g'(3) + g(3)f'(3)) = 2[(-1) \cdot 2] \cdot \left((-1) \cdot \dfrac{3}{2} + 2 \cdot \dfrac{1}{4}\right) = 4$

45. $g'(f(1)) \cdot f'(1) = g'(2) \cdot \left(-\dfrac{1}{2}\right) = \dfrac{3}{5} \cdot \left(-\dfrac{1}{2}\right) = -\dfrac{3}{10}$

47. $-\dfrac{1}{[2+f(2)]^2} \cdot [1 + f'(2)] = -\dfrac{1}{(2+5)^2} \cdot (1+1) = -\dfrac{2}{49}$

49. $\dfrac{dg}{dw} = \dfrac{(32w+10) \cdot 6w - 3w^2 \cdot 32}{(32w+10)^2} = \dfrac{96w^2 + 60w}{(32w+10)^2}$ and $\dfrac{dw}{dt} = 300t^{1/2}$

Therefore, $g'(t) = \dfrac{dg}{dw} \cdot \dfrac{dw}{dt} = \dfrac{96w^2 + 60w}{(32w+10)^2} \cdot 300t^{1/2}$.

At $t = 100$, we have $w = 200 \cdot 100^{3/2} = 200{,}000$, and

$$g'(100) = \frac{96 \cdot (200{,}000)^2 + 60 \cdot (200{,}000)}{(32 \cdot (200{,}000) + 10)^2} \cdot 3{,}000 = 281.25.$$

51. $h'(t) = 1 - \dfrac{1}{2\sqrt{2t^2 + 100}} \cdot \dfrac{d}{dt}(2t^2 + 100) = 1 - \dfrac{2t}{\sqrt{2t^2 + 100}}$

EXERCISES 3.8

1. Explicit: $y = x^{1/3}$, so that $y' = \dfrac{1}{3}x^{-2/3}$

Implicit: $1 - 3y^2 \dfrac{dy}{dx} = 0$, so that $\dfrac{dy}{dx} = \dfrac{1}{3y^2} = \dfrac{1}{3x^{2/3}} = \dfrac{1}{3}x^{-2/3}$

3. Explicit: $y = \sqrt{x^2 - 1}$, so that $y' = \dfrac{x}{\sqrt{x^2 - 1}}$

Implicit: $2x - 2y\dfrac{dy}{dx} = 0$, so that $\dfrac{dy}{dx} = \dfrac{x}{y} = \dfrac{x}{\sqrt{x^2 - 1}}$

5. Explicit: $y = \ln(x^2 + 1)$, so that $y' = \dfrac{2x}{x^2 + 1}$

Implicit: $2x = e^y \dfrac{dy}{dx}$, so that $\dfrac{dy}{dx} = 2xe^{-y} = \dfrac{2x}{x^2 + 1}$

7. $2x + y + xy' + 2yy' = 0 \Longrightarrow y'(x + 2y) = -2x - y \Longrightarrow y' = -\dfrac{2x + y}{x + 2y}$

9. $e^y y' + e^{-y} y' = 2 \Longrightarrow y'(e^y + e^{-y}) = 2 \Longrightarrow y' = \dfrac{2}{e^y + e^{-y}}$

11. $y' + \dfrac{1}{y}y' = 2x \Longrightarrow y'\left(\dfrac{y + 1}{y}\right) = 2x \Longrightarrow y' = \dfrac{2xy}{y + 1}$

13. (a) $2x + 8yy' = 0 \Longrightarrow y' = -x/4y$. Therefore, the slope at $(1, \sqrt{3}/2)$ is $-\dfrac{1}{4 \cdot \sqrt{3}/2} = -\dfrac{1}{2\sqrt{3}}$.

(b) Setting $1/2 = y' = -x/4y$ gives $x = -2y$. Substituting that into the equation of the curve yields $(-2y)^2 + 4y^2 = 4$. Therefore, $y^2 = 1/2$, so that $y = \pm 1/\sqrt{2} = \pm\sqrt{2}/2$ and $x = -2y = \mp\sqrt{2}$. The points are $(\sqrt{2}, -\frac{\sqrt{2}}{2})$ and $(-\sqrt{2}, \frac{\sqrt{2}}{2})$.

15. (a) The equation of the circle of radius r centered at 0 is $x^2 + y^2 = r^2$. Implicit differentiation gives $2x + 2yy' = 0$, so that $y' = -x/y$, which is the slope of the circle at any point of (x, y) with $y \neq 0$.

(b) A radial line that is vertical must go from $(0, 0)$ to $(0, r)$ or $(0, -r)$. In that case, the tangent is horizontal, so that the lines are perpendicular. Similarly, a radial line that is horizontal meets a tangent that is vertical either at the point $(r, 0)$ or at $(-r, 0)$. Any radial line that is neither vertical nor horizontal goes from $(0, 0)$ (the center of the circle) to a point (x, y) on the circumference, with $x \neq 0$ and $y \neq 0$. Its slope equals y/x, and, as we just saw in (a), the slope of the tangent line at (x, y) equals $-x/y$. Therefore, the product of the slopes of the radial and tangent lines is -1, which means that the lines are orthogonal.

17. Using the result of the previous exercise, we get

$$\frac{\sqrt{3/8}\left(1-\frac{6}{8}-\frac{2}{8}\right)}{\sqrt{1/8}\left(1+\frac{6}{8}+\frac{2}{8}\right)} = \frac{0}{2\sqrt{1/8}} = 0.$$

19. (a) The points $(1,2)$ and $(-1,-2)$ satisfy both equations. The rest follows from part (b) as we explain next.

(b) Suppose the curves intersect at a point (x_1, y_1). Since $x_1 y_1 = b > 0$, we must have $y_1 \neq 0$ and $x_1 \neq 0$. Using implicit differentiation, we can find the slopes of the two curves as follows:

$$y^2 - x^2 = a \Longrightarrow 2yy' - 2x = 0 \Longrightarrow y' = x/y$$

$$xy = b \Longrightarrow xy' + y = 0 \Longrightarrow y' = -y/x.$$

It follows that at the intersection point (x_1, y_1) the curves have slopes x_1/y_1 and $-y_1/x_1$, respectively. Since the product of these slopes is -1, the tangents are perpendicular.

21. Taking the derivative of both sides with respect to t gives $\frac{dq}{dt} = -400e^{-0.08p}\frac{dp}{dt}$. According to the given information, $p = 10$ and $dp/dt = -0.5$, Therefore, $dq/dt = -400e^{-0.8}(-0.5) \approx 90$ barrels per day.

23. Since $r = 0.5$, the volume of water is $V = \pi r^2 h = 0.25\pi h$. Taking the derivative of both sides with respect to t gives $\frac{dV}{dt} = 0.25\pi \frac{dh}{dt}$, and solving for dh/dt yields $\frac{dh}{dt} = \frac{1}{0.25\pi}\frac{dV}{dt}$. The volume of water is decreasing by 0.1 cubic feet per minute, so that $dV/dt = -0.1$. Therefore, $\frac{dh}{dt} = \frac{1}{0.25\pi} \cdot (-0.1) \approx -0.13$ feet per minute.

25. Taking the derivative of both sides of the equation $c = 2\pi r$ with respect to t gives $\frac{dc}{dt} = 2\pi \frac{dr}{dt}$. The given information says that $dr/dt = 0.02$ centimeters per second. Therefore, $dc/dt = 2\pi \cdot 0.02 = 0.04\pi \approx 0.126$ centimeters per second.

27. Let x be the distance of the boat from the pier and z be the length of rope, both functions of time t. By the Pythagorean theorem, they satisfy the equation $z^2 = x^2 + 4^2$. Taking the derivative of both sides with respect to t, we get $2z\frac{dz}{dt} = 2x\frac{dx}{dt}$, so that $\frac{dx}{dt} = \frac{z}{x}\frac{dz}{dt}$. At $x = 12$, we have

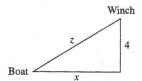

$$z = \sqrt{12^2 + 4^2} = \sqrt{160} = 4\sqrt{10}.$$

Therefore, $\frac{dx}{dt} = \frac{4\sqrt{10}}{12} \cdot 2 = \frac{2}{3}\sqrt{10} \approx 2.1$ feet per minute.

29. (a) $\dfrac{dA}{dt} = 2\pi r\,\dfrac{dr}{dt}$

(b) We are given that $r = 3$ centimeters and $dr/dt = 0.2$ centimeters per second. Therefore, $A = \pi \cdot 3^2 = 9\pi$ square centimeters and $dA/dt = 2\pi \cdot 3 \cdot 0.2 = 1.2\pi$ square centimeters per second.

31. Let x be the coordinate of car A and y the coordinate of car B along their respective lines of motion, and let s be the distance between them. Then $s^2 = x^2 + y^2$.

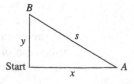

(a) After 1 hour, $x = 40$ and $y = 30$. Therefore,
$s = \sqrt{40^2 + 30^2} = 50$ miles.

(b) Taking the derivative with respect to t gives $2s\frac{ds}{dt} = 2x\frac{dx}{dt} + 2y\frac{dy}{dt}$, so that

$$\frac{ds}{dt} = \frac{1}{s}\left(x\frac{dx}{dt} + y\frac{dy}{dt}\right).$$

At $t = 1$, $x = dx/dt = 40$, $y = dy/dt = 30$ and $s = 50$. Therefore,

$$\frac{ds}{dt} = \frac{1}{50}(40^2 + 30^2) = 50 \text{ mph.}$$

CHAPTER 3 REVIEW EXERCISES

1. $f'(x) = \lim_{h\to 0}\dfrac{f(x+h) - f(x)}{h} = \lim_{h\to 0}\dfrac{[(x+h)^2 - 3(x+h) + 2] - (x^2 - 3x + 2)}{h}$

$= \lim_{h\to 0}\dfrac{(x^2 + 2xh + h^2 - 3x - 3h + 2) - (x^2 - 3x + 2)}{h}$

$= \lim_{h\to 0}\dfrac{h(2x + h - 3)}{h} = \lim_{h\to 0}(2x + h - 3) = 2x - 3$

3. $f'(x) = \lim_{h\to 0}\dfrac{f(x+h) - f(x)}{h} = \lim_{h\to 0}\dfrac{\frac{2}{3(x+h)-1} - \frac{2}{3x-1}}{h}$

$= \lim_{h\to 0}\dfrac{1}{h}\left(\dfrac{2(3x - 1) - 2(3x + 3h - 1)}{(3x + 3h - 1)\cdot(3x - 1)}\right)$

$= \lim_{h\to 0}\dfrac{1}{h}\left(\dfrac{-6h}{(3x + 3h - 1)(3x - 1)}\right)$

$= \lim_{h\to 0}\dfrac{-6}{(3x + 3h - 1)(3x - 1)} = -\dfrac{6}{(3x - 1)^2}$

5. Writing $y = x^{1/2}$, we have $y' = \frac{1}{2}x^{-1/2} = 1/2\sqrt{x}$. The slope at $x = 9$ is $y'(9) = \frac{1}{6}$. Also, $y(9) = \sqrt{9} = 3$. The equation of the tangent line is $y - 3 = \frac{1}{6}(x - 9)$, or $y = \frac{1}{6}x + \frac{3}{2}$.

7. The derivative is $y' = 2x - \frac{8}{x^3}$, and the slope at $x = 2$ is $y'(2) = 2\cdot 2 - \frac{8}{2^3} = 3$. The equation of the tangent line is $y - 5 = 3(x - 2)$, or $y = 3x - 1$.

9. By the quotient rule, $y' = \frac{(x+1)-x}{(x+1)^2} = \frac{1}{(x+1)^2}$, so that the slope at $x = 1$ is $y'(1) = 1/4$. Also, $y(1) = 1/2$. The equation of the tangent line is $y - \frac{1}{2} = \frac{1}{4}(x - 1)$, or $y = \frac{1}{4}x + \frac{1}{4}$.

11. By the chain rule, $y' = 2xe^{x^2-1}$, so that the slope at $x = 1$ is $y'(1) = 2$. Also, $y(1) = e^0 = 1$. The equation of the tangent line is $y - 1 = 2(x - 1)$, or $y = 2x - 1$.

13. $f'(x) = 5x^4 - 12x^3 + 6x^2 + 2x - 5$

15. $f'(x) = e^{x^2-3x+1}\dfrac{d}{dx}(x^2 - 3x + 1) = (2x - 3)e^{x^2-3x+1}$

17. $f'(x) = \dfrac{1}{3}(3x + 1)^{-2/3}\dfrac{d}{dx}(3x + 1) = (3x + 1)^{-2/3}$

19. $f'(x) = 4(\ln x)^3\dfrac{d}{dx}\ln x = \dfrac{4}{x}(\ln x)^3$ **21.** $f'(x) = \dfrac{d}{dx}(1 + e^{-x}) = -e^{-x} = -\dfrac{1}{e^x}$

23. $f'(x) = 3x^2 \ln x + x^3 \cdot \dfrac{1}{x} = 3x^2 \ln x + x^2 = x^2(1 + 3\ln x)$

25. $f'(x) = \dfrac{1}{2\sqrt{e^{3x} + 1}} \cdot \dfrac{d}{dx}\left(e^{3x} + 1\right) = \dfrac{3e^{3x}}{2\sqrt{e^{3x} + 1}}$

27. Solving $y' = 2x + 3 = 9$, we find $x = 3$.

29. Since $y' = 1 + e^x > 0$ for all x, the equation $y' = 0$ has no solution. Therefore, the graph has no horizontal tangent line.

31. Setting $y' = \frac{3}{2}x^{1/2} = 1$, we get $x^{1/2} = \frac{2}{3}$. Therefore, $x = \frac{4}{9}$.

33. $f(x) = 2\ln x$, so that $f'(x) = \dfrac{2}{x}$ and $f''(x) = -\dfrac{2}{x^2}$.

35. $f'(x) = (2x + 1)^{-1/2}$, so that $f''(x) = -(2x + 1)^{-3/2}$.

37. $f'(x) = e^{2x} + 2xe^{2x}$, so that $f''(x) = 4e^{2x} + 4xe^{2x} = 4e^{2x}(1 + x)$

39. (a) $v(t) = s'(t) = \dfrac{15}{2}t^2 - 2t^{5/2}$ and $v(1) = \dfrac{11}{2}$ ft/sec.

   (b) $\dfrac{s(1) - s(0)}{1 - 0} = \dfrac{27}{14}$ ft/sec.

   (c) $a(t) = \dfrac{d^2 s}{dt^2} = 15t - 5t^{3/2}$ and $a(4) = 20$ ft/sec^2.

41. (a) $\dfrac{dq}{dp} = -400e^{-0.08p}$, so that $\dfrac{dq}{dp}(10) = -400e^{-0.8} \approx -179.73$.

   (b) $R(p) = pq = 5{,}000pe^{-0.08p}$, so that $\dfrac{dR}{dp} = 5{,}000(e^{-0.08p} - 0.08pe^{-0.08p})$. Therefore, $\dfrac{dR}{dp}(10) = 1{,}000e^{-0.8} \approx 449.33$.

43. $R'(x) = 220 - 8x$ and $C'(x) = 40$. Setting them equal gives $220 - 8x = 40$, whose solution is $x = 22.5$. That is not the same as the break even point, which is the solution of $R(x) = C(x)$. In this case, that equation takes the form $220x - 4x^2 = 900 + 40x$, whose (approximate) solutions are $x = 5.73$ and $x = 39.27$.

45. $f'(x) = \dfrac{1}{2\sqrt{x}}$, so that $\sqrt{x} \approx \sqrt{a} + \dfrac{1}{2\sqrt{a}}(x - a)$

   (a) With $a = 49$, we have $\sqrt{50} \approx 7 + \frac{1}{14} = \frac{99}{14}$

   (b) With $a = 81$, we have $\sqrt{83} \approx 9 + \frac{1}{18} \cdot 2 = \frac{82}{9}$

   (c) With $a = 100$, we have $\sqrt{98} \approx 10 - \frac{1}{10} = \frac{99}{10}$

47. Using $f(x) = 1/x$, with $f'(x) = -1/x^2$ and $a = 2$, we get

$$\frac{1}{2.108} \approx \frac{1}{2} - \frac{1}{4} \cdot 0.108 = 0.473.$$

49. Using $f(x) = \ln x$, with $f'(x) = 1/x$ and $a = 1$, we get

$$\ln(1.005) = \ln 1 + 0.005 = 0.005.$$

51. Using 5,568 as the half-life of C-14, we write the decay equation in the form $y = y_0 e^{-t(\ln 2)/5,568}$, where y_0 is the concentration of C-14 at $t = 0$. Then y_0 is also the concentration of C-14 in the fresh sample. Therefore, $0.445 y_0 = y_0 e^{-t(\ln 2)/5,568}$. Solving for t, we get $t \approx -5,568 \ln(0.445)/\ln 2 \approx$ 6,500 years.

53. $f'(g(3)) \cdot g'(3) = f'(2) \cdot \dfrac{3}{2} = 1 \cdot \dfrac{3}{2} = \dfrac{3}{2}$

55. $\dfrac{g(2)f'(2) - f(2)g'(2)}{(g(2))^2} = \dfrac{\left(-\frac{1}{3}\right) - 5 \cdot \frac{3}{5}}{\left(-\frac{1}{3}\right)^2} = -30$

57. $\left(-\dfrac{1}{f(1)^2}\right) \cdot f'(1) = \left(-\dfrac{1}{4}\right) \cdot \left(-\dfrac{1}{2}\right) = \dfrac{1}{8}$

59. (a) $f(x)$ is discontinuous at $x = -3$, $x = 3$, and $x = 6$. Explanation: At both $x \pm 3$, there are two one-sided limits that do not agree, so that $\lim_{x \to -3} f(x)$ and $\lim_{x \to 3} f(x)$ do not exist. At $x = 6$, the limit exists, but does not equal $f(6)$.

   (b) $f(x)$ is not differentiable at $x = -3$, $x = 0$, $x = 3$, $x = 5$, $x = 6$. Explanation: At $x = -3$, $x = 3$, and $x = 6$, $f(x)$ is not continuous, and cannot be differentiable. Although $f(x)$ is continuous at $x = 0$ and $x = 5$, the graph has a corner, and there is no well-determined tangent line.

61. Implicit differentiation gives $2x = \dfrac{2yy'}{1 + y^2}$, so that $y' = \dfrac{x(1 + y^2)}{y}$.

63. Implicit differentiation gives $e^y + xe^y y' - 2xy - x^2 y' = 2x$, and solving for y' yields

$$y' = \frac{2x + 2xy - e^y}{xe^y - x^2}.$$

65. First, $y' = -2cte^{-t^2}$. Second, $-2ty = -2tce^{-t^2}$. Thus, $y' = -2ty$.

67. By implicit differentiation, we get $2yy' + 2t = 0$, which means that $y' = -t/y$. Therefore, $yy' + t = y \cdot (-t/y) + t = 0$, which shows that y defined implicity by $y^2 + t^2 = c^2$ satisfies the differential equation $yy' + t = 0$.

69. First, we check the initial condition: $y(0) = (1 + e^0)^{-1} = \frac{1}{2}$. Next, taking the derivative gives $y' = e^{-t}/(1 + e^{-t})^2$. Finally, we check that

$$y(1 - y) = \frac{1}{1 + e^{-t}}\left(1 - \frac{1}{1 + e^{-t}}\right) = \frac{e^{-t}}{(1 + e^{-t})^2} = y'.$$

Therefore $y = (1 + e^{-t})^{-1}$ satisfies both the initial condition and the differential equation.

71. First, we check the initial condition: $y(1) = \frac{1}{3} + \frac{5}{3} = 2$. Next, taking the derivative gives $y' = \frac{2}{3}t - \frac{5}{3t^2}$. Finally,

$$ty' + y = \frac{2}{3}t^2 - \frac{5}{3t} + \frac{1}{3}t^2 + \frac{5}{3t} = t^2.$$

Thus, $y = \frac{1}{3}t^2 + \frac{5}{3t}$ satisfies both the initial condition and the differential equation.

73. $f(x)$ is differentiable at every x. The only questionable point is at $x = 0$, and there we have

$$\lim_{h \to 0^+} \frac{f(0+h) - f(0)}{h} = \lim_{h \to 0^+} \frac{h^2}{h} = \lim_{h \to 0^+} h = 0$$

and

$$\lim_{h \to 0^-} \frac{f(0+h) - f(0)}{h} = \lim_{h \to 0^-} \frac{h^3}{h} = \lim_{h \to 0^-} h^2 = 0,$$

so that $f'(0) = \lim_{h \to 0} \frac{f(0+h) - f(0)}{h} = 0$, as the graph shows.

On the other hand, $g(x)$ is not differentiable at $x = 0$. Observe that the graph has a corner at that point and no well-determined tangent. More precisely, we have

$$\lim_{h \to 0^+} \frac{g(0+h) - g(0)}{h} = \lim_{h \to 0^+} \frac{h}{h} = 1$$

and

$$\lim_{h \to 0^-} \frac{g(0+h) - g(0)}{h} = \lim_{h \to 0^-} \frac{h^2}{h} = \lim_{h \to 0^-} h = 0,$$

which shows that $\lim_{h \to 0} \frac{g(0+h) - g(0)}{h}$ does not exist.

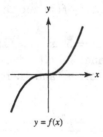

$y = f(x)$

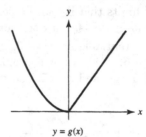

$y = g(x)$

75. (a) Taking the derivative with respect to t on both sides of the equation $V = \frac{4}{3}\pi r^3$ gives $\frac{dV}{dt} = 4\pi r^2 \frac{dr}{dt}$, which we can rewrite as $\frac{dr}{dt} = \frac{1}{4\pi r^2} \frac{dV}{dt}$. From the given information $r = 2$ and $dV/dt = 1$, we get $dr/dt = 1/16\pi$ inches per second.

(b) Taking the derivative with respect to t on both sides of the equation $S = 4\pi r^2$, we get $\frac{dS}{dt} = 8\pi r \frac{dr}{dt}$. Using the result of (a), we obtain

$$\frac{dS}{dt} = 8\pi \cdot 2 \cdot \frac{1}{16\pi} = 1 \text{ square inch per second.}$$

CHAPTER 3 EXAM

1. (c) because

$$\lim_{h \to 0} \frac{f(4+h) - f(4)}{h} = \lim_{h \to 0} \frac{\sqrt{2(4+h)+1} - \sqrt{2 \cdot 4 + 1}}{h} = \lim_{h \to 0} \frac{\sqrt{9 + 2h} - 3}{h}$$

3. Writing $f(x) = 5 + 2x^{1/2} - 3x^{-3}$, we obtain

$$f'(x) = x^{-1/2} + 9x^{-4} = \frac{1}{\sqrt{x}} + \frac{9}{x^4}.$$

5. Solving $y' = 12x^2 - 5 = 43$ yields $x = \pm 2$, and $y(2) = 23$ and $y(-2) = -21$. Thus, the graph of $y = 4x^3 - 5x + 1$ has slope 43 at $(2, 23)$ and $(-2, -21)$.

7. The rate at which the population is changing is $P'(t) = -400e^{-0.04t}$, and $P'(100) = -400e^{-4}$ organisms per hour.

9. With $f(x) = x^{1/3}$ and $a = 1,000$, we have $f'(x) = \frac{1}{3}x^{-2/3}$, so that $f(1,000) = 10$ and $f'(1,000) = 1/300$. Therefore,

$$\sqrt[3]{1,000 + h} \approx f(1,000) + f'(1,000)h = 10 + \frac{1}{300}h.$$

11. The true statement is (a). First, $f(x)$ is continuous at $x = 2$ since $|2 - 2| = 0$ and $\lim_{x \to 2} |x - 2| = 0$. Second, $f'(2)$ does not exist since

$$\lim_{x \to 2^+} \frac{|x - 2|}{x - 2} = 1 \quad \text{but} \quad \lim_{x \to 2^-} \frac{|x - 2|}{x - 2} = -1.$$

13. $f(x) = x^{4/5}$ is not differentiable at $x = 0$ since

$$\lim_{h \to 0^+} \frac{f(0 + h) - f(0)}{h} = \lim_{h \to 0^+} \frac{h^{4/5}}{h} = \lim_{h \to 0^+} h^{-1/5} = \infty$$

and

$$\lim_{h \to 0^-} \frac{f(0 + h) - f(0)}{h} = \lim_{h \to 0^-} \frac{h^{4/5}}{h} = \lim_{h \to 0^-} h^{-1/5} = -\infty.$$

The one-sided limits are different and neither is finite.

15. Since $y' = 2x \ln x + x$, the slope at $x = 1$ is $y'(1) = 2 \ln 1 + 1 = 1$. Also, $y(1) = 1 \ln 1 = 0$. The equation of the tangent line is $y = x - 1$.

17. Since both $y' = -ce^{-t} + 2$ and $2t - y = -ce^{-t} + 2$, we conclude that $y = ce^{-t} + 2t - 2$ is a solution to the differential equation $y' = 2t - y$.

19. According to Newton's law of cooling, the temperature at the end of t minutes is given by $H(t) = 20 + Ae^{kt}$, and $H'(t) = kAe^{kt}$. Since $H(15) = 48$ and $H'(15) = -3.5 = -7/2$, we get $28 = Ae^{15k}$ and $-\frac{7}{2} = kAe^{15k}$. Taking the quotient, we get $-\frac{1}{8} = k$, so that $28 = Ae^{-15/8}$ and $A = 28e^{15/8}$. Therefore, the initial temperature was $H(0) = 20 + 28e^{15/8} \approx 202.6°C$.

21. By the chain rule, $V' = 4\pi r^2 r'$. Substituting $V' = 8\pi$ and $r' = \frac{1}{2}$ yields $8\pi = 4\pi r^2 \cdot \frac{1}{2}$. Therefore, $r = 2$ and $V = \frac{4}{3}\pi \cdot 2^3 = \frac{32}{3}\pi$ cubic feet.

23. Implicit differentiation gives $3x^2 + xy' + y - 3y^2 y' = 0$. By rearranging terms and factoring, we get $3x^2 + y = y'(3y^2 - x)$. Therefore, $y' = \frac{3x^2 + y}{3y^2 - x}$.

CHAPTER 4
Using the Derivative

EXERCISES 4.1

1. Set $f'(x) = 4x - 7 = 0$. There is a critical point at $x = \frac{7}{4}$.

x	$(-\infty, \frac{7}{4})$	$(\frac{7}{4}, \infty)$
$f'(x)$	$-$	$+$

Thus, $f(x)$ is increasing on $(\frac{7}{4}, \infty)$ and decreasing on $(-\infty, \frac{7}{4})$. There is a local minimum at $x = \frac{7}{4}$.

3. Set $f'(x) = 6x^2 + 6x - 36 = 6(x+3)(x-2) = 0$. There are critical points at $x = -3$ and $x = 2$.

x	$(-\infty, -3)$	$(-3, 2)$	$(2, \infty)$
$f'(x)$	$+$	$-$	$+$

Thus, $f(x)$ is increasing on $(-\infty, -3)$ and $(2, \infty)$ and decreasing on $(-3, 2)$. There is a local maximum at $x = -3$ and a local minimum at $x = 2$.

5. Set $f'(x) = 14x(x^2 - 4)^6 = 0$. There are critical points at $x = -2, x = 0$, and $x = 2$.

x	$(-\infty, -2)$	$(-2, 0)$	$(0, 2)$	$(2, \infty)$
$f'(x)$	$-$	$-$	$+$	$+$

Thus, $f(x)$ is increasing on $(0, \infty)$ and decreasing on $(-\infty, 0)$. There is a local minimum at $x = 0$.

7. Set $f'(x) = xe^x + e^x = e^x(x+1) = 0$. There is a critical point at $x = -1$.

x	$(-\infty, -1)$	$(-1, \infty)$
$f'(x)$	$-$	$+$

Thus, $f(x)$ is increasing on $(-1, \infty)$ and decreasing on $(-\infty, -1)$. There is a local minimum of $f(x)$ at $x = -1$.

9. There is a critical point at $x = 0$ since $f'(x) = \frac{2}{5}x^{-3/5}$ does not exist there, and no other critical points since $\frac{2}{5}x^{-3/5} = 0$ has no solution.

x	$(-\infty, 0)$	$(0, \infty)$
$f'(x)$	$-$	$+$

Thus, $f(x)$ is increasing on $(0, \infty)$ and decreasing on $(-\infty, 0)$. There is a local minimum of $f(x)$ at $x = 0$.

11. Set $f'(x) = e^{-x} - xe^{-x} = e^{-x}(1 - x) = 0$. There is a critical point at $x = 1$.

x	$(-\infty, 1)$	$(1, \infty)$
$f'(x)$	$+$	$-$

Thus, $f(x)$ is increasing on $(-\infty, 1)$ and decreasing on $(1, \infty)$. There is a local maximum of $f(x)$ at $x = 1$.

13. The derivative is $f'(x) = \begin{cases} -1 & \text{if } x < 1 \\ 1 & \text{if } x > 1 \end{cases}$ and does not exist at $x = 1$. The only critical point is at $x = 1$.

x	$(-\infty, 1)$	$(1, \infty)$
$f'(x)$	$-$	$+$

Thus, $f(x)$ is increasing on $(1, \infty)$ and decreasing on $(-\infty, 1)$. There is a local minimum at $x = 1$.

15. Since $f'(x) = 3x^2 - 4x + 3 = 3(x - \frac{2}{3})^2 + \frac{5}{3} > 0$ for all x, there is no critical point, and $f(x)$ is increasing over the entire x-axis.

17. Set $f'(x) = \frac{x \cdot \frac{1}{x} - (1 + \ln x)}{x^2} = -\frac{\ln x}{x^2} = 0$. There is a critical point at $x = 1$.

x	$(0, 1)$	$(1, \infty)$
$f'(x)$	$+$	$-$

Thus, $f(x)$ is increasing on $(0, 1)$ and $f(x)$ is decreasing on $(1, \infty)$. There is a local maximum at $x = 1$.

19. Set $f'(x) = \frac{(x^2 + 1)2x - x^2 \cdot 2x}{(x^2 + 1)^2} = \frac{2x}{(x^2 + 1)^2} = 0$. There is a critical point at $x = 0$.

x	$(-\infty, 0)$	$(0, \infty)$
$f'(x)$	$-$	$+$

Thus, $f(x)$ is increasing on $(0, \infty)$ and decreasing on $(-\infty, 0)$. There is a local minimum of $f(x)$ at $x = 0$.

21. **(a)** Solving $P'(y) = -2y + 20 = 0$ yields $y = 10$.

(b) Testing one point, such as $x = 5$, on $(0, 10)$ and one point, such as $x = 15$, on $(10, 20)$ gives

y	$(0, 10)$	$(10, 20)$
$P'(y)$	$+$	$-$

(c) There is a local maximum at $y = 10$, and, by the single critical point principle, it is a global maximum, with $P(10) = 180$. Since $P(y)$ is increasing on $(0, 10)$, its smallest value on $[0, 10]$ is at $y = 0$, with $P(0) = 80$. And since $P(y)$ is decreasing on $(10, 20)$, its smallest value on $[10, 20]$ is at $y = 20$ with $P(20) = 80$. Thus, the greatest amount of rain was at $y = 10$ years, and the least at $y = 0$ and $y = 20$.

23. Since $S'(t) = \frac{1}{3}t^{-2/3}$ is positive for $t > 0$, $S(t)$ is an increasing function on $(0, \infty)$, which means the daily sales will not reach a maximum.

25. The marginal cost is $MC(x) = C'(x) = 2(5x + 2)^{-3/5}$, and its derivative is $MC'(x) = C''(x) = -6(5x + 2)^{-8/5}$. Since $MC'(x) < 0$ for all $x > 0$ (the natural domain of MC), the marginal cost decreases with increased production.

27. **(a)**

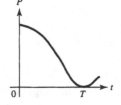

   (b)   The derivative is negative in $(0, T)$ and positive in (T, ∞).

   (c)   The population had a maximum at $t = 0$ and a minimum at $t = T$.

29. b

31. c

33. **(a)** $(-3, -1)$ and $(1, 3)$

   (b) $(-1, 1)$

   (c) Two points, $x = 1$ and -1

   (d) At $x = -1$

   (e) At $x = 1$

35.

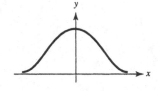

37.

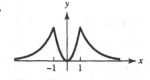

39.

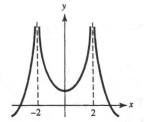

41. **(c)** Reason: The graph of the derivative shows that the derivative is negative for $x < 0$ and positive for $x > 0$, which means that the function is decreasing on $(-\infty, 0)$ and increasing on $(0, \infty)$.

43. **(d)** Reason: The graph of the function shows that it is decreasing over its entire domain, which means that the derivative is always negative.

45.

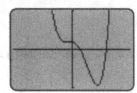

The function has a local maximum at $x \approx 0.5387$ and a local minimum at $x \approx 2.361$, found by solving $f'(x) = 0$ using the equation solver.

47. Set $f'(x) = 2 - 6x = 0$. There is a critical point of $f(x)$ at $x = 1/3$.

x	$(-\infty, 1/3)$	$(1/3, \infty)$
$f'(x)$	$+$	$-$

Thus, $f(x)$ has a local maximum at $x = \frac{1}{3}$, with value $f(\frac{1}{3}) = \frac{16}{3}$. By the single critical point principle, it is also a global maximum.

49. Set $f'(x) = (1 - x)e^{-x} = 0$. There is a critical point of $f(x)$ at $x = 1$.

x	$(-\infty, 1)$	$(1, \infty)$
$f'(x)$	$+$	$-$

Thus, $f(x)$ has a local maximum at $x = 1$, with value $f(1) = 1/e$. By the single critical point principle, it is also a global maximum.

51. Set $f'(x) = 2x/(x^2 + 1) = 0$. There is a critical point of $f(x)$ at $x = 0$.

x	$(-\infty, 0)$	$(0, \infty)$
$f'(x)$	$-$	$+$

Thus, $f(x)$ has a local minimum of $f(x)$ at $x = 0$, with value $f(0) = 0$. By the single critical point principle, it is also a global minimum.

53. Set $f'(x) = -2xe^{-x^2} = 0$. There is a critical point of $f(x)$ at $x = 0$.

x	$(-\infty, 0)$	$(0, \infty)$
$f'(x)$	$+$	$-$

Thus, $f(x)$ has a local maximum of $f(x)$ at $x = 0$, with value $f(0) = 1$. By the single critical point principle, it is also a global maximum.

55. Set $f'(x) = (2\ln x)/x = 0$. There is a critical point at $x = 1$.

x	$(0, 1)$	$(1, \infty)$
$f'(x)$	$-$	$+$

Thus, $f(x)$ has a local minimum of $f(x)$ at $x = 1$, with value $f(1) = 0$. By the single critical point principle, it is also a global minimum.

57. Set $f'(x) = 10x(x^2 - 1)^4 = 0$. There are critical points of $f(x)$ at $x = -1, x = 0$ and $x = 1$.

x	$(-\infty, -1)$	$(-1, 0)$	$(0, 1)$	$(1, \infty)$
$f'(x)$	$-$	$-$	$+$	$+$

Thus, $f(x)$ has a local minimum of $f(x)$ at $x = 0$, with value $f(0) = -1$. Since $f(x)$ is decreasing on $(-\infty, 0)$ and increasing on $(0, \infty)$, we conclude that there is also a global minimum at $x = 0$.

59. **(a)** $f(-2) = -27 < 0$ and $f(-1) = 5 > 0$. Since $f(x)$ is a continuous function, it follows from the intermediate value theorem that $f(x) = 0$ has a solution between $x = -2$ and $x = -1$.

   (b) Since $f'(x) = 5x^4 + 1 \geq 1$ for all x, it is never zero, and $f(x)$ has no critical points. Moreover, $f(x)$ is always increasing, and the graph cannot cross the x-axis more than once. In other words, $f(x) = 0$ cannot have two solutions.

61. Set $f'(x) = e^x - 1 = 0$. There is a critical point of $f(x)$ at $x = 0$, and checking signs gives

x	$(-\infty, 0)$	$(0, \infty)$
$f'(x)$	$-$	$+$

Thus, $f(x)$ has a local minimum at $x = 0$, which (by the single critical point principle) is also a global minimum. The global minimum value is $f(0) = 1$. In other words, $f(x) \geq 1$ for all x, so that $f(x)$ cannot equal zero.

63. If $0 < p < 12$, then $\frac{dp}{dt} = 0.03p(1 - \frac{p}{12}) > 0$, which means that $p(t)$ is an increasing function for all $t > 0$ and does not achieve a maximum.

EXERCISES 4.2

In exercises 1 and 3, we use the fact that $f'(x)$ is increasing (resp., decreasing) if and only if $f(x)$ is concave up (resp., down).

1. B   **3.** A   **5.** (a), (c), (f), (g), (k)

7. (b), (f), (i)   **9.** (f)   **11.** (i)

13. (a), (g)   **15.** (c)   **17.** (h)

19. Set $f''(x) = 12x + 12 = 0$. The solution is $x = -1$.

x	$(-\infty, -1)$	$(-1, \infty)$
$f''(x)$	$-$	$+$

Thus, the graph of $f(x)$ is concave down on $(-\infty, -1)$ and concave up on $(-1, \infty)$. There is an inflection point at $(-1, 10)$.

21. Set $f''(x) = 2x^2 - 6x - 8 = 0$. Solutions are $x = -1$ and $x = 4$.

x	$(-\infty, -1)$	$(-1, 4)$	$(4, \infty)$
$f''(x)$	$+$	$-$	$+$

Thus, the graph of $f(x)$ is concave up on $(-\infty, -1)$ and $(4, \infty)$ and concave down on $(-1, 4)$. There are inflection points at $(-1, -\frac{65}{6})$ and $(4, -\frac{235}{3})$.

23. The second derivative, $f''(x) = 6/(x - 2)^4$, $x \neq 2$, is never zero, but it is undefined at $x = 2$. Checking signs gives

x	$(-\infty, 2)$	$(2, \infty)$
$f''(x)$	$+$	$+$

The graph of $f(x)$ is concave up on both $(-\infty, 2)$ and $(2, \infty)$. There is no inflection point.

25. The second derivative, $f''(x) = 2/(x - 1)^3$, $x \neq 1$, is never zero, but it is undefined at $x = 1$.

x	$(-\infty, 1)$	$(1, \infty)$
$f''(x)$	$-$	$+$

The graph of $f(x)$ is concave down on $(-\infty, 1)$ and $f(x)$ is concave up on $(1, \infty)$. Since 1 is not in the domain of f, there is no inflection point.

27. Set $f''(x) = (2 - 2\ln x)/x^2 = 0$. The solution is $x = e$.

x	$(0, e)$	(e, ∞)
$f''(x)$	$+$	$-$

Thus, the graph of $f(x)$ is concave up on $(0, e)$ and concave down on (e, ∞). There is an inflection point at $(e, 1)$.

29. Set $f''(x) = -18(x^2 - 1)/(x^2 + 3)^3 = 0$. The solutions are $x = -1$ and $x = 1$.

x	$(-\infty, -1)$	$(-1, 1)$	$(1, \infty)$
$f''(x)$	$-$	$+$	$-$

Thus, the graph of $f(x)$ is concave down on $(-\infty, -1)$ and $(1, \infty)$ and concave up on $(-1, 1)$. There are inflection points at $(-1, \frac{1}{4})$ and $(1, \frac{1}{4})$.

31. Set $f''(x) = (6x^2 - 8)/(x^2 + 4)^3 = 0$. The solutions are $x = -2/\sqrt{3}$ and $x = 2/\sqrt{3}$.

x	$(-\infty, -2/\sqrt{3})$	$(-2/\sqrt{3}, 2/\sqrt{3})$	$(2/\sqrt{3}, \infty)$
$f''(x)$	$+$	$-$	$+$

Thus, the graph of $f(x)$ is concave up on $(-\infty, -\frac{2}{\sqrt{3}})$ and $(\frac{2}{\sqrt{3}}, \infty)$ and concave down on $(-\frac{2}{\sqrt{3}}, \frac{2}{\sqrt{3}})$. There are inflection points at $(-\frac{2}{\sqrt{3}}, \frac{3}{16})$ and $(\frac{2}{\sqrt{3}}, \frac{3}{16})$.

33. Set $f'(x) = 3x^2 - 6 = 0$. There are critical points of $f(x)$ at $x = \sqrt{2}$ and $x = -\sqrt{2}$. Since $f''(x) = 6x$, we have $f''(\sqrt{2}) = 6\sqrt{2} > 0$ and $f''(-\sqrt{2}) = -6\sqrt{2} < 0$. Thus, $f(x)$ has a local minimum at $x = \sqrt{2}$ and a local maximum at $x = -\sqrt{2}$. The local minimum is $f(\sqrt{2}) = 1 - 4\sqrt{2}$, and the local maximum is $f(\sqrt{2}) = 1 + 4\sqrt{2}$.

35. Set $f'(x) = (3x + 1)e^{3x} = 0$. There is a critical point of $f(x)$ at $x = -1/3$. Since $f''(x) = e^{3x}(6 + 9x)$, we have $f''(-\frac{1}{3}) = 3e^{-1} > 0$. Thus, $f(x)$ has a local minimum at $x = -1/3$ with value $f(-\frac{1}{3}) = -\frac{1}{3}e^{-1}$.

37. Set $f'(x) = 1 - (4/x^2) = 0$. There are critical points of $f(x)$ at $x = -2$ and $x = 2$. Since $f''(x) = 8/x^3$, we have $f''(2) = 1 > 0$ and $f''(-2) = -1 < 0$. Thus, $f(x)$ has a local minimum at $x = 2$ with value $f(2) = 4$, and a local maximum at $x = -2$ with value $f(-2) = -4$.

39. Set $f'(x) = (-2x + 2)e^{-x^2+2x} = 0$. There is a critical point of $f(x)$ at $x = 1$. Since $f''(x) = (4x^2 - 8x + 2)e^{-x^2+2x}$, we have $f''(1) = -2e < 0$. Thus, $f(x)$ has a local maximum at $x = 1$, with value $f(1) = e$.

41. Set $f'(x) = 4x^3 - 4x = 0$. There are three critical points of $f(x)$, at $x = -1, x = 0$, and $x = 1$. Since $f''(x) = 12x^2 - 4$, we have $f''(1) = f''(-1) = 8$ and $f''(0) = -4$. Thus, $f(x)$ has a local minimum at both $x = -1$ and $x = 1$, with value $f(-1) = f(1) = 2$, and a local maximum at $x = 0$ with value $f(0) = 3$.

43. Set $f'(x) = 1 - \frac{1}{2}x^{-1/2} = 0$. There is a critical point of $f(x)$ at $x = 1/4$. Since $f''(x) = \frac{1}{4}x^{-3/2}$, we have $f''(\frac{1}{4}) = 2$. Thus $f(x)$ has a local minimum at $x = 1/4$ with value $f(\frac{1}{4}) = -\frac{1}{4}$.

45. We first compute

$$D'(t) = \frac{3.2e^{4-0.4t}}{(1 + e^{4-0.4t})^2} \quad \text{and} \quad D''(t) = \frac{2.56e^{8-0.8t}}{(1 + e^{4-0.4t})^3} - \frac{1.28e^{4-0.4t}}{(1 + e^{4-0.4t})^2}.$$

Since we want to maximize $D'(t)$, we solve $D''(t) = 0$ to obtain $t = 10$. Checking the sign of $D''(t)$ yields

t	$(0, 10)$	$(10, \infty)$
$D''(t)$	$+$	$-$

Thus, $D'(t)$ is increasing on $(0, 10)$ and decreasing on $(10, \infty)$ and has a local maximum at $t = 10$. By the single critical point principle, it also has a global maximum at $t = 10$, and we conclude that the rate of depth increase was highest on the tenth day.

47. Solving $f'(x) = 12x^3 + 24x^2 + 12x = 12x(x + 1)^2 = 0$, we find the critical points of $f(x)$ are at $x = -1$ and $x = 0$. Next, we compute $f''(x) = 36x^2 + 48x + 12 = 12(3x^2 + 4x + 1)$. At $x = 0$, we have $f''(0) = 12$, and there is a local minimum with value $f(0) = -1$. At $x = -1$, we have $f''(-1) = 0$, and we cannot use the second derivative test. Checking the sign of $f'(x)$ shows that it is negative on both $(-\infty, -1)$ and $(-1, 0)$. Therefore, there is no local extremum at $x = -1$.

49. Solving $f'(x) = 8x(x^2 - 1)^3 = 0$, we find the critical points are at $x = -1, x = 0$, and $x = 1$. Next, we compute $f''(x) = 8(x^2 - 1)^2(7x^2 - 1)$. At $x = 0$, we have $f''(0) = -8$, and there is a local maximum. At $x = \pm 1$, we have $f''(\pm 1) = 0$, and we cannot use the second derivative test. Checking the sign of $f'(x)$ gives

x	$(-\infty, -1)$	$(-1, 0)$	$(0, 1)$	$(1, \infty)$
$f'(x)$	$-$	$+$	$-$	$+$

Thus, $f(x)$ has a local minimum at $x = -1$, with value $f(-1) = 0$ and a local minimum at $x = 1$ wit value $f(1) = 0$.

51. **(a)** The graph of the total enrollment is concave down. Recall that the graph of $f(x)$ is concave down if and only $f'(x)$ is decreasing.

   (b) Change *smaller* to *larger*.

53. **(a)** *Employment*

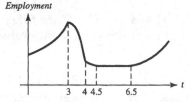

   (b) There are local minima at $t = 0$ and at all t satisfying $4.5 \le t \le 6.5$. There is a local maximum at $t = 3$.

t	$(0, 3)$	$(3, 4)$	$(4, 4.5)$	$(4.5, 6.5)$	$(6.5, \infty)$
derivative	$+$	$-$	$-$	0	$+$

   (c)

t	$(0, 3)$	$(3, 4)$	$(4, 4.5)$	$(4.5, 6.5)$	$(6.5, \infty)$
second derivative	$+$	$-$	$+$	0	$+$
concavity	up	down	up		up

   (d) Inflection points occur at $t = 3$, $t = 4$.

   (e) *ever-increasing rate* and *more and more* and *slowed down the layoff rate*.

55. Using the ZOOM function as shown, we see that there is an inflection point at approximately $x = -1.835$, where the graph of $f(x)$ changes from concave down to concave up.

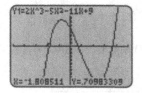 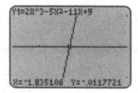

Similar use of the ZOOM function shows that the graph of $f(x)$ is concave down on $(-\infty, -1.835)$ and $(0.669, 3.665)$, concave up on $(-1.835, 0.669)$ and $(3.665, \infty)$. Inflection points at $x = -1.835$, 0.669, and 3.665. (All numbers are approximations.)

57. **(a)** $f'(x) = 0$ at $x \approx -0.5$, where the graph of $f(x)$ has a local maximum.

   (b) $f'(x)$ has a maximum at $x \approx -1.26$, where the graph of $f(x)$ changes from concave up to concave down.

   (c) $f'(x)$ has a minimum at $x \approx 0.26$, where the graph of $f(x)$ changes from concave down to concave up.

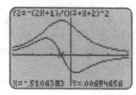 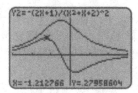 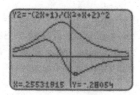

59. Since $0 < p(t) < 1$, we know that $dp/dt = 0.02p(1-p) > 0$, so that $p(t)$ is increasing. Next, by taking the derivative on both sides of this equation, we get

$$\frac{d^2p}{dt^2} = 0.02\frac{dp}{dt}(1-p) - 0.02p\frac{dp}{dt} = 0.02(1-2p)\frac{dp}{dt}.$$

Since $dp/dt > 0$, we see that $d^2p/dt^2 > 0$ for $0 < p < 0.5$ and $d^2p/dt^2 < 0$ for $0.5 < p < \infty$. Therefore, the graph of $p(t)$ is concave up when $0 < p < 0.5$, and concave down on when $0.5 < p < 1$.

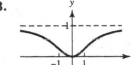

EXERCISES 4.3

1.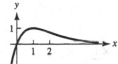

3.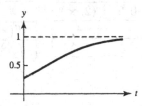

5.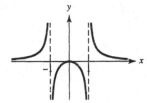

7.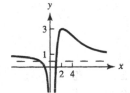

9. Domain: $-\infty < x < \infty$
Symmetry: none
Asymptotes: none
Local maximum at $(-1, 7)$
Local minimum at $(1, 3)$
Inflection point at $(0, 5)$

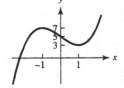

x	$(-\infty, -1)$	$(-1, 1)$	$(1, \infty)$
$f'(x)$	$+$	$-$	$+$
$f(x)$	↗	↘	↗

$$f'(x) = 3x^2 - 3$$

x	$(-\infty, 0)$	$(0, \infty)$
$f''(x)$	$-$	$+$
$f(x)$	∩	∪

$$f''(x) = 6x$$

11. Domain: $-\infty < x < \infty$
Symmetry: about the y-axis
Asymptotes: none
Local maximum at $(0, 10)$
Global minima at $(-2, -6)$ and $(2, -6)$
Inflection points at
$(-\frac{2}{\sqrt{3}}, \frac{10}{9})$ and $(\frac{2}{\sqrt{3}}, \frac{10}{9})$

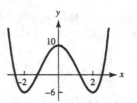

x	$(-\infty, -2)$	$(-2, 0)$	$(0, 2)$	$(2, \infty)$
$f'(x)$	$-$	$+$	$-$	$+$
$f(x)$	↘	↗	↘	↗

$$f'(x) = 4x^3 - 16x$$

x	$(-\infty, -\frac{2}{\sqrt{3}})$	$(-\frac{2}{\sqrt{3}}, \frac{2}{\sqrt{3}})$	$(\frac{2}{\sqrt{3}}, \infty)$
$f''(x)$	$+$	$-$	$+$
$f(x)$	∪	∩	∪

$$f''(x) = 12x^2 - 16$$

13. Domain: $x \neq 2$
Vertical asymptote: $x = 2$
Horizontal asymptote: $y = 0$
Local or global extrema: none
Inflection point: none

x	$(-\infty, 2)$	$(2, \infty)$
$f'(x)$	$-$	$-$
$f(x)$	↘	↘

$$f'(x) = -\frac{1}{(x-2)^2}$$

x	$(-\infty, 2)$	$(2, \infty)$
$f''(x)$	$-$	$+$
$f(x)$	∩	∪

$$f''(x) = \frac{2}{(x-2)^3}$$

15. Domain: $x \neq -1/2$
Vertical asymptote: $x = -1/2$
Horizontal asymptote: $y = 1/2$
Local or global extrema: none
Inflection point: none

x	$(-\infty, -\frac{1}{2})$	$(-\frac{1}{2}, \infty)$
$f'(x)$	$+$	$+$
$f(x)$	↗	↗

$$f'(x) = \frac{1}{(2x+1)^2}$$

x	$(-\infty, -\frac{1}{2})$	$(-\frac{1}{2}, \infty)$
$f''(x)$	$+$	$-$
$f(x)$	∪	∩

$$f''(x) = -\frac{4}{(2x+1)^3}$$

17. Domain: $x \neq 0$
Symmetry: about $(0,0)$
Vertical asymptote: $x = 0$
Slant asymptote: $y = x$
Local maximum at $(-2,-4)$
Local minimum at $(2,4)$
Inflection point: none

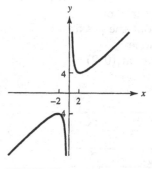

x	$(-\infty,-2)$	$(-2,0)$	$(0,2)$	$(2,\infty)$
$f'(x)$	$+$	$-$	$-$	$+$
$f(x)$	↗	↘	↘	↗

x	$(-\infty,0)$	$(0,\infty)$
$f''(x)$	$-$	$+$
$f(x)$	∩	∪

$$f'(x) = 1 - \frac{4}{x^2}$$

$$f''(x) = \frac{8}{x^3}$$

19. Domain: $-\infty < x < \infty$
Symmetry: about $(0,0)$
Horizontal asymptote: $y = 0$
Global maximum at $(2, \frac{1}{4})$
Global minimum at $(-2, -\frac{1}{4})$
Inflection points at
$$\left(-2\sqrt{3}, -\tfrac{\sqrt{3}}{8}\right), \left(2\sqrt{3}, \tfrac{\sqrt{3}}{8}\right), (0,0)$$

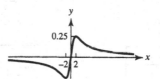

x	$(-\infty,-2)$	$(-2,2)$	$(2,\infty)$
$f'(x)$	$-$	$+$	$-$
$f(x)$	↘	↗	↘

x	$(-\infty,-2\sqrt{3})$	$(-2\sqrt{3},0)$	$(0,2\sqrt{3})$	$(2\sqrt{3},\infty)$
$f''(x)$	$-$	$+$	$-$	$+$
$f(x)$	∩	∪	∩	∪

$$f'(x) = \frac{-x^2 + 4}{(x^2+4)^2}$$

$$f''(x) = \frac{2x^3 - 24x}{(x^2+4)^3}$$

21. Domain: $-\infty < x < \infty$
Symmetry: none
Horizontal asymptote: $y = 0$
Global maximum at $(1, e^{-1})$
Inflection point: $(2, 2e^{-2})$

x	$(-\infty,1)$	$(1,\infty)$
$f'(x)$	$+$	$-$
$f(x)$	↗	↘

x	$(-\infty,2)$	$(2,\infty)$
$f''(x)$	$-$	$+$
$f(x)$	∩	∪

$$f'(x) = (1-x)e^{-x}$$

$$f''(x) = (x-2)e^{-x}$$

23. Domain: $-\infty < x < \infty$
Symmetry: about the y-axis
Asymptotes: none
Global minimum at $(0, 2)$
Inflection point: none

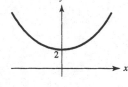

x	$(-\infty, 0)$	$(0, \infty)$
$f'(x)$	$-$	$+$
$f(x)$	$\searrow$	$\nearrow$

$$f'(x) = e^x - e^{-x}$$

x	$(-\infty, \infty)$
$f''(x)$	$+$
$f(x)$	$\cup$

$$f''(x) = e^x + e^{-x}$$

25. Domain: $-\infty < x < \infty$
Symmetry: about the y-axis
Asymptotes: none
Global minimum at $(0, 2\ln 2)$
Inflection points:
  $(-2, 3\ln 2), (2, 3\ln 2)$

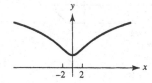

x	$(-\infty, 0)$	$(0, \infty)$
$f'(x)$	$-$	$+$
$f(x)$	$\searrow$	$\nearrow$

$$f'(x) = \frac{2x}{x^2 + 4}$$

x	$(-\infty, -2)$	$(-2, 2)$	$(2, \infty)$
$f''(x)$	$-$	$+$	$-$
$f(x)$	$\cap$	$\cup$	$\cap$

$$f''(x) = \frac{-2x^2 + 8}{(x^2 + 4)^2}$$

27. $f'(x) = 5x^4 + 12x^2 - 24x - 18$ and $f''(x) = 20x^3 + 24x - 24$

(a) Since $f''(0) < 0$ and $f''(1) > 0$, there is an inflection point between $x = 0$ and $x = 1$. And since $\frac{d}{dx} f''(x) = 60x^2 + 24 > 0$ for all x, $f''(x)$ is increasing and cannot cross the x-axis more than once. Therefore, there is exactly one inflection point. Using the equation solver with initial guess 0.5 gives the approximate solution $x_1 \approx 0.7063$ to $f''(x) = 0$.

(b) Since $f''(x) < 0$ on $(-\infty, x_1)$, with x_1 as in (a), $f'(x)$ is decreasing throughout that interval and can have at most one root there. Similarly, $f'(x)$ is increasing on (x_1, ∞) and can have at most one root there. Moreover, $f'(x_1) \neq 0$. Therefore, $f'(x) = 0$ has at most two solutions. Using the equation solver with initial guess 0 gives the approximate solution $x \approx -0.567$, and using it with initial guess 1 gives $x \approx 1.52$.

(c) $\lim\limits_{x \to \infty} f(x) = \infty$

  $\lim\limits_{x \to -\infty} f(x) = -\infty$

(d)

[−3, 3] × [−80, 90] window

29.

x	$(0, 80)$	$(80, 120)$
$R'(x)$	$+$	$-$
$R(x)$	$\nearrow$	$\searrow$

x	$(0, 40)$	$(40, 120)$
$R''(x)$	$+$	$-$
$R(x)$	$\cup$	$\cap$

$R'(x) = 0.00064(240x - 3x^2)$ $R''(x) = 0.00064(240 - 6x)$

The infection is spreading fastest when 80 trees are infected, since $R(x)$ takes a maximum value of $R(80) \approx 164$ at $x = 80$.

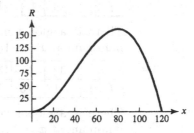

31. (a) $P'(x) = -0.1e^{-0.1x} < 0$ for all x
$P''(x) = 0.01e^{-0.1x} > 0$ for all x
Therefore, the graph is decreasing and concave up.

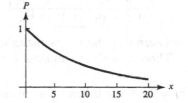

(b) $e^{-0.1x} \geq 0.50 \iff x \leq 10 \ln 2 \approx 6.9$ Therefore, the greatest depth at which the plant can survive is about 6.9 feet.

33. (a)

s	$(-\infty, 0)$	$(0, \infty)$
dN/ds	$+$	$-$
N	$\nearrow$	$\searrow$

s	$(-\infty, -10)$	$(-10, 10)$	$(10, \infty)$
d^2N/ds^2	$+$	$-$	$+$
N	$\cup$	$\cap$	$\cup$

$\dfrac{dN}{ds} = -18.27se^{-0.005s^2}$ $\dfrac{d^2N}{dS^2} = 0.1827(s^2 - 100)e^{-0.005s^2}$

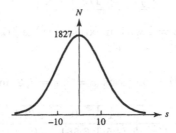

(b) According to this model, the adjusted score 0 was achieved by the largest number of students. That means that the score achieved by the largest number of students was the same as the average score.

EXERCISES 4.4

1. $f'(x) = 3x^2 - 6x = 3x(x-2) = 0$ has the solutions $x = 0$ and $x = 2$.

   (a)

x	-1	0	1
$f(x)$	-3	1	-1

   There is a global maximum at $x = 0$ with maximum value $f(0) = 1$. There is a global minimum at $x = -1$ with minimum value $f(-1) = -3$.

   (b)

x	1	2	3
$f(x)$	-1	-3	1

   There is a global maximum at $x = 3$ with maximum value $f(3) = 1$. There is a global minimum at $x = 2$ with minimum value $f(2) = -3$.

   (c)

x	-2	0	2	3
$f(x)$	-19	1	-3	1

   There is a global maximum at both $x = 0$ and $x = 3$ with maximum value $f(0) = f(3) = 1$. There is a global minimum at $x = -2$ with minimum value $f(-2) = -19$.

3. $f'(x) = 2x + 1 = 0$ has the solution $x = -1/2$.

x	0	2
$f(x)$	3	9

   There is a global maximum at $x = 2$ with maximum value $f(2) = 9$. There is a global minimum at $x = 0$ with minimum value $f(0) = 3$.

5. $f'(x) = (-x^2 + 4)/(x^2 + 4)^2 = 0$ has the solutions $x = -2$ and $x = 2$.

x	1	2	3
$f(x)$	1/5	1/4	3/13

   There is a global maximum at $x = 2$ with maximum value $f(2) = 1/4$. There is a global minimum at $x = 1$ with minimum value $f(1) = 1/5$.

7. $f'(x) = 32x(4x^2 - 9)^3 = 0$ has the solutions $x = 0, x = 3/2$ and $x = -3/2$.

x	-1	0	3/2	2
$f(x)$	625	6,561	0	2,401

   There is a global maximum at $x = 0$ with maximum value $f(0) = 6,561$. There is a global minimum at $x = \frac{3}{2}$ with minimum value $f(\frac{3}{2}) = 0$.

9. $f'(x) = (x-1)/\sqrt{x^2 - 2x + 2} = 0$ has the solution $x = 1$.

x	0	1	2
$f(x)$	$\sqrt{2}$	1	$\sqrt{2}$

   There is a global maximum at both $x = 0$ and $x = 2$ with maximum value $f(0) = f(2) = \sqrt{2}$. There is a global minimum at $x = 1$ with minimum value $f(1) = 1$.

11. $f'(x) = 2x - 2 = 0$ has the solution $x = 1$.

x	0	1	3
$f(x)$	-3	-4	0

There is a global maximum at $x = 3$ with maximum value $f(3) = 0$. There is a global minimum at $x = 1$ with minimum value $f(1) = -4$.

13. $f'(x) = 1 - \frac{4}{x^2} = 0$ has solutions $x = \pm 2$. Also, $\lim_{x \to 0+} f(x) = \infty$ and $\lim_{x \to \infty} f(x) = \infty$.

x	$(0, 2)$	$(2, \infty)$
$f'(x)$	$-$	$+$
$f(x)$	$\searrow$	$\nearrow$

There is a global minimum at $x = 2$ with minimum value $f(2) = 4$. There is no global maximum.

15. $f'(x) = (1 - 2x)e^{-2x} = 0$ has the solution $x = 1/2$. Also, $\lim_{x \to -\infty} f(x) = -\infty$ and $\lim_{x \to \infty} f(x) = 0$.

x	$(-\infty, 1/2)$	$(1/2, \infty)$
$f'(x)$	$+$	$-$
$f(x)$	$\nearrow$	$\searrow$

There is a global maximum at $x = \frac{1}{2}$ with maximum value $f(\frac{1}{2}) = \frac{1}{2e}$. There is no global minimum.

17. $f'(x) = 3x^2 + 1 > 0$ for all x. Therefore, $f(x)$ is always increasing, and takes its minimum at the left-hand endpoint and its maximum at the right-hand endpoint. There is a global maximum at $x = 1$ with the maximum value $f(1) = 1$. There is a global minimum at $x = -1$ with the minimum value $f(-1) = -3$.

19. $f'(x) = x(2 \ln x + 1) = 0$ has the positive solution $e^{-1/2}$. Also, $\lim_{x \to 0+} f(x) = 0$ and $\lim_{x \to \infty} f(x) = \infty$.

x	$(0, e^{-1/2})$	$(e^{-1/2}, \infty)$
$f'(x)$	$-$	$+$
$f(x)$	$\searrow$	$\nearrow$

There is a global minimum at $x = e^{-1/2}$ with minimum value $f(e^{-1/2}) = -1/2e$. There is no global maximum.

21. $f'(x) = e^x(x - 2)/x^3 = 0$ has the positive solution $x = 2$. Also, $\lim_{x \to 0+} f(x) = \infty$ and $\lim_{x \to \infty} f(x) = \infty$.

x	$(0, 2)$	$(2, \infty)$
$f'(x)$	$-$	$+$
$f(x)$	$\searrow$	$\nearrow$

There is a global minimum at $x = 2$ with minimum value $f(2) = e^2/4$. There is no global maximum.

23. $f'(x) = 4x^3 - 12x = 0$ has the solutions $x = -\sqrt{3}, x = 0$, or $x = \sqrt{3}$.

x	-2	$-\sqrt{3}$	0	$\sqrt{3}$	2
$f(x)$	-8	-9	0	-9	-8

There is a global maximum at $x = 0$ with maximum value $f(0) = 0$. There is a global minimum at both $x = \sqrt{3}$ and $x = -\sqrt{3}$ with minimum value $f(-\sqrt{3}) = f(\sqrt{3}) = -9$.

25. $f'(x) = 2e^{2x} - 6e^x = 2e^x(e^x - 3) = 0$ has the solution $x = \ln 3$. Also, $\lim_{x \to \infty} f(x) = \lim_{x \to \infty}$
$e^x(e^x - 6) = \infty$.

x	$(-\infty, \ln 3)$	$(\ln 3, \infty)$
$f'(x)$	$-$	$+$
$f(x)$	$\searrow$	$\nearrow$

There is a global minimum at $x = \ln 3$ with minimum value $f(\ln 3) = -9$. There is no global maximum.

27. $f'(x) = 4x/(x^2 + 1)^2 = 0$ has the solution $x = 0$. Also $f(x) < 1$ for all x, and $\lim_{x \to \infty} f(x) = 1$ and $\lim_{x \to -\infty} f(x) = 1$.

x	$(-\infty, 0)$	$(0, \infty)$
$f'(x)$	$-$	$+$
$f(x)$	$\searrow$	$\nearrow$

There is a global minimum at $x = 0$ with minimum value $f(0) = -1$. There is no global maximum.

29. To maximimze $h(t)$, we solve $h'(t) = -32t + 160 = 0$, which yields $t = 5$. Since $h''(5) = -32 < 0$, there is a local maximum at $t = 5$, and, by the single critical point principle, it is also a global maximum. Since $h(5) = 600$, we conclude that the projectile will reach a maximal height of 600 feet at the end of 5 seconds.

31. Set $A'(x) = 10 - \frac{4,000}{(x+1)^2} = 0$. The only positive solution is $x = 19$. Checking the sign of $A'(x)$, we get

x	$(0, 19)$	$(19, \infty)$
$A'(x)$	$-$	$+$
$A(x)$	$\searrow$	$\nearrow$

Thus, there is a local minimum at $x = 19$, and, by the single critical point principle, it is a global minimum relative to the interval $(0, \infty)$. To minimize the average cost, 19 units should be produced, and the minimal value is $A(19) = 690$.

EXERCISES 4.5

1. The revenue function is given by

$$R(p) = pq = 6,000p - 200p^{3/2}, \quad 0 \le p \le 900,$$

and $R'(p) = 6,000 - 300\sqrt{p} = 0$ has the solution $p = 400$. Checking $R(p)$ at the critical point and endpoints yields

p	0	400	900
$R(p)$	0	800,000	0

which shows that R has a maximum if $p = 400$ dollars.

3. The average cost function is given by

$$A(q) = \frac{C(q)}{q} = \frac{3,200}{q} - 15 + 2q, \quad q > 0,$$

and $A'(q) = -\frac{3,200}{q^2} + 2 = 0$ has the solution $q = 40$. Checking the sign of $A'(q)$ yields

q	$(0, 40)$	$(40, \infty)$
$A'(q)$	$-$	$+$
$A(q)$	$\searrow$	$\nearrow$

which shows that the global minimum over $(0, \infty)$ occurs at $q = 40$.

5. (a) Assume $q = mp + b$ for some real numbers m and b. The given data say that $48 = 9m + b$ and $42 = 12m + b$, whose solutions are $m = -2$ and $b = 66$. Therefore, the demand equation is $q = -2p + 66$.

   (b) The revenue function is given by $R(p) = p \cdot q = -2p^2 + 66p$, $0 \leq p \leq 33$, and solving $R'(p) = -4p + 66$ gives the critical point $p = 16.50$. Since $R''(p) = -4 < 0$, there is a local maximum at that point, which is also a global one by the single critical point principle. Thus, the restaurant should charge \$16.50 per appetizer to maximize its revenue.

   (c) The profit function is given by

   $$P(p) = R(p) - 4q = -2p^2 + 66p - 4(-2p + 66) = -2p^2 + 74p - 264.$$

   Solving $P'(p) = -4p + 74 = 0$ gives the single critical point $p = 18.50$, and using the second derivative test shows there is a global maximum at that point. Thus, the restaurant should charge \$18.50 per appetizer to maximize its profit.

7. We want to optimize the caloric "profit," given by

   $$P(w) = G(w) - E(w) = 0.3w^2 - 0.1w^3,$$

   where $0 \leq w \leq 2.5$. To find any critical points, we solve $P'(w) = 0.6w - 0.3w^2 = 0$, obtaining $w = 0$ and $w = 2$.

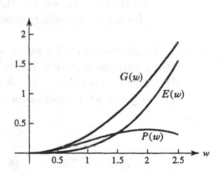

   Checking the value of $P(w)$ at the critical and end points leads to the following:

w	0	2	2.5
$P(w)$	0	0.4	0.3125

   Therefore, the optimum prey weight is 2 pounds.

9. (a) $P(w) = G(w) - R(w) = 1 - e^{-0.032w} - 0.01w$, $w \geq 0$.

   (b) Solving $P'(w) = 0.032e^{-0.032w} - 0.01 = 0$ yields the critical point $w = 31.25 \ln \frac{16}{5}$. Since $P''(w) = -(0.032)^2 e^{-0.032w} < 0$ for all w, the function has a global maximum at the single critical point. We conclude that the optimal rack width for the deer is $31.25 \ln \frac{16}{5} \approx 36.35$ inches.

11. The derivative is $B'(t) = 3t^2 - 12t$, whose roots are $t = 0$ and 4. Comparing $B(t)$ at the critical and end points gives

t	0	4	8
$B(t)$	36	4	164

   Thus, during the first 8 hours, the minimum number of bacteria is 4 million and the maximum is 164 million.

13. Let x and y be the dimensions of the page in inches and A be the size of the printed area (in square inches). Then $xy = 150$, or $y = 150/x$, and

$$A = (x - 2)(y - 3) = xy - 3x - 2y + 6 = 156 - 3x - \frac{300}{x}.$$

To maximize A, we first solve $A'(x) = -3 + (300/x^2) = 0$, whose only positive solution is $x = 10$. Since $A''(10) = -0.6 < 0$, there is a local maximum at $x = 10$, which is also a global maximum over $(0, \infty)$ by the single critical point principle. Then $y = 150/10 = 15$, so that the dimensions are 10 inches (wide) by 15 inches (high).

15. Let x be the width of the kennel, and let y be its length and also the length of each interior fence. The total amount of fencing is $2x + 5y = 320$, so that $y = 64 - \frac{2}{5}x$. Therefore, the area is given by

$$A = xy = x\left(64 - \frac{2}{5}x\right) = 64x - \frac{2}{5}x^2$$

and $A'(x) = -\frac{4}{5}x + 64$, whose only root is $x = 80$. The second derivative test and single critical point principle show that it gives a global maximum, and $y = 32$.

17. **(a)** $r(y) = ky(M - y)$, $0 \le y \le M$

   (b) We solve $r'(y) = k(M - 2y) = 0$ to get $y = M/2$. Since $r(0) = r(M) = 0$ and $r(M/2) = k(M/2)^2 > 0$, we see that the maximum of r over $[0, M]$ occurs at $y = M/2$. Thus, the disease is spreading most rapidly when exactly half the community is infected.

19. At a steady speed of v miles per hour, a 1,600 mile trip will take $1,600/v$ hours, which means the driver's wages will total $22.50(\frac{1,600}{v}) = \frac{36,000}{v}$ dollars. In addition, the total fuel cost is $1,600(\frac{v}{140}) = \frac{80v}{7}$, and the total cost for depreciation and wear is $1,600 \cdot (0.27) = 432$ dollars. Thus, the total cost is given as a function of v by

$$C(v) = \frac{36,000}{v} + 432 + \frac{80v}{7}, \quad 45 \le v \le 60.$$

Solving $C'(v) = -\frac{36,000}{v^2} + \frac{80}{7} = 0$ gives $v = \sqrt{3,150} = 15\sqrt{14} \approx 56.1$, and comparing the values of $C(v)$ at the critical and end points yields the following (rounded to 2 decimal places):

v	45	$15\sqrt{14}$	60
$C(v)$	1,746.29	1,714.85	1,717.71

Thus the cost is minimized if $v = 15\sqrt{14} \approx 56$ mph.

21. Suppose the company places x orders per year, each for $270,000/x$ boxes. The average number of boxes in stock over the year will be $135,000/x$, which means the storage cost will be $0.02(135,000/x) = 2,700/x$ dollars. The total shipping cost will be $100x$ dollars. Therefore, the cost of shipping and storage is given by $C(x) = \frac{2,700}{x} + 100x$. The cost of the merchandise itself is constant. To find the minimum, we solve $C'(x) = -\frac{2,700}{x^2} + 100 = 0$, whose only positive solution is $x = \sqrt{27} = 3\sqrt{3}$. The second derivative test and the single critical point principle show that $C(x)$ has a minimum there. Thus, to minimize its costs, the company should place $3\sqrt{3} \approx 5.2$ orders per year, each for $30,000\sqrt{3} \approx 51,961.5$. Rounding to the nearest integers gives 5 orders per year, each for 54,000 boxes.

23. The distance of any point (x, y) to $(1, 0)$ is $\sqrt{(x-1)^2 + y^2}$. If the point is on the graph of $y = \sqrt{x}$, then $y^2 = x$, so that the distance is given by

$$f(x) = \sqrt{(x-1)^2 + x} = \sqrt{x^2 - x + 1}.$$

Then $f'(x) = (2x-1)/2\sqrt{x^2 - x + 1}$, whose only root is $x = 1/2$. The sign of $f'(x)$ satisfies

x	$(0, 1/2)$	$(1/2, \infty)$
$f'(x)$	$-$	$+$
$f(x)$	$\searrow$	$\nearrow$

which shows that $f(x)$ has a local minimum (and, by the single critical point principle, a global one) at $x = 1/2$. We conclude that $(\frac{1}{2}, \frac{\sqrt{2}}{2})$ is the point on the graph of $y = \sqrt{x}$ that is closest to the point $(1, 0)$.

25. With x and y as in the preceding exercise, the volume satisfies $V = x^2 y = 20$, so that $y = 20/x^2$. The total cost is given by

$$C = 3x^2 + 2x^2 + 4xy = 5x^2 + 4xy = 5x^2 + \frac{80}{x}.$$

Then $C'(x) = 10x - \frac{80}{x^2}$, whose only root is $x = 2$. The second derivative test and single critical point principle show that $C(2) = 60$ is a global minimum for $x > 0$.

27. Let $2x$ be the length of the rectangle's base, and y be its height. The x, y, and 1 are the lengths of the sides of a right triangle, as shown. Then $y = \sqrt{1 - x^2}$, and the area of the rectangle is $A(x) = 2x\sqrt{1 - x^2}$. Solving

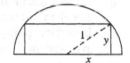

$$A'(x) = 2\sqrt{1-x^2} - \frac{2x^2}{\sqrt{1-x^2}} = \frac{2(1-2x^2)}{\sqrt{1-x^2}}, \quad x > 0,$$

whose only positive root is $x = \sqrt{2}/2$. The first derivative test and the single critical point principle show that the maximum value of $A(\sqrt{2}/2) = 1$ is the maximum value of $A(x)$ for $x > 0$. Thus, the area of the largest rectangle equals 1.

29. (a) The height of the box is x, and each edge of the base has length $30 - 2x$. Since neither of these quantities can be negative, we must have $x \geq 0$ and $2x \leq 30$, or, more simply, $0 \leq x \leq 15$. The volume of the box is given as a function of x by $V(x) = x(30 - 2x)^2 = 4x^3 - 120x^2 + 900x$.

(b) To maximize $V(x)$, $0 \leq x \leq 15$, we solve

$$V'(x) = 12x^2 - 240x + 900 = 0.$$

The solutions are $x = 5$ and $x = 15$. Since $V(0) = V(15) = 0$ and $V(5) = 2,000$, we conclude that the maximum volume is 2,000 cubic inches, achieved for $x = 5$.

EXERCISES 4.6

Most answers is this section are approximations, obtained using the TRACE function and either the ZOOM function (repeatedly) or the equation solver.

1. Local minima at $(-1.25723, 10.92465)$ and $(1.38725, -19.59187)$, local maximum at $(-0.43002, 15.13373)$. Inflection points at $(0.68103, -4.85644)$ and $(-0.88102, 12.888)$. The x-intercepts are at $x = 0.519303$ and 1.93300.

Several sample screens are shown below: The local maximum, first with the TRACE function and then the equation solver; the leftmost x-intercept, first with the TRACE and then with repeated use of the ZOOM function.

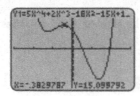 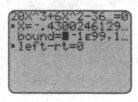 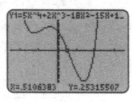 

3. Local minimum at $(0.96242, -0.65120)$, local maximum at $(-0.96242, 0.65120)$. The x-intercepts are at $x = -1.62213$, $x = 0$, and $x = 1.62213$. (Observe the symmetry about the origin. Also $(0,0)$ is an inflection point.)

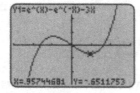 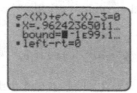 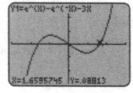 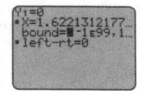

5. Local maximum: $(0.12808, -0.87297)$
   Local minimum: $(5.20526, -66.31221)$
   Inflection point: $(2.66667, -33.59359)$
   x-intercept: $x = 7.758841$

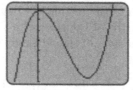

[−3, 9] × [−70, 5] window

7. Local maximum: $(0.57590, 3.94630)$
   Local minimum: $(-2.13805, -17.28567)$ and $(0.812150, 3.92140)$.
   Inflection points: $(-1.19648, -8.60727)$ and $(0.69648, 3.93366)$
   x-intercepts: $x = -3$ (exact) and $x = -0.46557$

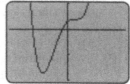

[−5, 5] × [−20, 10] window

9. Local maximum: $(-1.93045, -19.86606)$ and $(1.03603, 1.01796)$
   Local minimum: $(-1.03603, -25.01796)$ and $(1.93045, -4.13394)$.
   Inflection points: $(0, -12)$ (exact), $(-1.54919, -22.16007)$
      and $(1.54919, -1.83993)$
   x-intercepts: $x = 0.76756485$, $x = 1.32116$ and $x = 2.26633$

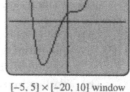

[−3, 3] × [−30, 5] window

11. We first compute $f'(x) = e^x - e^{-x} - 4$ and $f''(x) = e^x + e^{-x} > 0$. Therefore, the graph is always concave up, with no inflection point. Also, $f'(x)$ is always increasing and can have at most one root. The equation solver finds it to be at $x = 1.44364$, and the graph has a local (and global) minimum at $(1.44364, -1.30241)$. Again using the equation solver, we find the x-intercepts: $x = 0.58939$ and $x = 2.12680$

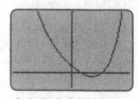

[−4, 4] × [−5, 20] window

13. We first observe that the graph is symmetric about the y-axis, and $0 < f(x) < 1$. So there are no x-intercepts. Next, we compute

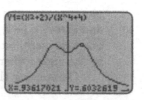

$$f'(x) = \frac{2x(4 - 4x^2 - x^4)}{(x^4 + 4)^2}.$$

Therefore $f'(0) = 0$. In addition, $f'(0.5) > 0$ and $f'(1) < 0$, so that there is a second critical point in the interval $(0.5, 1)$, and, by symmetry, another in $(-1, -0.5)$. Drawing the graph of f in a $[-4, 4] \times [0, 1]$ window, we see there is are a local minimum at $(0, 0.5)$ and (using the TRACE function) local maxima at $(\pm 0.936, 0.603)$. A more accurate estimate, found using the equation solver to solve $f'(x) = 0$ is $(\pm 0.91018, 0.60355)$.

The graph shows there are four inflection points. To find them, we graph $f'(x)$ (shown below in a $[-3, 3] \times [-1, 1]$ window) and use the TRACE and ZOOM functions to estimate its local maxima and minima (the positive ones as shown below). We conclude that the inflection points are approximately at $x = \pm 0.52$ and $x = \pm 1.55$, and by evaluating $f(x)$ we find the (approximate) inflection points $(\pm 0.52, 0.56)$ and $(\pm 1.55, 0.45)$.

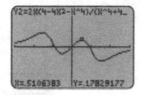 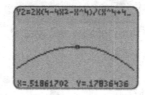 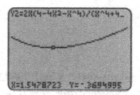

15. Local minimum: $(0, 0)$ (exact)
x-intercept: $x = 0$
Local maxima: $(\pm \sqrt{e-1}, 0.36788) \approx (\pm 1.3108325, 0.36788)$
Inflection points: $(\pm 2.13986, 0.30812)$ and $(\pm 0.42999, 0.14318)$

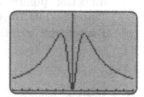

$[-7, 7] \times [-0, 0.5]$ window

17. We first observe that the x-intercepts are at $x = \pm 1$, since $f(\pm 1) = 0$.

Next, we compute $f'(x) = \frac{e^x(1 + 4x^3 - x^4) + 4x^3}{(e^x + 1)^2}$. Graphing $y = f'(x)$ is a $[-2, 15] \times [-3, 3]$ window, as shown, we see that it crosses the axis at two points, one slightly to the left of zero where $f'(x)$ changes from negative to positive (a local minimum of f) and one near 4 where $f'(x)$ changes from positive to negative (a local maximum of f). By using the equation solver to find these x and then evaluating $f(x)$, we find a local minimum of f at $(-0.453156, -0.58561)$ and a local maximum at $(4.0822, 4.5906)$.

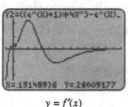 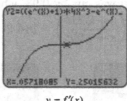 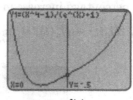

$y = f'(x)$ $y = f'(x)$ $y = f(x)$

The inflection points occur where $f'(x)$ changes from increasing to decreasing or vice-versa, and the graph of $y = f'(x)$ shows two such points, one near $x = 2$ and the other near $x = 6$. The picture also suggests there is at least one more, between $x = -1$ and $x = 1$. In order to get a better picture, we zoom in twice, as shown. In addition, we graph $y = f(x)$, choosing (after some trial and error) a $[-1, 1] \times [0, 0.6]$ window, as shown. It suggests there are two more inflection points, one near zero, where the graph changes from concave up to concave down, and a second slighlty to the right, where it changes back to concave up.

Computing the second derivative, we get

$$f''(x) = \frac{e^{2x}(x^4 - 8x^3 + 12x^2 - 1) - e^x(x^4 + 8x^3 - 24x^2 - 1) + 12x^2}{(1 + e^x)^3}.$$

We see directly that $f''(0) = 0$. In addition, by using the equation solver, we find that $f''(x) = 0$ at $x \approx 0.02120$, $x \approx 2.088511$, and $x = 6.04255$. Computing $f(x)$ for each of these, we get the inflection points $(0, -1/2)$ (exact), $(2.08851, 1.98680)$, $(6.04255, 3.15703)$, and $(0.02120, -0.49470)$.

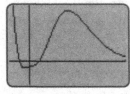

[−2, 10] × [−2, 5] window

19. Vertical asymptote: $x = 1.79632$
    Horizontal asymptote: $y = 0$ (exact)
    Local maximum: $(-1.2599, 0.53160)$
    Inflection points: $(0.40324, -0.18592)$, $(-0.41368, 0.22623)$,
        and $(-1.91275, 0.42107)$
    x-intercept: $x = 0$ (exact)

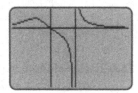

[−3, 6] × [−3, 1] window

21. Vertical asymptotes: $x = \pm 2$ (exact)
    Horizontal asymptote: $y = 0$ (exact)
    No local maximum or minimum.
    Inflection point: $(0, 0)$ (exact)
    x-intercept: $x = 0$ (exact)

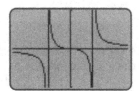

[−5, 5] × [−10, 10] window

23. (a) The efforts appear to be successful, since the graphs indicate that the percentage increase in the year 2000 was cut by about one-third. In fact, $D_p(50) = 2.58$ and $D_a(50) = 1.65$, so that the percentage increase in 2000 was cut by 0.93, which is a little more than one-third.

    (b) $\lim\limits_{t \to \infty} |D_p(t) - D_a(t)| = \lim\limits_{t \to \infty} \ln\left(\frac{e + 0.21t}{e + 0.05t}\right) = \ln 4.2 \approx 1.435$

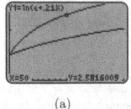

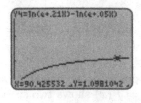

(a)   (b)

CHAPTER 4 REVIEW EXERCISES

1. Solving $f'(x) = 6x - 6 = 0$ gives $x = 1$, and the sign of $f'(x)$ is as follows:

x	$(-\infty, 1)$	$(1, \infty)$
$f'(x)$	$-$	$+$
$f(x)$	↘	↗

    Conclusion: $f(x)$ is decreasing on $(-\infty, 1)$ and increasing on $(1, \infty)$. There is a local minimum at $x = 1$ with minimum value $f(1) = -12$.

3. Solving $f'(x) = 3x^2 - 8x - 3 = 0$ gives $x = -1/3$ and $x = 3$, and

x	$(-\infty, -1/3)$	$(-1/3, 3)$	$(3, \infty)$
$f'(x)$	$+$	$-$	$+$
$f(x)$	↗	↘	↗

    Thus, $f(x)$ is increasing on $(-\infty, -1/3)$ and $(3, \infty)$ and decreasing on $(-1/3, 3)$. There is a local maximum at $x = -1/3$ with maximum value $f(-\frac{1}{3}) = -\frac{121}{27}$. There is a local minimum at $x = 3$ with minimum value $f(3) = -23$.

5. Solving $f'(x) = -(x + 2)/x^3 = 0$ gives $x = -2$. In addition, $x = 0$ is a vertical asymptote. The sign of $f'(x)$ is as follows:

x	$(-\infty, -2)$	$(-2, 0)$	$(0, \infty)$
$f'(x)$	$-$	$+$	$-$
$f(x)$	↘	↗	↘

    Thus, $f(x)$ is decreasing on $(-\infty, -2)$ and $(0, \infty)$, and increasing on $(-2, 0)$. There is a local minimum at $x = -2$ with minimum value $f(-2) = -1/4$.

7. Solving $f'(x) = 2xe^{-3x} - 3x^2 e^{-3x} = xe^{-3x}(2 - 3x) = 0$ gives $x = 0$ and $x = 2/3$.

x	$(-\infty, 0)$	$(0, 2/3)$	$(2/3, \infty)$
$f'(x)$	$-$	$+$	$-$
$f(x)$	↘	↗	↘

    Thus, $f(x)$ is decreasing on $(-\infty, 0)$ and $(\frac{2}{3}, \infty)$, and increasing on $(0, \frac{2}{3})$. There is a local minimum at $x = 0$ with minimum value $f(0) = 0$. There is a local maximum at $x = \frac{2}{3}$ with maximum value $f(\frac{2}{3}) = \frac{4}{9}e^{-2}$.

9. Solving $f'(x) = e^x - 2e^{-2x} = 0$ is equivalent to solving $e^{3x} = 2$, which gives $x = \frac{1}{3} \ln 2$.

x	$(-\infty, \frac{1}{3} \ln 2)$	$(\frac{1}{3} \ln 2, \infty)$
$f'(x)$	$-$	$+$
$f(x)$	↘	↗

    Thus, $f(x)$ is decreasing on $(-\infty, \frac{1}{3} \ln 2)$ and increasing on $(\frac{1}{3} \ln 2, \infty)$. There is a local minimum at $x = \frac{1}{3} \ln 2$ with minimum value $f(\frac{1}{3} \ln 2) = \frac{3}{2} \sqrt[3]{2}$.

11. Solving $f'(x) = (x^2 - 1)^{1/3} + \frac{2}{3}x^2(x^2 - 1)^{-2/3} = (\frac{5}{3}x^2 - 1)/(x^2 - 1)^{2/3} = 0$ gives $x = \pm\frac{\sqrt{15}}{5}$. Also, $f'(x)$ does not exist at $x = \pm 1$.

x	$(-\infty, -1)$	$(-1, -\frac{\sqrt{15}}{5})$	$(-\frac{\sqrt{15}}{5}, \frac{\sqrt{15}}{5})$	$(\frac{\sqrt{15}}{5}, 1)$	$(1, \infty)$
$f'(x)$	+	+	−	+	+
$f(x)$	↗	↗	↘	↗	↗

Thus, $f(x)$ is increasing on $(-\infty, -\frac{\sqrt{15}}{5})$, $(\frac{\sqrt{15}}{5}, \infty)$. It is decreasing on $(-\frac{\sqrt{15}}{5}, \frac{\sqrt{15}}{5})$. There is a local maximum at $x = -\frac{\sqrt{15}}{5}$ with maximum value $f(-\frac{\sqrt{15}}{5}) = \frac{\sqrt{15}}{5} \cdot (\frac{2}{5})^{1/3}$, and a local minimum at $x = \frac{\sqrt{15}}{5}$ with the minimum value $f(\frac{\sqrt{15}}{5}) = -\frac{\sqrt{15}}{5} \cdot (\frac{2}{5})^{1/3}$.

13. Solving $f''(x) = 12x^2 - 12 = 0$ gives $x = \pm 1$.

x	$(-\infty, -1)$	$(-1, 1)$	$(1, \infty)$
$f''(x)$	+	−	+
$f(x)$	∪	∩	∪

The graph of $f(x)$ is concave up on $(-\infty, -1)$ and $(1, \infty)$, and concave down on $(-1, 1)$. The inflection points are $(-1, -7)$ and $(1, -13)$.

15. Since $f''(x) = 4e^{-2x} > 0$ for all x, the graph is concave up on $(-\infty, \infty)$.

17. Solving $f''(x) = 2 - \frac{1}{x} = 0$ gives $x = 1/2$.

x	$(0, 1/2)$	$(1/2, \infty)$
$f''(x)$	−	+
$f(x)$	∩	∪

The graph of $f(x)$ is concave down on $(0, \frac{1}{2})$ and concave up on $(\frac{1}{2}, \infty)$. The inflection point is $(\frac{1}{2}, \frac{1}{4} + \frac{1}{2}\ln 2)$.

19. Solving $f''(x) = (4x^3 - 6x)e^{-x^2} = 0$ gives $x = 0$ and $x = \pm\frac{\sqrt{6}}{2}$.

x	$(-\infty, -\frac{\sqrt{6}}{2})$	$(-\frac{\sqrt{6}}{2}, 0)$	$(0, \frac{\sqrt{6}}{2})$	$(\frac{\sqrt{6}}{2}, \infty)$
$f''(x)$	−	+	−	+
$f(x)$	∩	∪	∩	∪

The graph of $f(x)$ is concave down on $(-\infty, -\frac{\sqrt{6}}{2})$ and $(0, \frac{\sqrt{6}}{2})$. It is concave up on $(-\frac{\sqrt{6}}{2}, 0)$ and $(\frac{\sqrt{6}}{2}, \infty)$. The inflection points are $(-\frac{\sqrt{6}}{2}, -\frac{\sqrt{6}}{2}e^{-3/2})$, $(0, 0)$, and $(\frac{\sqrt{6}}{2}, \frac{\sqrt{6}}{2}e^{-3/2})$.

21. The domain is the entire x-axis. Solving $f''(x) = 6x(1 + x)/(1 + x + x^2)^3 = 0$ gives $x = 0$ and $x = -1$.

x	$(-\infty, -1)$	$(-1, 0)$	$(0, \infty)$
$f''(x)$	+	−	+
$f(x)$	∪	∩	∪

The graph of $f(x)$ is concave up on $(-\infty, -1)$ and $(0, \infty)$ and concave down on $(-1, 0)$. The inflection points are $(-1, 1)$ and $(0, 1)$.

23. (d) $f'(x)$ goes from very negative to less negative, and is therefore increasing.

25. **(a)** $f(x)$ is increasing on $(0, \infty)$, since $f'(x) > 0$; and $f(x)$ is decreasing on $(-\infty, 0)$, since $f'(x) < 0$.

(b) $f(x)$ is concave up on $(-\infty, \infty)$, since $f'(x)$ is increasing.

27. (d)   **29.** (f)

31. Solving $f'(x) = 1 - \frac{1}{x}$ gives the critical point $x = 1$. The second derivative is $f''(x) = 1/x^2$. Then, $f''(1) = 1$, so that there is a local minimum at $x = 1$.

33. Solving $f'(x) = \frac{1}{x} - \frac{2\ln x}{x}$ gives the critical $x = e^{1/2}$. The second derivative is $f''(x) = \frac{2\ln x - 3}{x^2}$. Then, $f''(e^{1/2}) = -2e^{-1}$, so that there is a local maximum at $x = e^{1/2}$.

35.

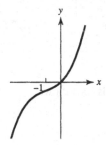

37. Domain: $-\infty < x < \infty$.
Asymptotes: none
Increasing on $(-\infty, -\frac{2}{3})$ and $(\frac{2}{3}, \infty)$
Decreasing on $(-\frac{2}{3}, \frac{2}{3})$
Local maximum at $(-\frac{2}{3}, \frac{28}{3})$
Local minimum at $(\frac{2}{3}, -\frac{4}{3})$
Concave up on $(0, \infty)$
Concave down on $(-\infty, 0)$
Inflection point at $(0, 4)$

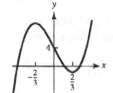

x	$(-\infty, -\frac{2}{3})$	$(-\frac{2}{3}, \frac{2}{3})$	$(\frac{2}{3}, \infty)$
$f'(x)$	$+$	$-$	$+$
$f(x)$	↗	↘	↗

$$f'(x) = 27x^2 - 12$$

x	$(-\infty, 0)$	$(0, \infty)$
$f''(x)$	$-$	$+$
$f(x)$	∩	∪

$$f''(x) = 54x$$

39. Domain: $x \neq \pm 2$
Vertical asymptote: $x = -2$, $x = 2$
Horizontal asymptote: $y = 1$
Increasing on $(-\infty, -2)$ and $(-2, 0)$
Decreasing on $(0, 2)$ and $(2, \infty)$
Local maximum at $x = 0$.
Concave up on $(-\infty, -2)$ and $(2, \infty)$
Concave down on $(-2, 2)$
Symmetry: about the y-axis

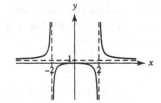

x	$(-\infty, -2)$	$(-2, 0)$	$(0, 2)$	$(2, \infty)$
$f'(x)$	$+$	$+$	$-$	$-$
$f(x)$	↗	↗	↘	↘

$$f'(x) = \frac{-8x}{(x^2 - 4)^2}$$

x	$(-\infty, -2)$	$(-2, 2)$	$(2, \infty)$
$f''(x)$	$+$	$-$	$+$
$f(x)$	∪	∩	∪

$$f''(x) = \frac{24x^2 + 32}{(x^2 - 4)^3}$$

41. Domain: $-\infty < x < \infty$
Increasing on $(-\infty, -27)$ and $(27, \infty)$
Decreasing on $(-27, 27)$
Local maximum at $(-27, 162)$
Local minimum at $(27, -162)$
Concave up on $(0, \infty)$
Concave down on $(-\infty, 0)$
Inflection point: $(0, 0)$
Symmetry: about $(0, 0)$

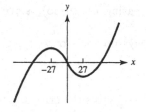

x	$(-\infty, -27)$	$(-27, 27)$	$(27, \infty)$
$f'(x)$	$+$	$-$	$+$
$f(x)$	↗	↘	↗

x	$(-\infty, 0)$	$(0, \infty)$
$f''(x)$	$-$	$+$
$f(x)$	∩	∪

$$f'(x) = \frac{5}{3}x^{2/3} - 15 \qquad f''(x) = \frac{10}{9}x^{-1/3}$$

43. Solving $f'(x) = 3x^2 - 3 = 0$ gives critical points $x = \pm 1$. The sign of $f'(x)$ is as follows:

x	$(-\infty, -1)$	$(-1, 1)$	$(1, \infty)$
$f'(x)$	$+$	$-$	$+$
$f(x)$	↗	↘	↗

Therefore, there is a local maximum at $x = -1$ and a local minimum at $x = 1$. However, there is no global maximum or minimum, since

$$\lim_{x \to \infty} f(x) = \infty \quad \text{and} \quad \lim_{x \to -\infty} f(x) = -\infty.$$

45. Solving $f'(x) = \ln x + 1 = 0$ gives the critical point $x = e^{-1}$. Since $f''(x) = 1/x$, we have $f''(e^{-1}) = e > 0$, which means there is a local minimum at $x = e^{-1}$ with value $f(e^{-1}) = -e^{-1}$. By applying the single critical point principle, we conclude it is also a global minimum. There is no global maximum, since $\lim_{x \to \infty} f(x) = \infty$.

47. $f'(x) = x^2 - 4x + 3 = 0$ has the solutions $x = 1$ and $x = 3$. For each of the intervals listed, we compare the value at the endpoints and any critical points inside the interval.

(a)

x	0	1	2
$f(x)$	-1	$1/3$	$-1/3$

There is a global minimum at $x = 0$ with the minimum value $f(0) = -1$. There is a global maximum at $x = 1$ with the maximum value $f(1) = \frac{1}{3}$.

(b)

x	-1	1
$f(x)$	$-19/3$	$1/3$

There is a global minimum at $x = -1$ with the minimum value $f(-1) = -\frac{19}{3}$. There is a global maximum at $x = 1$ with the maximum value $f(1) = \frac{1}{3}$.

(c)

x	2	3	4
$f(x)$	$-1/3$	-1	$1/3$

There is a global minimum at $x = 3$ with the minimum value $f(3) = -1$. There is a global maximum at $x = 4$ with the maximum value $f(4) = \frac{1}{3}$.

49. The demand function is given by an equation of the form $q = mp + b$, with p being the price of a ticket and q the number sold. Using the given data, we obtain $27{,}000 = 10m + b$ and $33{,}000 = 8m + $

b, and solving these for m and b give $m = -3{,}000$ and $b = 57{,}000$. Therefore, $q = -3{,}000p + 57{,}000$. The revenue function and its derivative are

$$R(p) = pq = -3{,}000p^2 + 57{,}000p, \quad p > 0,$$
$$R'(p) = -6{,}000p + 57{,}000.$$

Solving $R'(p) = 0$ gives $p = 9.50$ dollars, and the second derivative test and the single critical point principle show that the revenue is maximized at that price.

51. There are three interior fences, each y feet long. At \$4 per foot, their total cost is $12y$. In addition there are two outside fences of length y and two of length x. At \$12 per foot their total cost is $12(2x + 2y) = 24x + 24y$. Therefore, the total cost is given by $C = 24x + 36y$. Since the area satisfies $A = xy = 2{,}400$, we have $y = 2{,}400/x$, and making that substitution leads to the cost function

$$C(x) = 24x + \frac{86{,}400}{x}, \quad x > 0,$$

with derivative $C'(x) = 24 - \frac{86{,}400}{x^2}$, whose only positive root is $x = 60$. The sign of $C'(x)$ is as follows:

x	$(0, 60)$	$(60, \infty)$
$C'(x)$	$-$	$+$
$C(x)$	↘	↗

The first derivative test and the single critical point principle show there is a global minimum over $(0, \infty)$ at $x = 60$. It follows that $y = 2{,}400/60 = 40$.

53. Let r be the distance between the origin and a point (x, y) on the line $x + 3y = 1$ is

$$r = \sqrt{x^2 + y^2} = \sqrt{(1 - 3y)^2 + y^2} = \sqrt{1 - 6y + 10y^2},$$

and the derivative with respect to y is $r'(y) = (10y - 3)/\sqrt{1 - 6y + 10y^2}$, whose only root is $y = 3/10$. Then

y	$(-\infty, 3/10)$	$(3/10, \infty)$
$r'(y)$	$-$	$+$
r	↘	↗

The first derivative test and the single critical point principle show there is a global minimum at $y = 3/10$, and $x = 1 - 3y = 1/10$. Therefore, $(\frac{1}{10}, \frac{3}{10})$ is the point on the line $x + 3y = 1$ that is closest to the origin.

55. Since $y'(t) = -e^{-y(t)} < 0$ for all t, $y(t)$ is a decreasing function.

CHAPTER 4 EXAM

1. Solving $f'(x) = 24x(x^2 - 4)^{11} = 0$ gives the critical points $x = 0, 2$, and -2. The sign of $f'(x)$ is as follows:

x	$(-\infty, -2)$	$(-2, 0)$	$(0, 2)$	$(2, \infty)$
$f'(x)$	$-$	$+$	$-$	$+$
$f(x)$	↘	↗	↘	↗

There is a local minimum at $x = -2$ with the minimum value $f(-2) = 0$. There is a local maximum at $x = 0$ with the maximum value $f(0) = 4^{12}$. There is a local minimum at $x = 2$ with the minimum value $f(2) = 0$.

3. Solving $f'(x) = (-x^3 + 3x^2)e^{-x} = x^2(3-x)e^{-x} = 0$ gives $x = 0$ and $x = 3$. The sign of $f'(x)$ is as follows:

x	$(-\infty, 0)$	$(0, 3)$	$(3, \infty)$
$f'(x)$	$+$	$+$	$-$
$f(x)$	↗	↗	↘

Thus, the only local extreme point is at $x = 3$, and it is a local maximum. It is also a global maximum. To see why, we first observe that since it is the only critical point on $(0, \infty)$ it is a global maximum relative to that interval, with value $f(3) = 27e^{-3}$. And, since $f(x) < 0$ on $(-\infty, 0)$, there can be no greater value of $f(x)$ over the entire x-axis.

5. (e) The second derivative test is inconclusive if $f'(x) = f''(x) = 0$.

7.

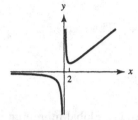

9. Solving $f'(x) = \frac{(2x-1)e^{2x}}{x^2} = 0$ gives $x = 1/2$.

(a) Comparing the values of $f(x)$ at the critical point and endpoints gives

x	$1/4$	$1/2$	1
$f(x)$	$4\sqrt{e} \approx 6.6$	$2e \approx 5.4$	$e^2 \approx 7.4$

The maximum value of $f(x)$ on $[\frac{1}{4}, 1]$ is equal to e^2 and the minimum value is equal to $2e$.

(b) Since there is no critical point in $(1, 4)$, we need only consider the values at the endpoints:

x	1	4
$f(x)$	$e^2 \approx 7.4$	$e^8/4 \approx 745.2$

The minimum value of $f(x)$ on $[1, 4]$ is equal to e^2 and the maximum value is equal to $e^8/4$.

11. The revenue function is given as a function of the price by

$$R(p) = pq = 3{,}000p - 100p^{3/2}, \quad 0 \le p \le 900,$$

and its derivative is $R'(p) = 3{,}000 - 150p^{1/2}$. Then $R'(p) = 0$ is equivalent to $p^{1/2} = 20$, or $p = 400$. Comparing the values of $R(p)$ at this critical point and the endpoints gives

p	0	400	900
$R(p)$	0	$400{,}000$	0

A continuous function on a closed, bounded interval must have a maximum value, and if the function is differentiable, the maximum can only occur at a critical point or endpoint. Therefore, we conclude that revenue is maximized if $p = 400$.

13. Since $y'(t) = e^{y(t)} > 0$ for all t, y is an increasing function.

15. Since $y'(t) = e^{y(t)}$, taking the derivative with respect to t on both sides of the given equation gives $y''(t) = y'(t)e^{y}(t)$ (by the chain rule). Substituting for $y'(t)$, we get $y''(t) = e^{2y(t)} > 0$ for all t. Therefore, the graph of $y(t)$ is concave up.

17. (a) First, we compute $p'(t)$, using the chain rule:

$$p'(t) = -\frac{22{,}400}{(28 + 52e^{-0.02t})^2} \cdot (-1.04e^{-0.02t}) = \frac{23{,}296e^{-0.02t}}{(28 + 52e^{-0.02t})^2}$$

Next, we compute the right-hand side of the differential equation. Since

$$\frac{p}{800} = \frac{28}{28 + 52e^{-0.02t}},$$

we obtain

$$0.02p\left(1 - \frac{p}{800}\right) = 0.02\left(\frac{22{,}400}{28 + 52e^{-0.02t}}\right)\left(1 - \frac{28}{28 + 52e^{-0.02t}}\right)$$

$$= 0.02\left(\frac{22{,}400}{28 + 52e^{-0.02t}}\right)\left(\frac{52e^{-0.02t}}{28 + 52e^{-0.02t}}\right) = \frac{23{,}296e^{-0.02t}}{(28 + 52e^{-0.02t})^2}.$$

Therefore, $\frac{dp}{dt} = 0.02p(1 - \frac{p}{800})$. Finally, $p(0) = \frac{22{,}400}{28+52} = 280$.

(b) $p(50) = \dfrac{22{,}400}{28 + 52e^{-1}} \approx 475.3$ million

(c) Since $\lim_{t\to\infty} e^{-0.02t} = 0$, we have $\lim_{t\to\infty} p(t) = \frac{22{,}400}{28} = 800$.

(d) In part (a) we saw that $p'(t) = \frac{23{,}290e^{-0.02t}}{(28+52e^{-0.02t})^2}$, which is positive for all t. Therefore, $p(t)$ is increasing.

(e) Taking the derivative with respect to t on both sides of the differential equation, we obtain

$$p'' = 0.02\left[p'\left(1 - \frac{p}{800}\right) - \frac{pp'}{800}\right] = 0.02p'\left(1 - \frac{p}{400}\right).$$

Since, as we remarked in (d), $p' > 0$, the sign of p'' is the same as that of $1 - \frac{p}{400}$, which is positive for $p < 400$ and negative for $p > 400$. Therefore, the graph of p is concave up for $p < 400$ and concave down for $p > 400$.

(f) Referring to (e), we see that $p'' = 0$ if $p = 400$, and the sign of p'' changes from negative (for $p < 400$) to positive (for $p > 400$).

(g) If $p = 400$, then $dp/dt = 0.02 \cdot 400(1 - \frac{400}{800}) = 4$.

(h)

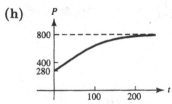

CHAPTER 5

Integration

EXERCISES 5.1

1. $P(x) = \int MP(x)\,dx = \int(-0.8x + 2{,}000)\,dx = -0.4x^2 + 2{,}000x + c$. Setting $x = 30$ and $P(30) = 15{,}000$ gives $-0.4(30)^2 + 2{,}000(30) + c = 15{,}000$, from which we get $c = -44{,}640$. Therefore, $P(x) = -0.4x^2 + 2{,}000x - 44{,}640$.

3. Since $\int \sqrt{x}\,dx = \int x^{1/2}\,dx = \frac{2}{3}x^{3/2} + c$, an antiderivative of $f(x)$ is $\frac{2}{3}x^{3/2}$.

5. Since $\int(s - \frac{1}{s})\,ds = \int s\,ds - \int \frac{1}{s}\,ds = \frac{1}{2}s^2 - \ln|s| + c$, an antiderivative of $h(s)$ is $\frac{1}{2}s^2 - \ln|s|$.

7. Since $\int(x^5 - \sqrt[3]{x} + x^{-4})\,dx = \int x^5\,dx - \int x^{1/3}\,dx + \int x^{-4}\,dx = \frac{1}{6}x^6 - \frac{3}{4}x^{4/3} - \frac{1}{3}x^{-3} + c$, an antiderivative of $f(x)$ is $\frac{1}{6}x^6 - \frac{3}{4}x^{4/3} - \frac{1}{3}x^{-3}$.

9. Since $\int(\pi r^2 - 2\pi r)\,dr = \pi \int r^2\,dr - 2\pi \int r\,dr = \frac{\pi}{3}r^3 - \pi r^2 + c$, an antiderivative of $A(r)$ is $\frac{\pi}{3}r^3 - \pi r^2$.

11. Since $F'(x) = \frac{d}{dx}(1 + e^{x^2}) = 2xe^{x^2} = f(x)$, $F(x)$ is an antiderivative of $f(x)$.

13. Since $F'(x) = \frac{d}{dx}(\ln|\frac{x-1}{x+1}|) = \frac{d}{dx}(\ln|x-1| - \ln|x+1|) = \frac{1}{x-1} - \frac{1}{x+1} = \frac{(x+1)-(x-1)}{(x-1)(x+1)} = \frac{2}{x^2-1} = f(x)$, $F(x)$ is antiderivative of $f(x)$.

15. We want $F'(x) = f(x)$. Since $F'(x) = \frac{5A}{5x+1}$, we get $\frac{5A}{5x+1} = \frac{1}{5x+1}$, or $5A = 1$. So $A = 1/5$.

17. Since $F'(x) = \frac{2Ax}{x^2+1}$, we get $\frac{2Ax}{x^2+1} = \frac{x}{x^2+1}$, or $2A = 1$. So, $A = 1/2$.

19. Since $\frac{d}{dt}[A(2t-1)^6 + c] = 12A(2t-1)^5$, we get $12A(2t-1)^5 = (2t-1)^5$, or $12A = 1$. So $A = 1/12$.

21. Since $\frac{d}{dt}(A\ln|\frac{t}{t-1}| + c) = A\frac{d}{dt}[\ln|t| - \ln|t-1| + c] = A(\frac{1}{t} - \frac{1}{t-1}) = \frac{-A}{t^2-t}$, we get $\frac{-A}{t^2-t} = \frac{1}{t^2-t}$. Therefore, $A = -1$.

23. Since $\frac{d}{dx}(A\ln|x-2| + B\ln|x+3| + c) = \frac{A}{x-2} + \frac{B}{x+3}$, we get

$$\frac{1}{x^2+x-6} = \frac{A}{x-2} + \frac{B}{x+3} = \frac{(A+B)x + (3A-2B)}{x^2+x-6}.$$

Therefore, we must have $A + B = 0$ and $3A - 2B = 1$. The solutions are $A = 1/5$ and $B = -1/5$.

25. Since $\int 2\,dx = 2x + c$, we can take $F_1(x) = 2x - 2$, $F_2(x) = 2x - 1$, $F_3(x) = 2x$, $F_4(x) = 2x + 1$, and $F_5(x) = 2x + 2$.

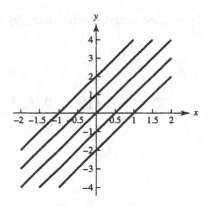

27. Since $\int \frac{1}{2}x\,dx = \frac{1}{4}x^2 + c$, we can take $F_1(x) = \frac{x^2}{4} - 2$, $F_2(x) = \frac{x^2}{4} - 1$, $F_3(x) = \frac{x^2}{4}$, $F_4(x) = \frac{x^2}{4} + 1$, and $F_5(x) = \frac{x^2}{4} + 2$.

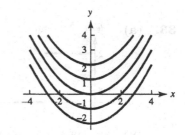

29. Since $\int 2e^{-x}\,dx = -2e^{-x} + c$, we can take $F_1(x) = -2e^{-x}$, $F_2(x) = -2e^{-x} + 1$, $F_3(x) = -2e^{-x} + 2$, $F_4(x) = -2e^{-x} + 3$, and $F_5(x) = -2e^{-x} + 4$.

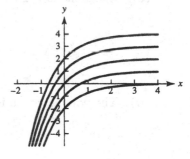

31. From the graph of f, we get $f(x) = \begin{cases} -2 & \text{if } 0 \le x \le 2, \\ x - 4 & \text{if } 2 < x \le 6. \end{cases}$ Taking antiderivatives gives

$$F(x) = \begin{cases} -2x + c & \text{if } 0 \le x \le 2, \\ \frac{1}{2}x^2 - 4x + b & \text{if } 2 < x \le 6, \end{cases}$$

for some constants c and b. In order for F to be continuous, the two branches must coincide at $x = 2$, which means that $-4 + c = -6 + b$, or $b = 2 + c$. Thus, we get

$$F(x) = \begin{cases} -2x + c & \text{if } 0 \le x \le 2, \\ \frac{1}{2}x^2 - 4x + 2 + c & \text{if } 2 < x \le 6. \end{cases}$$

The initial condition $F_1(0) = 2$ gives $c = 2$, and

$$F_1(x) = \begin{cases} -2x + 2 & \text{if } 0 \le x \le 2, \\ \frac{1}{2}x^2 - 4x + 4 & \text{if } 2 < x \le 6. \end{cases}$$

The initial condition $F_2(0) = 1$ gives $c = 1$ and

$$F_2(x) = \begin{cases} -2x + 1 & \text{if } 0 \le x \le 2, \\ \frac{1}{2}x^2 - 4x + 3 & \text{if } 2 < x \le 6. \end{cases}$$

Therefore, $F_1(1) = 0$, $F_2(1) = -1$, and $F_1(3) - F_2(3) = 1$.

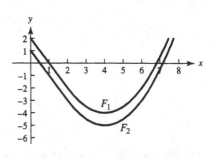

33. (a)

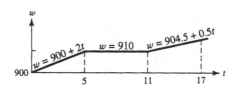

(b) No. The rate graph is discontinuous at $x = 5$ and $x = 11$.

(c) Using the initial condition $w(0) = 900$, we obtain

$$w(t) = \begin{cases} 2t + 900 & \text{if } 0 \le t \le 5, \\ 910 & \text{if } 5 < t \le 11, \\ 0.5t + 904.5 & \text{if } 11 < t \le 12. \end{cases}$$

(d) The volume graph is continuous.

35. $\displaystyle \int \frac{2}{x^3}\, dx = 2 \int x^{-3}\, dx = -x^{-2} + c = -\frac{1}{x^2} + c$

Check: $\dfrac{d}{dx}(-x^{-2} + c) = 2x^{-3} = \dfrac{2}{x^3}$

37. $\displaystyle \int e^{3t}\, dt = \frac{1}{3}e^{3t} + c$

Check: $\dfrac{d}{dt}\left(\dfrac{1}{3}e^{3t} + c\right) = e^{3t}$

39. $\displaystyle \int \frac{1}{2}(e^x + e^{-x})\, dx = \frac{1}{2}\left(\int e^x\, dx + \int e^{-x}\, dx \right) = \frac{1}{2}(e^x - e^{-x}) + c$

Check: $\dfrac{d}{dx}\left(\dfrac{1}{2}(e^x - e^{-x}) + c\right) = \dfrac{1}{2}(e^x + e^{-x})$

41. $\int y^2 \left(5y^3 + \dfrac{1}{y}\right) dy = \int (5y^5 + y)\, dy = \dfrac{5}{6}y^6 + \dfrac{1}{2}y^2 + c$

Check: $\dfrac{d}{dy}\left(\dfrac{5}{6}y^6 + \dfrac{1}{2}y^2 + c\right) = 5y^5 + y = y^2\left(5y^3 + \dfrac{1}{y}\right)$

43. $F(x) = x^3 - 12x$. $F(x)$ is decreasing when $f(x) < 0$, increasing when $f(x) > 0$. $F(x)$ is concave up when $f(x)$ is increasing, concave down when $f(x)$ is decreasing.

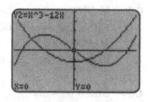

45. $F(x) = \frac{3}{4}x^{4/3} - \frac{1}{4}x^4$. $F(x)$ is decreasing when $f(x) < 0$, increasing when $f(x) > 0$. $F(x)$ is concave up when $f(x)$ is increasing, concave down when $f(x)$ is decreasing.

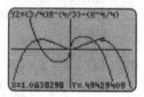

47. $y = \int (x^4 - 4x^{2/3})\, dx = \int x^4\, dx - 4\int x^{2/3}\, dx = \dfrac{1}{5}x^5 - \dfrac{12}{5}x^{5/3} + c$

49. $y = -\dfrac{100}{0.06}e^{-0.06t} + 350t + c = -\dfrac{5{,}000}{3}e^{-0.06t} + 350t + c$

51. Since $dy/dx = 1 + 2x$, we have $y = \int(1 + 2x)\, dx = x + x^2 + c$. The initial condition $y(0) = -1$ gives $0 + (0)^2 + c = -1$, or $c = -1$. Therefore, $y = x + x^2 - 1$.

53. Since $dy/dx = x + e^{-x}$, we have $y = \int(x + e^{-x})\, dx = \frac{1}{2}x^2 - e^{-x} + c$. The initial condition $y(0) = 1$ gives $\frac{1}{2}(0)^2 - e^0 + c = 1$, or $-1 + c = 1$, or $c = 2$. So, $y = \frac{1}{2}x^2 - e^{-x} + 2$.

55. Since $\frac{dQ}{dt} = e^{0.08t}$, we have $Q = \int e^{0.08t}\, dt = \frac{1}{0.08}e^{0.08t} + c = 12.5e^{0.08t} + c$. The initial condition $Q(0) = 80{,}000$ gives $12.5 + c = 80{,}000$, or $c = 79{,}987.5$. Therefore, $Q(t) = 12.5e^{0.08t} + 79{,}987.5$.

57. $y = \int (6x^2 - 4x + 1)\, dx = 2x^3 - 2x^2 + x + c$

The initial values give $c = -2, -1, 0, 1, 2$.

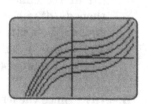

59. $y = \int \left(x + \dfrac{2}{x}\right) dx = \dfrac{1}{2}x^2 + 2\ln|x| + c$

The initial values give $c = -\frac{1}{2}, \frac{1}{2}, \frac{3}{2}, \frac{5}{2}, \frac{7}{2}$.

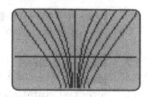

61. **(a)** Since $P'(x) = 15 - 0.01x$, we have $P(x) = 15x - 0.005x^2 + c$. The initial condition $P(0) = -1{,}000$ gives $c = -1{,}000$. Therefore,

$$P(x) = 15x - 0.005x^2 - 1{,}000.$$

(b) $P(1,000) = 15 \cdot (1,000) - 0.005 \cdot (1,000)^2 - 1,000 = 9,000.$

(c) Solving $P'(x) = 15 - 0.01x = 0$ gives $x = 1,500$. Since $P''(x) = -0.01 < 0$, the graph of $P(x)$ is concave down. The maximum profit that the company can earn in a single week is $P(1,500) = 15 \cdot (1,500) - 0.005 \cdot (1,500)^2 - 1,000 = 10,250$ dollars.

63. If $S(t)$ is the savings generated at the end of t years, $S'(t) = 300,000e^{-0.03t}$, and $S(t) = \int 300,000$ $e^{-0.03t}\, dt = -10,000,000e^{-0.03t} + c.$ The initial condition $S(0) = 0$ gives $c = 10,000,000$. So, $S(t) = 10,000,000(1 - e^{-0.03t})$. Setting $10,000,000(1 - e^{-0.03t}) = 2,000,000$ gives $e^{-0.03t} = \frac{4}{5}$, or $-0.03t = \ln \frac{4}{5}$, or $t = -\frac{1}{0.03} \ln \frac{4}{5} \approx 7.44$. The system will pay for itself after 7.44 years.

65. Since $y' = 2x - 3$, we have $y = \int (2x - 3)\, dx = x^2 - 3x + c.$ The graph passes through $(0, 4)$, so we have $c = y(0) = 4$. Therefore, $y = x^2 - 3x + 4$. Solving $y' = 2x - 3 = 0$ gives $x = 3/2$, at which there is a minimum. The minimum value is $y(3/2) = 7/4$, which is the minimum height of the graph.

67. **(a)** $y = \int -0.005t\, dt = -0.0025t^2 + c$

(b) Taking the initial population as 100%, the initial condition is $y(0) = 1$. Therefore, $c = 1$ and $y = -0.0025t^2 + 1.$

(c) $y(14) = -0.0025(14)^2 + 1 = 0.51$. Therefore, 49% of the bat population has died off after 14 years.

(d) The population will be eliminated when $-0.0025t^2 + 1 = 0$, or $t^2 = 1/(0.0025) = 400$, or $t = 20$ years.

69. Integrating with respect to t gives $J(t) = e^t - 0.9t + 7.5t^2 + c.$ Since $J(0) = 36$, we have $1 + c = 36$, or $c = 35$. Therefore, $J(t) = e^t - 0.9t + 7.5t^2 + 35.$

71. Write $P_{BN}(t)$ and $P_A(t)$ for the profits earned by Book Nook and Adams Books at the time t, respectively. Since $P'_{BN}(t) = 0.92P'_A(t)$, we see $P_{BN} = 0.92P_A + c.$ Now $P_{BN}(0) = -2,000$ and $P_A(0) = -2,400$. Thus, $c = -2,000 + 0.92 \cdot 2,400 = 208$ and $P_{BN} = 0.92P_A + 208 = 0.92 \cdot 3,100 + 208 = 3,060.$

73. Write $v(t)$ for the velocity of the car at the time t and $s(t)$ for the distance travelled by the car at the time t. If the car decelerates at the constant rate a, where a is a negative number, then $dv/dt = a$, and integrating both sides gives $v(t) = at + c.$ Since $v(0) = 80$, we have $c = 80$, and $ds/dt = v(t) = at + 80.$ Thus, $s(t) = \frac{1}{2}at^2 + 80t + c.$ Since $s(0) = 0$ we have $c = 0$, and so $s(t) = \frac{1}{2}at^2 + 80t.$ Setting $240 = s(4) = \frac{1}{2}a \cdot 4^2 + 80 \cdot 4$, we obtain $a = -10$ ft/sec^2. Thus, $v(4) = (-10) \cdot 4 + 80 = 40$ ft/sec.

75. **(a)** Since the fertilizer is leaving the tank, the volume $V(t)$ remaining in the tank is decreasing and satisfies the differential equation $dV/dt = -20e^{-0.2t}$, with the initial condition $V(0) = 100.$

(b) Integrating gives $V(t) = 100e^{-0.2t} + c.$ Since $V(0) = 100$, we must have $100 + c = 100$, or $c = 0$. Thus, $V(t) = 100e^{-0.2t}.$

77. First, we check the initial condition: $c(0) = (c_0 - C)e^{-0} + C = c_0.$ Next, taking the derivative gives $c'(t) = \frac{-kA}{V}(c_0 - C)e^{-kAt/V}.$ Finally, substituting for $c(t)$ on the right-hand side of the differential equation gives

$$\frac{kA}{V}[C - c(t)] = \frac{kA}{V}[C - (c_0 - C)e^{-kAt/V} - C] = -\frac{kA}{V}(c_0 - C)e^{-kAt/V}.$$

Thus, the given function satisfies the differential equation.

79. Following the method of the previous exercise with $v(0) = 0$ (starting from a position of rest), we obtain the following:

D3	▼	=	=C3+(0.25)*C2										
								acceleration					
A	B	C	D	E	F	G	H	I	J	K	L	M	N
t	0	0.25	0.5	0.75	1	1.25	1.5	1.75	2	2.25	2.5	2.75	3
a	0.00	0.34	0.76	1.30	2.04	2.92	4.15	5.80	8.12	10.80	14.26	16.82	18.05
v	0.00	0.00	0.09	0.28	0.60	1.11	1.84	2.88	4.33	6.36	9.06	12.62	16.83

81. The estimated and actual values of $F(x)$ (to 3 decimal places) are as follows:

A	B	C	D	E	F	G	H	I	J	K	L
x	0.000	0.100	0.200	0.300	0.400	0.500	0.600	0.700	0.800	0.900	1.000
Estimated	0.000	0.100	0.197	0.290	0.381	0.468	0.553	0.635	0.714	0.791	0.865
Actual to 3 decimal places	0.000	0.098	0.193	0.285	0.374	0.461	0.544	0.624	0.702	0.778	0.850

The estimated values are obtained as in the previous exercise. The actual values are found by using the indefinite integral $F(x) = \int e^{-x/3}\, dx = -3e^{-x/3} + c$ and the initial condition $F(0) = 0$, which gives $c = 3$.

83. (a) With the command **Integrate [f(x), x]** Mathematica gives an antiderivative of $f(x)$, while with the command **D[f(x), x]** gives the derivative of $f(x)$. In our case $f(x) = 1/(24 - 10x - 3x^2 + x^3)$ and Mathematica gives:

```
In[1] := F = Integrate[1 / (24 - 10 x - 3 x² + x³), x]

Out[1] = 1/14 Log[-4 + x] - 1/10 Log[-2 + x] + 1/35 Log[3 + x]

In[2] := f = D[F, x]

Out[2] =      1                1              1
         ─────────── - ─────────── + ───────────
         14 (-4 + x)   10 (-2 + x)   35 (3 + x)

In[3] := Simplify[f]

Out[3] =             1
          ─────────────────────
          24 - 10 x - 3 x² + x³
```

Here we used the command **Simplify** to show that indeed the function in the fourth line equals to $1/(24 - 10x - 3x^2 + x^3)$.

(b) In Mathematica, to solve a differential equation of the form $y' = f(x)$, we use the command **DSolve[y'[x]==f(x), y[x], x]**. In our case we have $f(x) = 3x^2 + 1$. So we have

```
In[1] := solution = DSolve[y'[x] == 3 x ^ 2 + 1, y[x], x]

Out[1] = {{y[x] → x + x³ + C[1]}}
```

As expected the output contains a constant C[1]. The next command will define the solution as a function of x and the constant c, where c is the replacement of C[1]. Here we denote these solutions by $y1(x, c)$.

```
In[2] := y1[x_, c_] = y[x] / . First[solution] / . C[1] -> c

Out[2] = c + x + x³
```

To plot the solutions $y1(x, c)$, for $-1 < x < 1$, and for $c = -4, -3, -2, -1, 0, 1, 2, 3, 4$, we use the following command.

In[3] := Plot[Evaluate[Table[y1[x,c],{c,-4,4,1}]],
        {x,-1,1}, AspectRatio->1]

This way we obtain the following graphs of the solutions.

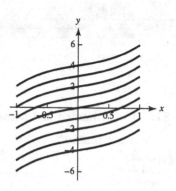

EXERCISES 5.2

1. Let $u = x^2 + 1$. Then $du = 2x\,dx$, and

$$\int \frac{2x}{x^2 + 1}\,dx = \int \frac{du}{u} = \ln|u| + c = \ln|x^2 + 1| + c = \ln(x^2 + 1) + c.$$

3. Let $u = -t$. Then $du = -dt$, or $dt = -du$, and

$$\int e^{-t}\,dt = \int e^u(-du) = -\int e^u\,du = -e^u + c = -e^{-t} + c.$$

5. Let $u = 3x + 4$. Then $du = 3\,dx$, or $dx = \frac{1}{3}\,du$, and

$$\int \sqrt{3x + 4}\,dx = \int \sqrt{u}\left(\frac{1}{3}\,du\right) = \frac{1}{3}\int u^{1/2}\,du = \frac{2}{9}u^{3/2} + c = \frac{2}{9}(3x + 4)^{3/2} + c.$$

7. Let $u = \ln(x^2 + 1)$. Then $du = \frac{2x}{x^2+1}\,dx$, or $\frac{x}{x^2+1}\,dx = \frac{1}{2}\,du$, and

$$\int \frac{x\ln(x^2 + 1)}{x^2 + 1}\,dx = \int u\left(\frac{1}{2}\,du\right) = \frac{1}{4}u^2 + c = \frac{1}{4}(\ln(x^2 + 1))^2 + c.$$

9. Let $u = 1 + \sqrt{x}$, so that $\sqrt{x} = u - 1$ and $1 - \sqrt{x} = 2 - u$. Also, $du = \frac{1}{2\sqrt{x}}\,dx$, so that $dx = 2\sqrt{x}\,du = 2(u - 1)\,du$. Therefore,

$$\int \frac{1 - \sqrt{x}}{1 + \sqrt{x}}\,dx = 2\int \frac{(2 - u)(u - 1)}{u}\,du = 2\int\left(-u + 3 - \frac{2}{u}\right)du$$

$$= -u^2 + 6u - 4\ln|u| + c$$

$$= -(1 + \sqrt{x})^2 + 6(1 + \sqrt{x}) - 4\ln|1 + \sqrt{x}| + c$$

$$= 4\sqrt{x} - x - 4\ln(1 + \sqrt{x}) + C. \text{ (Here } C = c + 5.)$$

11. (a) Let $u = \sqrt{x}$. Then $du = \frac{1}{2\sqrt{x}}\,dx$, or $dx = 2\sqrt{x}\,du = 2u\,du$. Thus,

$$\int \frac{1}{1 + \sqrt{x}}\,dx = \int \frac{2u}{u + 1}\,du$$

$$= \int\left(2 - \frac{2}{u + 1}\right)du = 2u - 2\ln|u + 1| + c = 2\sqrt{x} - 2\ln(\sqrt{x} + 1) + c.$$

(b) The two answers differ only by a constant. That is,

$$2(1 + \sqrt{x}) - 2\ln(1 + \sqrt{x}) + c = 2\sqrt{x} - 2\ln(1 + \sqrt{x}) + c + 2.$$

13. If $u = 1 + e^t$, then $du = e^t \, dt$, and $dt = \frac{1}{e^t} \, du = \frac{1}{u-1} \, du$. Therefore,

$$\int \sqrt{1 + e^t} \, dt = \int \frac{\sqrt{u}}{u - 1} \, du \neq \int \sqrt{u} \, du.$$

15. Let $u = x^2 + 4$. Then $du = 2x \, dx$, and

$$\int \frac{2x}{\sqrt{x^2 + 4}} \, dx = \int \frac{1}{\sqrt{u}} \, du = \int u^{-1/2} \, du = 2u^{1/2} + c = 2\sqrt{x^2 + 4} + c.$$

Differentiating gives

$$\frac{d}{dx}\left(2\sqrt{x^2 + 4} + c\right) = 2 \cdot \frac{1}{2}(x^2 + 4)^{-1/2} \cdot 2x + 0 = \frac{2x}{\sqrt{x^2 + 4}}.$$

17. Let $u = x^2 - 3$. Then $du = 2x \, dx$ or $x \, dx = \frac{1}{2} \, du$, and

$$\int (x^2 - 3)^5 x \, dx = \int u^5 \cdot \frac{1}{2} \, du = \frac{1}{12}u^6 + c = \frac{1}{12}(x^2 - 3)^6 + c.$$

Differentiating gives

$$\frac{d}{dx}\left(\frac{1}{12}(x^2 - 3)^6 + c\right) = \frac{1}{12} \cdot 6(x^2 - 3)^5 \cdot 2x + 0 = (x^2 - 3)^5 x.$$

19. Let $u = x^2 + x + 1$. Then $du = (2x + 1) \, dx$, and

$$\int \frac{2x + 1}{(x^2 + x + 1)^3} \, dx = \int \frac{1}{u^3} \, du = \int u^{-3} \, du = -\frac{1}{2}u^{-2} + c$$

$$= -\frac{1}{2}(x^2 + x + 1)^{-2} + c = -\frac{1}{2(x^2 + x + 1)^2} + c.$$

Differentiating gives

$$\frac{d}{dx}\left(-\frac{1}{2}(x^2 + x + 1)^{-2} + c\right) = -\frac{1}{2} \cdot (-2)(x^2 + x + 1)^{-3} \cdot (2x + 1) = \frac{2x + 1}{(x^2 + x + 1)^3}.$$

21. Let $u = -0.1t^4$. Then $du = -0.4t^3 \, dt$ or $t^3 \, dt = -2.5 \, du$, and

$$\int t^3 e^{-0.1t^4} \, dt = -2.5 \int e^u \, du = -2.5e^u + c = -2.5e^{-0.1t^4} + c.$$

Differentiating gives

$$\frac{d}{dt}\left(-2.5e^{-0.1t^4} + c\right) = -2.5e^{-0.1t^4} \cdot (-0.1 \cdot 4t^3) = t^3 e^{-0.1t^4}.$$

23. Let $u = 2x + 1$. Then $du = 2 \, dx$, or $dx = \frac{1}{2} \, du$, and

$$\int \frac{3}{2x + 1} \, dx = \int \frac{3}{u} \cdot \frac{1}{2} \, du = \frac{3}{2}\ln|u| + c = \frac{3}{2}\ln|2x + 1| + c.$$

Differentiating gives

$$\frac{d}{dx}\left(\frac{3}{2}\ln|2x + 1| + c\right) = \frac{3}{2} \cdot \frac{1}{2x + 1} \cdot 2 = \frac{3}{2x + 1}.$$

25. Let $u = \ln t$. Then $du = \frac{1}{t} dt$, and

$$\int \frac{\ln(t^2)}{t} dt = \int 2\ln t \cdot \frac{1}{t} dt = 2\int u \, du = u^2 + c = (\ln t)^2 + c.$$

Differentiating gives

$$\frac{d}{dt}\left((\ln t)^2 + c\right) = 2 \cdot \ln t \cdot \frac{1}{t} + 0 = \frac{2\ln t}{t} = \frac{\ln(t^2)}{t}.$$

27. Let $u = \ln t$. Then $du = \frac{1}{t} dt$, and

$$\int \frac{dt}{t\ln t} = \int \frac{1}{\ln t} \cdot \frac{1}{t} dt = \int \frac{1}{u} du = \ln|u| + c = \ln|\ln t| + c.$$

Differentiating gives

$$\frac{d}{dt}\left(\ln|\ln t| + c\right) = \frac{1}{\ln t} \cdot \frac{1}{t} = \frac{1}{t\ln t}.$$

29. Let $u = 1 + \sqrt{x}$. Then $\sqrt{x} = u - 1$ and $du = \frac{1}{2\sqrt{x}} dx$, so that $dx = 2\sqrt{x}\, du = 2(u-1)\, du$. Therefore,

$$\int \frac{\sqrt{x}}{1 + \sqrt{x}} dx = \int \frac{u-1}{u} \cdot 2(u-1)\, du = 2\int \left(u - 2 + \frac{1}{u}\right) du$$

$$= u^2 - 4u + 2\ln|u| + c$$

$$= 1 + 2\sqrt{x} + x - 4 - 4\sqrt{x} + 2\ln|1 + \sqrt{x}| + c$$

$$= x - 2\sqrt{x} + 2\ln(1 + \sqrt{x}) + C. \text{ (Here } C = c - 3.)$$

Differentiating gives

$$\frac{d}{dx}\left(x - 2\sqrt{x} + 2\ln(1 + \sqrt{x}) + C\right) = 1 - 2 \cdot \frac{1}{2}\frac{1}{\sqrt{x}} + 2 \cdot \frac{1}{1 + \sqrt{x}} \cdot \frac{1}{2}\frac{1}{\sqrt{x}}$$

$$= 1 - \frac{1}{\sqrt{x}} + \frac{1}{(1 + \sqrt{x})\sqrt{x}}$$

$$= \frac{\sqrt{x} - 1}{\sqrt{x}} + \frac{1}{(1 + \sqrt{x})\sqrt{x}}$$

$$= \frac{(\sqrt{x} - 1)(\sqrt{x} + 1) + 1}{(1 + \sqrt{x})\sqrt{x}}$$

$$= \frac{x - 1 + 1}{(1 + \sqrt{x})\sqrt{x}} = \frac{\sqrt{x}}{1 + \sqrt{x}}.$$

31. To compute $y = \int \sqrt{4t + 1}\, dt$, we let $u = 4t + 1$. Then $du = 4\, dt$ or $dt = \frac{1}{4} du$, and

$$y = \int u^{1/2} \cdot \frac{1}{4} du = \frac{1}{6} u^{3/2} + c = \frac{1}{6}(4t + 1)^{3/2} + c.$$

Since $3 = y(0) = \frac{1}{6} + c$, we have $c = \frac{17}{6}$, so that $y = \frac{1}{6}(4t + 1)^{3/2} + \frac{17}{6}$.

33. To compute $r = \int \frac{e^x - e^{-x}}{e^x + e^{-x}}\, dx$, we let $u = e^x + e^{-x}$. Then $du = (e^x - e^{-x})\, dx$, and $r = \int \frac{1}{u}\, du = \ln|u| + c = \ln(e^x + e^{-x}) + c$. The condition $r(0) = 0$ gives $0 = \ln(1 + 1) + c$, or $c = -\ln 2$. Thus, $r = \ln(e^x + e^{-x}) - \ln 2 = \ln \frac{e^x + e^{-x}}{2}$.

35. Since $\frac{dv}{dt} = a(t) = 5 - t(t^2 + 1)^{-1}$, we have $v(t) = \int (5 - \frac{t}{t^2+1})\, dt$. Using the substitution $u = t^2 + 1$, we obtain $v(t) = 5t - \frac{1}{2}\ln(t^2 + 1) + c$. Since the boat starts from a position of rest, $v(0) = 0$, which gives $c = 0$ and $v(t) = 5t - \frac{1}{2}\ln(t^2 + 1)$. So, $v(1) = 5 - \frac{1}{2}\ln 2 \approx 4.65$ miles per hour.

37. Integrating gives $y = \int (70e^{0.14t} - 120e^{0.03t})\, dt = 70 \int e^{0.14t}\, dt - 120 \int e^{0.03t}\, dt$. Using the substitutions $u = 0.14t$ for the first integral and $u = 0.03t$ for the second, we obtain

$$y = \frac{70}{0.14}e^{0.14t} - \frac{120}{0.03}e^{0.03t} + c = 500e^{0.14t} - 4,000e^{0.03t} + c.$$

From the initial condition $y(0) = 3,700$ we obtain $500 - 4,000 + c = 3,700$, or $c = 7,200$. Therefore, the fish population after t years is $y(t) = 500e^{0.14t} - 4,000e^{0.03t} + 7,200$ and $y(10) = 500e^{1.4} - 4,000e^{0.3} + 7,200 \approx 3,828$.

39. (a) Substituting $u = 0.4h^2 + h + 1$ and $du = (0.8h + 1)\, dh$, we obtain

$$R(h) = \int 37\frac{0.8h + 1}{0.4h^2 + h + 1}\, dh = 37 \int \frac{1}{u}\, du = 37\ln|u| + c = 37\ln(0.4h^2 + h + 1) + c.$$

The initial condition $R(0) = 168$ gives $c = 168$. Thus,

$$R(h) = 37\ln(0.4h^2 + h + 1) + 168.$$

(b) $R(1) = 37\ln(0.4 + 1 + 1) + 168 = 37\ln 2.4 + 168 \approx 200$.

41. To compute $E(y) = \int -2.52y(0.1y^2 + 1)^{-4}\, dy$, we let $u = 0.1y^2 + 1$. Then $du = 0.2y\, dy$ or $y\, dy = 5\, du$. Therefore,

$$E(y) = -12.6 \int u^{-4}\, du = 4.2u^{-3} + c = 4.2(0.1y^2 + 1)^{-3} + c.$$

Since $E(0) = 3.6$, we have $4.2 + c = 3.6$, or $c = -0.6$. Thus $E(y) = 4.2(0.1y^2 + 1)^{-3} - 0.6$. Solving $E(y) = 0$, or $(0.1y^2 + 1)^3 = 7$, we find $y = \sqrt{10(\sqrt[3]{7} - 1)} \approx 3$ years.

43. (a) The amount x of the drug in the body at the time t satisfies the equation $t = \int \frac{dx}{r - kx}$. Let $u = r - kx$. Then $du = -k\, dx$ or $dx = -\frac{1}{k}\, du$, and

$$t = -\frac{1}{k} \int \frac{1}{u}\, du = -\frac{1}{k}\ln|u| + c = -\frac{1}{k}\ln|r - kx| + c.$$

Since $x = 0$ at $t = 0$, we have $-\frac{1}{k}\ln r + c = 0$, or $c = \frac{1}{k}\ln r$. Thus, $t = \frac{1}{k}(\ln r - \ln|r - kx|)$ or

$$t = \frac{1}{k}\ln\left|\frac{r}{r - kx}\right| = \frac{1}{k}\ln\left(\frac{r}{r - kx}\right) \qquad \text{for } x < \frac{r}{k}.$$

(b) Multiplying the last formula by k gives $\ln \frac{r}{r - kx} = kt$. Then exponentiating gives $\frac{r}{r - kx} = e^{kt}$, and taking reciprocal gives $\frac{r - kx}{r} = e^{-kt}$, or $1 - \frac{k}{r}x = e^{-kt}$, or $\frac{k}{r}x = 1 - e^{-kt}$. Therefore, $x = \frac{r}{k}(1 - e^{-kt})$.

(c) Since $e^{-kt} \to 0$ as $t \to \infty$, we have $\lim_{x \to \infty} x(t) = \frac{r}{k}$.

In exercise 45, we expect $F(x)$ to be increasing when $f(x)$ is positive, decreasing when $f(x)$ is negative, concave up when $f(x)$ is increasing, and concave down when $f(x)$ is decreasing.

45. Substituting $u = e^x + e^{-x}$, we have $du = (e^x - e^{-x})\,dx$. Therefore,

$$F(x) = \int \frac{du}{u^2} = -\frac{1}{u} + c = -\frac{1}{e^x + e^{-x}} + c.$$

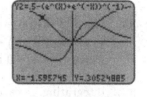

The initial condition $F(0) = 0$ gives $c = 0.5$, and

$$F(x) = 0.5 - \frac{1}{e^x + e^{-x}}.$$

47. Substituting $u = x^2 + 1$, we have $du = 2x\,dx$, or $x\,dx = \frac{1}{2}\,du$. Therefore,

$$y = \int \frac{1}{2\sqrt{u}}\,du = \sqrt{u} + c = \sqrt{x^2 + 1} + c.$$

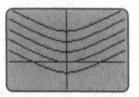

The initial conditions give $c = -2, -1, 0, 1, 2$.

49. Substituting $u = \ln x$, we have $du = \frac{1}{x}\,dx$. Therefore,

$$y = \int 2u\,du = u^2 + c = (\ln x)^2 + c.$$

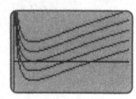

The initial conditions give $c = -2, -1, 0, 1, 2$.

EXERCISES 5.3

1. Let $u = x$ and $dv = e^{2x}\,dx$. Then $du = dx$ and $v = \frac{1}{2}e^{2x}$. Therefore,

$$\int xe^{2x}\,dx = \frac{1}{2}xe^{2x} - \int \frac{1}{2}e^{2x}\,dx = \frac{1}{2}xe^{2x} - \frac{1}{4}e^{2x} + c = \frac{1}{2}e^{2x}\left(x - \frac{1}{2}\right) + c.$$

3. Let $u = t$ and $dv = e^{-t/2}\,dt$. Then $du = dt$ and $v = -2e^{-t/2}$, and

$$\int te^{-t/2}\,dt = -2te^{-t/2} + 2\int e^{-t/2}\,dt = -2te^{-t/2} - 4e^{-t/2} + c.$$

5. Let $u = t^2$ and $dv = e^{-t}\,dt$. Then $du = 2t\,dt$ and $v = -e^{-t}$, and

$$\int t^2 e^{-t}\,dt = -t^2 e^{-t} + 2\int te^{-t}\,dt.$$

Next, we apply integration by parts again, this time with $u = t$ and $dv = e^{-t}\,dt$. Then $du = dt$ and $v = -e^{-t}$, and

$$\int t^2 e^{-t}\,dt = -t^2 e^{-t} + 2\left[-te^{-t} - \int(-e^{-t})\,dt\right]$$

$$= -t^2 e^{-t} - 2te^{-t} + 2\int e^{-t}\,dt = -t^2 e^{-t} - 2te^{-t} - 2e^{-t} + c.$$

7. Let $u = \ln x$ and $dv = x^2\,dx$. Then $du = \frac{1}{x}\,dx$ and $v = \frac{1}{3}x^3$. Therefore,

$$\int x^2 \ln x\,dx = \frac{1}{3}x^3 \ln x - \frac{1}{3}\int x^2\,dx = \frac{1}{3}x^3 \ln x - \frac{1}{9}x^3 + c.$$

9. Let $u = x$ and $dv = \sqrt{3x+2}\,dx$. Then $du = dx$ and $v = \frac{2}{9}(3x+2)^{3/2}$. ($v$ is found by using substitution, with $t = 3x + 2$ and $dt = 3\,dx$.) Integrating by parts gives

$$\int x\sqrt{3x+2}\,dx = \frac{2}{9}x(3x+2)^{3/2} - \int \frac{2}{9}(3x+2)^{3/2}\,dx,$$

and, again using the substitution $t = 3x + 2$, we get

$$\int x\sqrt{3x+2}\,dx = \frac{2}{9}x(3x+2)^{3/2} - \frac{2}{9}\cdot\frac{1}{3}\cdot\frac{2}{5}(3x+2)^{5/2} + c$$

$$= \frac{2}{9}x(3x+2)^{3/2} - \frac{4}{135}(3x+2)^{5/2} + c.$$

11. Let $u = x^2$ and $dv = x\sqrt{x^2+1}\,dx$. Then $du = 2x\,dx$ and $v = \int x\sqrt{x^2+1}\,dx$. Using the substitution
$t = x^2 + 1$, we find $v = \frac{1}{3}(x^2+1)^{3/2}$. Now using integration by parts gives

$$\int x^3\sqrt{x^2+1}\,dx = \frac{1}{3}x^2(x^2+1)^{3/2} - \int 2x\cdot\frac{1}{3}(x^2+1)^{3/2}\,dx$$

$$= \frac{1}{3}x^2(x^2+1)^{3/2} - \frac{1}{3}\int (x^2+1)^{3/2}(x^2+1)'\,dx$$

$$= \frac{1}{3}x^2(x^2+1)^{3/2} - \frac{2}{15}(x^2+1)^{5/2} + c,$$

where we again used the substitution $t = x^2 + 1$ in computing the last integral.

13. Let $\frac{1}{(x-1)(x+5)} = \frac{A}{x-1} + \frac{B}{x+5}$. By clearing denominators, we get

$$1 = A(x+5) + B(x-1).$$

Substituting $x = -5$ gives $1 = A(-5+5) + B(-5-1)$, or $B = -\frac{1}{6}$. Next, substituting $x = 1$ gives
$1 = A(1+5) + B(1-1)$ or $A = \frac{1}{6}$. Therefore,

$$\frac{1}{(x-1)(x+5)} = \frac{1}{6}\cdot\frac{1}{x-1} - \frac{1}{6}\cdot\frac{1}{x+5}.$$

15. Let $\frac{1}{x^2-9x+20} = \frac{1}{(x-4)(x-5)} = \frac{A}{x-4} + \frac{B}{x-5}$. After clearing denominators, we obtain the equation $1 =
A(x-5) + B(x-4)$, so $A = -1$ and $B = 1$. Thus the given integral equals

$$\int \frac{1}{x-5}\,dx - \int \frac{1}{x-4}\,dx = \ln|x-5| - \ln|x-4| + c = \ln\left|\frac{x-5}{x-4}\right| + c.$$

17. Let $\frac{1}{t^2+t} = \frac{1}{t(t+1)} = \frac{A}{t} + \frac{B}{t+1}$. After clearing denominators, we obtain the equation $1 = A(t+1) +
Bt$, so $A = 1$ and $B = -1$.

$$\int \frac{dt}{t^2+t} = \int \frac{1}{t}\,dt - \int \frac{1}{t+1}\,dt = \ln|t| - \ln|t+1| + c = \ln\left|\frac{t}{t+1}\right| + c.$$

19. Let $\frac{1}{(x-a)(x-b)} = \frac{A}{x-a} + \frac{B}{x-b}$. After clearing denominators, we obtain the equation $1 = A(x - b) + B(x - a)$. Substituting $x = a$ gives $1 = A(a - b)$, or $A = \frac{1}{a-b}$. Substituting $x = b$ gives $1 = B(b - a)$, or $B = -\frac{1}{a-b}$. Thus, the integral equals

$$\frac{1}{a-b}\left(\int \frac{1}{x-a}\,dx - \int \frac{1}{x-b}\,dx\right) = \frac{1}{a-b}\ln\left|\frac{x-a}{x-b}\right| + c.$$

21. Let $\frac{x}{(x-1)(x+2)} = \frac{A}{x-1} + \frac{B}{x+2}$. After clearing denominators, we obtain the equation $x = A(x + 2) + B(x - 1)$. Substituting $x = 1$ gives $1 = 3A$ or $A = \frac{1}{3}$, and substituting $x = -2$ gives $-2 = -3B$ or $B = \frac{2}{3}$. Thus, the integral equals

$$\frac{1}{3}\int \frac{1}{x-1}\,dx + \frac{2}{3}\int \frac{1}{x+2}\,dx = \frac{1}{3}\ln|x-1| + \frac{2}{3}\ln|x+2| + c.$$

23. Let $\frac{x-3}{x^2-3x-4} = \frac{x-3}{(x-4)(x+1)} = \frac{A}{x-4} + \frac{B}{x+1}$. After clearing denominators, we obtain the equation

$$x - 3 = A(x + 1) + B(x - 4).$$

Substituting $x = 4$ gives $1 = 5A$, so that $A = \frac{1}{5}$. Substituting $x = -1$ gives $-4 = -5B$, so that $B = \frac{4}{5}$. Thus, the given integral equals

$$\frac{1}{5}\int \frac{1}{x-4}\,dx + \frac{4}{5}\int \frac{1}{x+1}\,dx = \frac{1}{5}\ln|x-4| + \frac{4}{5}\ln|x+1| + c.$$

25. If $h(t)$ denotes the rocket's height at the time t, then $h'(t) = \frac{t+1}{\sqrt{2t+1}}$. Integrating gives $h(t) = \int (t+1)(2t+1)^{-1/2}\,dt$. We may compute this integral by using integration by parts with $u = t + 1$ and $dv = (2t + 1)^{-1/2}\,dt$. Then $du = dt$ and $v = (2t + 1)^{1/2}$. Thus,

$$h(t) = (t+1)(2t+1)^d 1/2 - \int (2t+1)^{1/2}\,dt = (t+1)\sqrt{2t+1} - \frac{1}{3}(2t+1)^{3/2} + c.$$

Since $h(0) = 0$, we have $1 - \frac{1}{3} + c = 0$, or $c = -\frac{2}{3}$. Thus,

$$h(t) = (t+1)\sqrt{2t+1} - \frac{1}{3}(2t+1)^{3/2} - \frac{2}{3},$$

and $h(12) = 13\sqrt{25} - \frac{1}{3}(25)^{3/2} - \frac{2}{3} = \frac{68}{3}$ ft.

27. Using the TI-83 Plus TABLE function, as shown below, gives the table of values for t from 0 to 12. Since the number of nematodes was reduced from 23,000 to a little over fifty in 12 days, the treatment was successful.

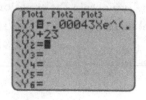 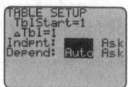  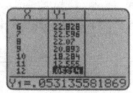

29. If t denotes time in days, then $t = \int \frac{10,000}{x(4,000-x)}\,dx$. To compute this integral, we use the method of partial fractions, writing

$$\frac{10,000}{x(4,000-x)} = \frac{A}{x} + \frac{B}{4,000-x}.$$

By multiplying both sides by $x(4{,}000 - x)$, we obtain $10{,}000 = A(4{,}000 - x) + Bx$. Setting $x = 0$ gives $A = 2.5$, and setting $x = 4{,}000$ gives $B = 2.5$. Therefore,

$$t = \int \frac{2.5}{x}\, dx + \int \frac{2.5}{4{,}000 - x}\, dx$$

$$= 2.5(\ln|x| - \ln|4{,}000 - x|) + c = 2.5 \ln\left|\frac{x}{4{,}000 - x}\right| + c.$$

Since $x = 8$ when $t = 0$, we must have $c = -2.5 \ln \frac{8}{3{,}992} = 2.5 \ln 499 \approx 15.53$. Thus $t = 2.5 \ln\left|\frac{x}{4{,}000 - x}\right| + 2.5 \ln 499$. If $x = 200$, then $t \approx 8.17$ days.

31. (a) The range is expanding, since $y' = \frac{9}{t^2 + 9t} > 0$ for $t \geq 1$.

(b) Let $\frac{9}{t^2 + 9t} = \frac{9}{t(t+9)} = \frac{A}{t} + \frac{B}{t+9}$. After clearing denominators, we obtain $9 = A(t + 9) + Bt$, whose solution is $A = 1$ and $B = -1$. Thus,

$$y = \int \frac{9}{t^2 + 9t}\, dt = \int \left(\frac{1}{t} - \frac{1}{t+9}\right) dt = \ln|t| - \ln|t+9| + c = \ln\left|\frac{t}{t+9}\right| + c.$$

Since $17.7 = y(1) = \ln \frac{1}{10} + c$, we have $c = 17.7 - \ln \frac{1}{10} = 17.7 + \ln 10$, and

$$y = \ln\left|\frac{t}{t+9}\right| + 17.7 + \ln 10 = \ln \frac{10t}{t+9} + 17.7, \quad t \geq 1.$$

(c) $\lim_{t \to \infty} y(t) = \ln 10 + 17.7 \approx 20$ acres.

In exercises 33 and 35, $F(x)$ is decreasing when $f(x)$ is negative, increasing when $f(x)$ is positive, concave up when $f(x)$ is increasing, and concave down when $f(x)$ is decreasing.

33. Integration by partial fractions gives

$$F(x) = \int \frac{1/2}{1+x}\, dx + \int \frac{1/2}{1-x}\, dx$$

$$= \frac{1}{2}[\ln(1+x) - \ln(1-x)] + c.$$

Setting $x = 0$ and $F(0) = 0$, we get $c = 0$.

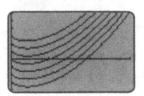

35. Integration by parts with $u = \ln x$ and $dv = dx$ gives

$$F(x) = x + \int \ln x\, dx$$

$$= x + x \ln x - x + c = x \ln x + c.$$

Setting $x = 1$ and $F(1) = 0$, we get $c = 0$.

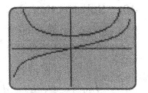

37. Using integration by parts with $u = \ln x$ and $dv = dx$, we get

$$y(x) = x^2 + \int \ln x\, dx$$

$$= x^2 + x \ln x - x + c.$$

The initial conditions give $c = -3, -2, -1, 0, 1, 2, 3$.

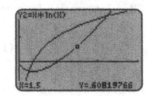

[0, 3] × [−4, 6] window

SUPPLEMENTARY EXERCISES 5.2 AND 5.3

39. Substitution. Let $u = 3x + 5$. Then $du = 3\,dx$ or $dx = \frac{1}{3}\,du$, and

$$\int (3x+5)^9\,dx = \frac{1}{3}\int u^9\,du = \frac{1}{30}u^{10} + c = \frac{1}{30}(3x+5)^{10} + c.$$

(In these exercises, checking the answer by differentiation is left for the reader.)

41. $\displaystyle \int \left(\frac{2}{x+4} - \frac{3}{x-1} \right) dx = 2\ln|x+4| - 3\ln|x-1| + c$

43. Substitution. Let $u = t^2 + 9$. Then $du = 2t\,dt$ or $t\,dt = \frac{1}{2}\,du$, and

$$\int t\sqrt{t^2+9}\,dt = \frac{1}{2}\int u^{1/2}\,du = \frac{1}{3}u^{3/2} + c = \frac{1}{3}(t^2+9)^{3/2} + c.$$

45. Partial fractions. Let $\frac{1}{t^2-1} = \frac{1}{(t-1)(t+1)} = \frac{A}{t-1} + \frac{B}{t+1}$. Clearing denominators gives $1 = A(t+1) + B(t-1)$, so that $A = \frac{1}{2}$ and $B = -\frac{1}{2}$. Thus,

$$\int \frac{1}{t^2-1}\,dx = \int \frac{1/2}{t-1}\,dt - \int \frac{1/2}{t+1}\,dt = \frac{1}{2}\ln|t-1| - \frac{1}{2}\ln|t+1| + c = \frac{1}{2}\ln\left|\frac{t-1}{t+1}\right| + c.$$

47. Substitution. Let $u = t - 1$. Then $du = dt$, and

$$\int \frac{1}{(t-1)^2}\,dt = \int \frac{1}{u^2}\,du = -\frac{1}{u} + c = -\frac{1}{t-1} + c.$$

49. Substitution. Let $u = x^4 - 4$. Then $du = 4x^3\,dx$ or $x^3\,dx = \frac{1}{4}\,du$, and

$$\int x^3(x^4-4)^5\,dx = \frac{1}{4}\int u^5\,du = \frac{1}{24}u^6 + c = \frac{1}{24}(x^4-4)^6 + c.$$

51. $\displaystyle \int (t + e^{4t})\,dt = \int t\,dt + \int e^{4t}\,dt = \frac{1}{2}t^2 + \frac{1}{4}e^{4t} + c$

53. Integration by parts. Let $u = 1 + \ln x$ and $dv = x\,dx$. Then $du = \frac{1}{x}\,dx$ and $v = \frac{1}{2}x^2$.

$$\int x(1 + \ln x)\,dx = \frac{1}{2}x^2(1 + \ln x) - \frac{1}{2}\int x\,dx = \frac{1}{2}x^2(1 + \ln x) - \frac{1}{4}x^2 + c.$$

55. Substitution. Let $u = 0.5t^2$. Then $du = t\,dt$, and

$$\int te^{0.5t^2}\,dt = \int e^u\,du = e^u + c = e^{0.5t^2} + c.$$

57. Partial fractions. Let $\frac{1}{x^2-3x+2} = \frac{1}{(x-1)(x-2)} = \frac{A}{x-1} + \frac{B}{x-2}$. Then $1 = A(x-2) + B(x-1)$, so $A = -1$ and $B = 1$, so that

$$\int \frac{1}{x^2 - 3x + 2}\,dx = -\int \frac{1}{x-1}\,dx + \int \frac{1}{x-2}\,dx = -\ln|x-1| + \ln|x-2| + c = \ln\left|\frac{x-2}{x-1}\right| + c.$$

59. Repeated use of integration by parts. First, let $u = (\ln x)^3$ and $dv = dx$. Then $du = \frac{3}{x}(\ln x)^2\, dx$ and $v = x$ so that $\int (\ln x)^3\, dx = x(\ln x)^3 - 3\int (\ln x)^2\, dx$. Next, let $u = (\ln x)^2$ and $dv = dx$. Then $du = \frac{2}{x}\ln x\, dx$ and $v = x$, so that

$$\int (\ln x)^2\, dx = x(\ln x)^2 - 2\int \ln x\, dx.$$

Finally, let $u = \ln x$ and $dv = dx$. Then $du = \frac{1}{x}\, dx$ and $v = x$, so that

$$\int \ln x\, dx = x \ln x - x + c.$$

Putting these steps together, we get

$$\int (\ln x)^3\, dx = x(\ln x)^3 - 3x(\ln x)^2 + 6x \ln x - 6x + c.$$

61. $\displaystyle \int \frac{2t}{t^2 - 3t}\, dt = \int \frac{2t}{t(t-3)}\, dt = \int \frac{2}{t-3}\, dt = 2\ln|t-3| + c$

63. First, we use a substitution, with $u = 1 + e^x$. Then $2 - u = 1 - e^x$ and $du = e^x\, dx$ or $dx = \frac{1}{(u-1)}\, du$. Therefore,

$$\int \frac{1 - e^x}{1 + e^x}\, dx = \int \frac{2 - u}{u(u-1)}\, du.$$

Next, use partial fractions, writing $\frac{2-u}{u(u-1)} = \frac{A}{u-1} + \frac{B}{u}$. Clearing denominators and setting $u = 1$ and then $u = 0$, we find that $A = 1$ and $B = -2$. Therefore,

$$\int \frac{2 - u}{u(u-1)}\, du = \int \frac{1}{u-1}\, du - 2\int \frac{1}{u}\, du + c = \ln|u-1| - 2\ln|u| + c.$$

Combining these steps, we get

$$\int \frac{1 - e^x}{1 + e^x}\, dx = \ln(e^x) - 2\ln(1 + e^x) + c = x - 2\ln(1 + e^x) + c.$$

65. First, we multiply both the numerator and denominator by e^x, to obtain

$$\int \frac{1}{e^x - e^{-x}}\, dx = \int \frac{e^x}{e^x(e^x - e^{-x})}\, dx = \int \frac{e^x}{e^{2x} - 1}\, dx.$$

Next we make the substitution $u = e^x$, so that $du = e^x\, dx$, and

$$\int \frac{e^x}{e^{2x} - 1}\, dx = \int \frac{1}{u^2 - 1}\, du.$$

Finally, we use the method of partial fractions to get

$$\int \frac{1}{u^2 - 1}\, du = \frac{1}{2}\int \frac{1}{u-1}\, du - \frac{1}{2}\int \frac{1}{u+1}\, du = \frac{1}{2}\ln|u-1| - \frac{1}{2}\ln|u+1| + c.$$

Putting these steps together, we obtain

$$\int \frac{1}{e^x - e^{-x}}\, dx = \frac{1}{2}\ln|e^x - 1| - \frac{1}{2}\ln|e^x + 1| + c = \frac{1}{2}\ln\left|\frac{e^x - 1}{e^x + 1}\right| + c.$$

67. Let $u = \sqrt{x}$, so that $du = \frac{1}{2\sqrt{x}} \, dx$. Then $dx = 2\sqrt{x} \, du = 2u \, du$. Thus,

$$\int \frac{u^2}{1+u} 2u \, du = 2 \int \frac{u^3}{u+1} \, du = 2 \int \left(u^2 - u + 1 - \frac{1}{u+1} \right) du,$$

where the last step comes from dividing $u+1$ into u^3 with remainder. Therefore,

$$\int \frac{x}{1+\sqrt{x}} \, dx = 2 \int \left(u^2 - u + 1 - \frac{1}{u+1} \right) du = \frac{2}{3}u^3 - u^2 + 2u - 2\ln|u+1| + c$$

$$= \frac{2}{3}x^{3/2} - x + 2\sqrt{x} - 2\ln(\sqrt{x} + 1) + c.$$

69. Let $u = \sqrt{x}$. Then $du = \frac{1}{2\sqrt{x}} \, dx$ or $\frac{1}{\sqrt{x}} \, dx = 2 \, du$. Therefore,

$$\int \frac{1}{\sqrt{x}\left(\sqrt{x}-1\right)\left(\sqrt{x}+2\right)} \, dx = 2 \int \frac{1}{(u-1)(u+2)} \, du.$$

Using partial fractions, we obtain

$$2 \int \frac{1}{(u-1)(u+2)} \, du = \frac{2}{3} \int \left(\frac{1}{u-1} - \frac{1}{u+2} \right) du = \frac{2}{3}(\ln|u-1| - \ln|u+2|) + c.$$

Therefore,

$$\int \frac{1}{\sqrt{x}\left(\sqrt{x}-1\right)\left(\sqrt{x}+2\right)} \, dx = \frac{2}{3}(\ln|\sqrt{x}-1| - \ln|\sqrt{x}+2|) + c = \frac{2}{3}\ln\left|\frac{\sqrt{x}-1}{\sqrt{x}+2}\right| + c.$$

71. Let $u = 2t + 1$. Then $du = 2 \, dt$, or $dt = \frac{1}{2} \, du$, and

$$y = \frac{1}{2} \int u^{1/2} \, du = \frac{1}{3}u^{3/2} + c = \frac{1}{3}(2t+1)^{3/2} + c.$$

The initial condition $y(0) = 1$ gives $\frac{1}{3} + c = 1$ or $c = \frac{2}{3}$. Thus,

$$y = \frac{1}{3}(2t+1)^{3/2} + \frac{2}{3}.$$

73. First, $y = \int (\ln t)^2 \, dt - \frac{1}{2}t^2 + c$. Next, we use integration by parts twice to obtain

$$\int (\ln t)^2 \, dt = t(\ln t)^2 - 2 \int \ln t \, dt$$

$$= t(\ln t)^2 - 2 \left(t\ln t - \int 1 \, dt \right) = t(\ln t)^2 - 2t\ln t + 2t.$$

Therefore, $y(t) = t(\ln t)^2 - 2t\ln t + 2t - \frac{1}{2}t^2 + c$. Setting $t = 1$ and $y(1) = 2$ gives $2 = 2 - \frac{1}{2} + c$, or $c = \frac{1}{2}$. Thus, $y = t(\ln|t|)^2 - 2t\ln|t| + 2t - \frac{1}{2}t^2 + \frac{1}{2}$.

75. We use integration by parts with $u = t$ and $dv = \sqrt{3t+1} \, dt$. Then $du = dt$ and $v = \frac{2}{9}(3t+1)^{3/2}$ (by substitution), and

$$y = \frac{2}{9}t(3t+1)^{3/2} - \frac{2}{9} \int (3t+1)^{3/2} \, dt = \frac{2}{9}t(3t+1)^{3/2} - \frac{4}{135}(3t+1)^{5/2} + c.$$

Since $0 = y(0) = -\frac{4}{135} + c$, we obtain $c = \frac{4}{135}$, and

$$y = \frac{2}{9}t(3t+1)^{3/2} - \frac{4}{135}(3t+1)^{5/2} + \frac{4}{135}.$$

EXERCISES 5.4

1. We use the Riemann sum formula with $\Delta x = \frac{b-a}{n} = \frac{2-1}{4} = \frac{1}{4}$ and $f(x) = \frac{1}{x}$. The values of $f(x)$ at the endpoints are as follows:

x	1	5/4	3/2	7/4	2
$f(x)$	1	4/5	2/3	4/7	1/2

(a) Using left-hand endpoints:

$$S_4 = \frac{1}{4} \cdot 1 + \frac{1}{4} \cdot \frac{4}{5} + \frac{1}{4} \cdot \frac{2}{3} + \frac{1}{4} \cdot \frac{4}{7} = \frac{1}{4} + \frac{1}{5} + \frac{1}{6} + \frac{1}{7} = \frac{319}{420} \approx 0.760.$$

(b) Using right-hand endpoints:

$$S_4 = \frac{1}{4} \cdot \frac{4}{5} + \frac{1}{4} \cdot \frac{2}{3} + \frac{1}{4} \cdot \frac{4}{7} + \frac{1}{4} \cdot \frac{1}{2} = \frac{1}{5} + \frac{1}{6} + \frac{1}{7} + \frac{1}{8} = \frac{533}{840} \approx 0.635.$$

(c) The midpoints of the four subintervals and the values of $f(x)$ at these points are given in the following table:

x	9/8	11/8	13/8	15/8
$f(x)$	8/9	8/11	8/13	8/15

Then $S_4 = \frac{1}{4} \cdot \frac{8}{9} + \frac{1}{4} \cdot \frac{8}{11} + \frac{1}{4} \cdot \frac{8}{13} + \frac{1}{4} \cdot \frac{8}{15} = \frac{2}{9} + \frac{2}{11} + \frac{2}{13} + \frac{2}{15} = \frac{4,448}{6,435} \approx 0.691.$

3. $S_8 = f(0.4) \cdot 0.2 + f(0.6) \cdot 0.2 + \cdots + f(1.8) \cdot 0.2$

$= 0.2(15 + 17 + 19 + 16 + 14 + 14 + 12 + 10) = 117(0.2) = 23.4$

5. In each of the following, $\Delta x = 0.25$ and we use left-hand endpoints.

(a) $\displaystyle\int_{-1}^{1} f(x)\,dx \approx [f(-1) + f(-0.75) + f(-0.5) + f(-0.25)$

$+ f(0) + f(0.25) + f(0.5) + f(0.75)] \cdot 0.25$

$\approx (3 + 3 + 2.5 + 2 + 1 + 0 - 1 - 2) \cdot 0.25 = 8.5(0.25) = 2.125$

(b) $\displaystyle\int_{-1}^{0.25} f(x)\,dx \approx [f(-1) + f(-0.75) + f(-0.5) + f(-0.25) + f(0)] \cdot 0.25$

$\approx (3 + 3 + 2.5 + 2 + 1) \cdot 0.25 = 11.5(0.25) = 2.875$

(c) $\displaystyle\int_{0.25}^{2} f(x)\,dx \approx [f(0.25) + f(0.5) + f(0.75) + f(1)$

$+ f(1.25) + f(1.50) + f(1.75)] \cdot 0.25$

$\approx [0 - 1 - 2 - 3 - 3.5 - 3.8 - 3.6] \cdot 0.25$

$= -16.9(0.25) = -4.225$

(d) $\displaystyle\int_{-1}^{2} f(x)\,dx = \int_{-1}^{0.25} f(x)\,dx + \int_{0.25}^{2} f(x)\,dx \approx 2.875 - 4.225 = -1.35$

Note: You can also estimate by counting squares. Area of each square is 1/4.

(a) $\displaystyle\int_{-1}^{1} f(x)\,dx \approx 10.5$ squares $- 4.5$ squares $= 6$ squares $= \frac{6}{4} = 1.5$

(b) $\displaystyle\int_{-1}^{0.25} f(x)\,dx \approx 10.5 \text{ squares} = (10.5)\cdot\frac{1}{4} = 2.625$

(c) $\displaystyle\int_{0.25}^{2} f(x)\,dx \approx -19 \text{ squares} = -\frac{19}{4} = -4.75$

(d) $\displaystyle\int_{-1}^{2} f(x)dx = \int_{-1}^{0.25} f(x)\,dx + \int_{0.25}^{2} f(x)\,dx = 2.625 - 4.75 \approx -2.125$

7. (a) By using midpoints with $\Delta x = 40$, we estimate the area of the upper region by $(60 + 86 + 94 + 108 + 100 + 44 + 41)\cdot 40 = 21{,}320$ square meters, and the area of the lower region by $(40 + 58 + 72 + 66 + 78 + 79 + 68)\cdot 40 = 18{,}440$ square meters. Therefore, the area of the lake is approximately 39,760 square meters.

(b) By using left endpoints with $\Delta x = 20$ we estimate the area of the upper region by $(60 + 70 + 86 + 90 + 94 + 98 + 108 + 105 + 100 + 82 + 44 + 40 + 41)\cdot 20 = 20{,}360$ square meters, and the area of the lower region by $(40 + 45 + 58 + 68 + 72 + 67 + 66 + 72 + 78 + 80 + 79 + 80 + 68)\cdot 20 = 17{,}460$ square meters. Therefore, the area of the lake is approximately equal to 37,820 square meters.

9. The integral is the area of the right-hand triangle, with vertices at $(0,0)$, $(1,0)$, and $(1,2)$, minus the area of the left-hand triangle, with vertices at $(-2,-4)$, $(-2,0)$, and $(0,0)$. Therefore,

$$\int_{-2}^{1} 2x\,dx = \frac{1\cdot 2}{2} - \frac{(-2)\cdot(-4)}{2} = -3.$$

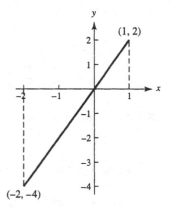

11. The integral is the negative of the area of the rectangle with vertices at $(0,0)$, $(0,-2)$, $(3,0)$, and $(3,-2)$. Therefore,

$$\int_{0}^{3} (-2)\,dx = -(\text{base})(\text{height}) = -3\cdot 2 = -6.$$

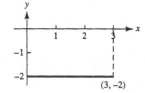

13. The integral is the area of the left-hand triangle, with vertices at $(0,0)$, $(1,0)$, and $(0,1)$, minus the area of the right-hand triangle, with vertices at $(1,0)$, $(3,0)$, and $(3,-2)$. Therefore,

$$\int_{0}^{3} (1-x)\,dx = \frac{1}{2} - 2 = -\frac{3}{2}.$$

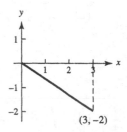

15. Dividing $[1,3]$ into 8 equal segments, we have $\Delta x = \frac{3-1}{8} = 0.25$.

(a) $\displaystyle\int_{1}^{3} f(x)\,dx \approx 0.25\cdot[f(1) + f(1.25) + \cdots + f(2.75)]$

$$= 0.25\cdot[0 + 0.22 + 0.41 + 0.56 + 0.69 + 0.81 + 0.92 + 1.01] = 1.155$$

(b) $\displaystyle\int_1^3 f(x)\,dx \approx 0.25 \cdot [f(1.25) + f(1.5) + \cdots + f(3.0)]$

$\qquad = 0.25 \cdot [0.22 + 0.41 + 0.56 + 0.69 + 0.81 + 0.92 + 1.01 + 1.10] = 1.43$

17. (a) $\displaystyle\int_0^7 r(t)\,dt$
(b) $\Delta t = 1$

(c) $\displaystyle\int_0^7 r(t)\,dt \approx 37 + 43 + 48 + 56 + 62 + 68 + 76 = 390$

19. $\displaystyle\int_0^2 v(t)\,dt = \int_0^2 (-32t - 50)\,dt$
21. $\displaystyle\int_0^{100} 0.05te^{-0.01t}\,dt$

23. Area $= -\displaystyle\int_{-2}^1 f(x)\,dx + \int_1^3 f(x)\,dx - \int_3^6 f(x)\,dx$

$\qquad = -\displaystyle\int_{-2}^1 f(x)\,dx + \int_1^3 f(x)\,dx - \left(\int_1^6 f(x)\,dx - \int_1^3 f(x)\,dx \right)$

$\qquad = 2.8 + 1.2 + 4.7 = 8.7$

25. $\displaystyle\int_{-1}^3 2f(x)\,dx = 2 \int_{-1}^3 f(x)\,dx = 2 \left(\int_{-1}^5 f(x)\,dx - \int_3^5 f(x)\,dx \right) = 2(-6 - 4) = -20$

27. A calculator gives $\ln 2 \approx 0.69314718$.

(a) With $n = 10$, $\Delta x = \frac{1}{10}$ and the midpoints are $\{ \frac{21}{20}, \frac{23}{20}, \frac{25}{20}, \ldots, \frac{37}{20}, \frac{39}{20} \}$.

$$\int_1^2 x^{-1}\,dx \approx \frac{1}{10} \cdot \left(\frac{20}{21} + \frac{20}{23} + \frac{20}{25} + \cdots + \frac{20}{37} + \frac{20}{39} \right)$$

$$= 2 \cdot \left(\frac{1}{21} + \frac{1}{23} + \frac{1}{25} + \cdots + \frac{1}{39} \right) \approx 0.69283536$$

Error is approximately 0.000312.

(b) With $n = 20$, $\Delta x = \frac{1}{20}$ and the midpoints are $\{ \frac{41}{40}, \frac{43}{40}, \frac{45}{40}, \ldots, \frac{77}{40}, \frac{79}{40} \}$.

$$\int_1^2 x^{-1}\,dx \approx \frac{1}{20} \cdot \left(\frac{40}{41} + \frac{40}{43} + \frac{40}{45} + \cdots + \frac{40}{77} + \frac{40}{79} \right)$$

$$= 2 \cdot \left(\frac{1}{41} + \frac{1}{43} + \frac{1}{45} + \cdots + \frac{1}{77} + \frac{1}{79} \right) \approx 0.69306910$$

Error is approximately 0.000078.

(c) With $n = 40$, $\Delta x = \frac{1}{40}$ and the midpoints are $\{ \frac{81}{80}, \frac{83}{80}, \frac{85}{80}, \ldots, \frac{157}{80}, \frac{159}{80} \}$.

$$\int_1^2 x^{-1}\,dx \approx \frac{1}{40} \cdot \left(\frac{80}{81} + \frac{80}{83} + \frac{80}{85} + \cdots + \frac{80}{157} + \frac{80}{159} \right)$$

$$= 2 \cdot \left(\frac{1}{81} + \frac{1}{83} + \frac{1}{85} + \cdots + \frac{1}{157} + \frac{1}{159} \right) \approx 0.693127652$$

Error is approximately 0.00002.

Note: these computations can be done using the LIST editor of a
TI-83 Plus. In part (c), for instance, we use the sequence operation,
seq(, to construct a list of numbers of the form $[80 + (2x - 1)]^{(-1)}$,
save it as a list L_1, and then use the sum function, **sum(**, and
multiply the result by 2, as shown.

29. With $n = 8$, $\Delta t = 0.25$, and the midpoints are
$\{-0.875, -0.625, \ldots 0.625, 0.875\}$.

$$\int_{-1}^{1} f(t)\, dt \approx -3.34375$$

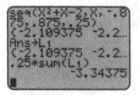

31. With $n = 10$, $\Delta t = 0.1$, and the left endpoints are
$\{0.5, 0.6, 0.7, \ldots 1.4, 1.5\}$.

$$\int_{0.5}^{1.5} f(t)\, dt \approx 1.35373$$

33. With $\Delta t = 0.25$, $n = 8$, and the left endpoints are
$\{-1, -0.75, -0.5, -0.25, 0, 0.25, 0.5, 0.75\}$.

$$\int_{-1}^{1} f(t)\, dt \approx 1.981955$$

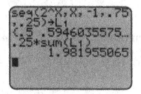

35. With $n = 20$, $\Delta x = 0.1$, and the midpoints are
$\{0.05, 0.15, 0.25, \ldots, 1.85, 1.95\}$.

$$\int_{0}^{2} f(x)\, dx \approx 3.995$$

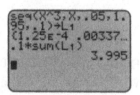

37. With $n = 30$, $\Delta x = 0.08$, and the midpoints are
$\{0.04, 0.12, 0.20, \ldots, 2.28, 2.36\}$.

$$\int_{0}^{2.4} f(x)\, dx \approx 1.232802$$

39. (a) $\quad 2 \cdot 1 + 3 \cdot 2 + 4 \cdot 3 + \cdots + n \cdot (n - 1)$

$$= 2 \cdot (2 - 1) + 3 \cdot (3 - 1) + 4 \cdot (4 - 1) + \cdots + n \cdot (n - 1)$$

$$= 2^2 - 2 + 3^2 - 3 + 4^2 - 4 + \cdots + n^2 - n$$

$$= 2^2 + 3^2 + 4^2 + \cdots + n^2 - (2 + 3 + 4 + \cdots + n)$$

$$= 1^2 + 2^2 + 3^2 + 4^2 + \cdots + n^2 - (1 + 2 + 3 + 4 + \cdots + n) = S - T$$

(b) $n^3 - 1$

$$= (2^3 - 1^3) + (3^3 - 2^3) + \cdots + [n^3 - (n-1)^3]$$
$$= (2^2 + 1 \cdot 2 + 1^2) + (3^2 + 2 \cdot 3 + 2^2) + \cdots + [n^2 + n(n-1)n + (n-1)^2]$$
$$= (2^2 + 3^2 + \cdots n^2) + (1^2 + 2^2 + 3^2 + \cdots + (n-1)^2) + (1 \cdot 2 + 2 \cdot 3 + \cdots + (n-1)n)$$
$$= (S - 1) + (S - n^2) + (S - T) = 3S - T - n^2 - 1$$

Therefore, $n^3 = 3S - T - n^2$ and $S = \frac{1}{3}(n^3 + n^2 + T)$.

(c) $S = \dfrac{1}{3}(n^3 + n^2 + T) = \dfrac{1}{3}\left[n^3 + n^2 + \dfrac{1}{2}n(n+1)\right]$

$$= \frac{1}{6}(2n^3 + 2n^2 + n^2 + n) = \frac{1}{6}(2n^3 + 3n^2 + n)$$

$$= \frac{1}{6}n(2n^2 + 3n + 1) = \frac{1}{6}n(n+1)(2n+1)$$

EXERCISES 5.5

1. The area is that of a rectangle with base 4 and height 1/2. Therefore,

$$A = (\text{base})(\text{height}) = 4 \cdot \frac{1}{2} = 2.$$

By the fundamental theorem,

$$A = \int_1^5 \frac{1}{2}\,dx = \frac{1}{2}x\,\Big|_1^5 = \frac{5}{2} - \frac{1}{2} = 2.$$

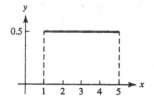

3. The area is that of a right rectangle with base 4 and height 2. Therefore,

$$A = \frac{1}{2}(\text{base})(\text{height}) = \frac{4 \cdot 2}{2} = 4.$$

By the fundamental theorem,

$$A = \frac{1}{2}\int_0^4 x\,dx = \frac{1}{4}x^2\,\Big|_0^4 = 4.$$

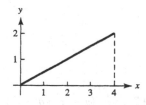

5. The area is the sum of two areas. The first is a triangle of base 1 and height 2, the second is a right triangle with base 1 and height 2. Therefore,

$$A = 1 \cdot 2 + \frac{1 \cdot 2}{2} = 3.$$

By the fundamental theorem,

$$A = \int_0^1 (2x + 2)\,dx = (x^2 + 2x)\,\Big|_0^1 = 3.$$

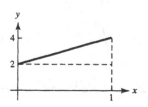

7. $A = \displaystyle\int_0^2 x^3 \, dx = \dfrac{x^4}{4}\bigg|_0^2 = \dfrac{2^4}{4} - \dfrac{0^4}{4} = 4$

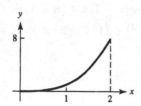

9. $A = \displaystyle\int_0^2 (2x - x^2) \, dx = x^2 - \dfrac{1}{3}x^3 \bigg|_0^2$

$$= 2^2 - \dfrac{1}{3}2^3 = \dfrac{4}{3}$$

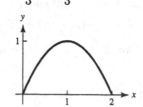

11. $\displaystyle\int_1^2 x \, dx + \int_1^2 \dfrac{1}{x} \, dx = \dfrac{x^2}{2}\bigg|_1^2 + \ln|x| \bigg|_1^2$

$$= \dfrac{4}{2} - \dfrac{1}{2} + (\ln 2 - \ln 1) = \dfrac{3}{2} + \ln 2$$

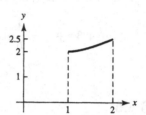

13. $A = \displaystyle\int_1^3 \dfrac{1}{x^2} \, dx = -\dfrac{1}{x}\bigg|_1^3 = -\dfrac{1}{3} - (-1) = \dfrac{2}{3}$

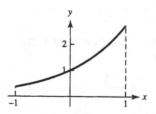

15. $A = \displaystyle\int_{-1}^1 e^x \, dx = e^x \bigg|_{-1}^1 = e - e^{-1}$

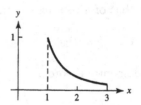

17. $\displaystyle\int_0^1 (x - x^3) \, dx = \left(\dfrac{1}{2}x^2 - \dfrac{1}{4}x^4 \right)\bigg|_0^1 = \left(\dfrac{1}{2}\cdot 1^2 - \dfrac{1}{4}\cdot 1^4 \right) - 0 = \dfrac{1}{4}$

19. $\displaystyle\int_1^3 (1 - x) \, dx = x - \dfrac{1}{2}x^2 \bigg|_1^3 = \left(3 - \dfrac{1}{2}\cdot 3^2 \right) - \left(1 - \dfrac{1}{2}\cdot 1^2 \right) = -2$

21. $\displaystyle\int_0^2 (x^3 - 4x^2 + 3x) \, dx = \dfrac{x^4}{4} - \dfrac{4x^3}{3} + \dfrac{3x^2}{2}\bigg|_0^2 = \dfrac{16}{4} - \dfrac{32}{3} + \dfrac{12}{2} = -\dfrac{2}{3}$

23. $\displaystyle\int_0^4 (x^4 - 4x^3) \, dx = \dfrac{1}{5}x^5 - x^4 \bigg|_0^4 = \dfrac{1}{5}4^5 - 4^4 = 4^4 \left(\dfrac{4}{5} - 1 \right) = -\dfrac{256}{5}$

25. $\displaystyle\int_{-1}^1 x^{1/3} \, dx = \dfrac{3}{4}x^{4/3}\bigg|_{-1}^1 = \dfrac{3}{4} - \dfrac{3}{4} = 0$

27. $\displaystyle\int_{-1}^{1} e^{-2x}\,dx = -\frac{1}{2}e^{-2x}\Big|_{-1}^{1} = -\frac{1}{2}(e^{-2} - e^2) = \frac{1}{2}(e^2 - e^{-2})$

29. Let $u = 2x + 1$. Then $du = 2\,dx$ or $dx = \frac{1}{2}\,du$. Therefore,

$$\int \sqrt{2x+1}\,dx = \frac{1}{2}\int \sqrt{u}\,du = \frac{1}{3}u^{3/2} + c = \frac{1}{3}(2x+1)^{3/2} + c,$$

and we obtain

$$\int_0^4 \sqrt{2x+1}\,dx = \frac{1}{3}(2x+1)^{3/2}\Big|_0^4 = \frac{1}{3}(9^{3/2} - 1^{3/2}) = \frac{26}{3}.$$

31. $\displaystyle\int_{-1}^{0} (e^x - e^{-x})\,dx = e^x + e^{-x}\Big|_{-1}^{0} = 2 - e^{-1} - e$

33. Let $u = \ln x$ and $dv = x\,dx$. Then $du = \frac{1}{x}\,dx$ and $v = \frac{1}{2}x^2$, so

$$\int x\ln x\,dx = \frac{1}{2}x^2\ln x - \frac{1}{2}\int x\,dx = \frac{1}{2}x^2\ln x - \frac{1}{4}x^2 + c.$$

Therefore,

$$\int_1^2 x\ln x\,dx = \frac{1}{2}x^2\ln x - \frac{1}{4}x^2\Big|_1^2 = (2\ln 2 - 1) - \left(-\frac{1}{4}\right) = 2\ln 2 - \frac{3}{4}.$$

35. First, $\int_0^2 (x^2 + x - 2)\,dx = \frac{x^3}{3} + \frac{x^2}{2} - 2x\big|_0^2 = \frac{8}{3} + 2 - 4 = \frac{2}{3}$. Next, since $f(x) = x^2 + x - 2 = (x + 2)(x - 1)$, we have $f(x) > 0$ in $(-\infty, -2)$ and $(1, \infty)$, and $f(x) < 0$ in $(-2, 1)$. Therefore, the area is

$$-\int_0^1 (x^2 + x - 2)\,dx + \int_1^2 (x^2 + x - 2)\,dx$$

$$= -\left(\frac{x^3}{3} + \frac{x^2}{2} - 2x\right)\Big|_0^1 + \left(\frac{x^3}{3} + \frac{x^2}{2} - 2x\right)\Big|_1^2$$

$$= -\left(\frac{1}{3} + \frac{1}{2} - 2\right) + \left(\frac{8}{3} + 2 - 4\right) - \left(\frac{1}{3} + \frac{1}{2} - 2\right) = 3.$$

37. Let $u = x^2 + 1$. Then $du = 2x\,dx$ or $x\,dx = \frac{1}{2}\,du$. Therefore,

$$\int \frac{x}{x^2+1}\,dx = \frac{1}{2}\int \frac{1}{u}\,du = \frac{1}{2}\ln|u| + c = \frac{1}{2}\ln(x^2+1) + c,$$

and we obtain $\int_0^1 \frac{x}{x^2+1}\,dx = \frac{1}{2}\ln(x^2+1)\big|_0^1 = \frac{1}{2}\ln 2$. Since $f(x) > 0$ in $(0,1)$, the area is also equal to $\frac{1}{2}\ln 2$.

39. Let $u = \ln x$ and $dv = dx$. Then $du = 1/x\,dx$ and $v = x$, so that

$$\int \ln x\,dx = x\ln x - \int \frac{1}{x}\cdot x\,dx = x\ln x - x + c = x(\ln x - 1) + c.$$

Therefore, $\int_1^2 \ln x\,dx = x(\ln x - 1)\big|_1^2 = 2(\ln 2 - 1) - (0 - 1) = 2\ln 2 - 1$. Since $f(x) > 0$ in $(1,2)$, the area is also equal to $2\ln 2 - 1$.

41. By the fundamental theorem of calculus,

$$\int_{-1}^{2} f(x)\,dx = F(2) - F(-1) = 3 - 5 = -2.$$

43. Since $A(t)$ is the area of a right triangle with base and height both equal to t,

$$A(t) = \frac{1}{2}(\text{base})(\text{height}) = \frac{1}{2}t \cdot t = \frac{1}{2}t^2,$$

and $A'(t) = t = f(t)$.

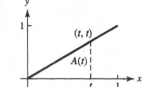

45. Since $A(t) = \int_1^t f(x)\,dx = \frac{t^3}{3} + 16\ln t - \frac{1}{3}$, it follows from Theorem 5.5.2 that $f(t) = A'(t) = t^2 + \frac{16}{t}$, so that $f(x) = x^2 + \frac{16}{x}$.

47. $G'(t) = \sqrt{t^2 - 3t + 5}$ by Theorem 5.5.2.

49. $G'(t) = \dfrac{d}{dt}\left(t + \int_0^t x^2 e^{x^4}\,dx\right) = 1 + \dfrac{d}{dt}\left(\int_0^t x^2 e^{x^4}\,dx\right) = 1 + t^2 e^{t^4}$

51. Let $F(u) = \int_1^u e^x\sqrt{x}\,dx$. By Theorem 5.5.2, we have $F'(u) = e^u\sqrt{u}$. Since $G(t) = F(t^2)$, the chain rule gives

$$G'(t) = \frac{d}{dt}F(t^2) = F'(t^2) \cdot \frac{d}{dt}t^2 = e^{t^2}\sqrt{t^2} \cdot 2t = 2t|t|e^{t^2}.$$

53. The change in temperature (degrees Celsius) is equal to

$$\int_0^5 -5.52e^{-0.07t}\,dt = \frac{5.52}{0.07}e^{-0.07t}\Big|_0^5 = \frac{552}{7}(e^{-0.35} - 1) \approx -23.3.$$

55. Let $S(t)$ be the amount of annual sales (in millions of dollars) t years after 2000. Then $S'(t) = r(t) = 1{,}000 - 10t^2$, which means that the change of sales from 2,000 to 2,010 is given by

$$S(10) - S(0) = \int_0^{10}(1{,}000 - 10t^2)\,dt = 1{,}000t - \frac{10}{3}t^3\Big|_0^{10} = 10{,}000 - \frac{10{,}000}{3} = \frac{20{,}000}{3}.$$

Since $S(0) = 800$, we have $S(10) = 800 + \frac{20{,}000}{3} \approx 7{,}466.67$ million dollars.

57. The profit function is of the form

$$P(x) = \int(-0.2x + 50)\,dx = -0.1x^2 + 50x + c.$$

The initial condition $P(100) = 1{,}500$ gives $-1{,}000 + 5{,}000 + c = 1{,}500$, or $c = -2{,}500$. Therefore, $P(x) = -0.1x^2 + 50x - 2{,}500$.

59. Recall that it is profitable to increase production if $P'(x) > 0$ and to decrease it if $P'(x) < 0$. In this case, $P'(x) = MP(x) = -0.6x + 420$, and $P'(500) = -300 + 420 = 120$. Therefore, increasing production is profitable.

To maximize $P(x)$, we set $P'(x) = -0.6x + 420 = 0$ and solve for x to get $x = 700$. Thus, the maximum profit should occurs at $x = 700$.

To find the maximum profit, we first note that the profit function is of the form $P(x) = \int(-0.6x + 420)\,dx = -0.3x^2 + 420x + c$. To find c, we set $x = 500$ and $P(500) = 150{,}000$, which gives $-0.3 \cdot 500^2 + 420 \cdot 500 + c = 150{,}000$, or $c = 15{,}000$. Therefore, $P(x) = -0.3x^2 + 420x + 15{,}000$, and $P(700) = -0.3 \cdot 700^2 + 420 \cdot 700 + 15{,}000 = 162{,}000$ dollars, which is the maximum profit.

61. The change in profit is

$$P(60) - P(40) = \int_{40}^{60} P'(x)\,dx = \int_{40}^{60} MP(x)\,dx$$

$$= \int_{40}^{60} [-0.006(x + 40)^2 + 500]\,dx = -0.002(x + 40)^3 + 500x \Big|_{40}^{60}$$

$$= -0.002 \cdot 100^3 + 500 \cdot 60 + 0.002 \cdot 80^3 - 500 \cdot 40 = 9{,}024.$$

63. According to the claim, the increase in egg production (in hundreds of eggs) from the first to the fifth week is

$$\int_{1}^{5} (0.08w + 0.3)\,dw = (0.04w^2 + 0.3w) \Big|_{1}^{5} = (0.04 \cdot 5^2 + 0.3 \cdot 5) - (0.04 \cdot 1^5 + 0.3 \cdot 1) = 2.16.$$

Thus, the predicted increase is 216 eggs, which is less than the actual increase of 300 eggs. Therefore, the new feed performed even better than promised.

65. If we let $a(t)$ denote the size (in centimeters) of type A fish after t years, and $b(t)$ that of type B, then we are given that $a'(t) = 0.072t^2$ and $b'(t) = 2.1t^{0.6}$.

(a) At 5 years, $a'(5) = 0.072 \cdot 5^2 = 1.8$ and $b'(5) = 2.1 \cdot 5^{0.6} \approx 5.52$. Thus, type B is growing faster.

At 12 years $a'(12) = 0.072 \cdot 12^2 = 10.37$ and $b'(12) = 2.1 \cdot 12^{0.6} \approx 9.33$. Thus, type A is growing faster.

(b) During the first 5 years, the growth of a type A fish is $\int_0^5 0.072t^2\,dt = 0.024t^3 \big|_0^5 = 3$, while that of a type B fish is $\int_0^5 2.1t^{0.6}\,dt = \frac{2.1}{1.6}t^{1.6}\big|_0^5 = 17.24$.

During the first 12 years, the type A grows a total of $\int_0^{12} 0.072t^2\,dt \approx 41.47$, while the type B grows $\int_0^{12} 2.1t^{0.6}\,dt = 69.95$.

During the first 20 years, the total growth of a type A is $\int_0^{20} 0.072t^2\,dt = 192$, while that of a type B is $\int_0^{20} 2.1t^{0.6}\,dt \approx 158.4$.

67. Using the Riemann sum with left endpoints and $\Delta = 0.25$ we get the estimate

$$\int_0^3 v(t)\,dt \approx 0.25 \cdot (2.11 + 2.18 + 2.24 + 2.29 + 2.33 + 2.37 + 2.40 + 2.42 + 2.45 + 2.47 + 2.48)$$

$$= 6.435.$$

Using right endpoints gives

$$\int_0^3 v(t)\,dt \approx 0.25 \cdot (2.18 + 2.24 + 2.29 + 2.33 + 2.37 + 2.40 + 2.42 + 2.45 + 2.47 + 2.48 + 2.5)$$

$$= 6.5325.$$

EXERCISES 5.6

1. $\displaystyle\int_0^{10} e^{-0.1t}\, dt = -10e^{-0.1t}\Big|_0^{10} = -10e^{-1} - (-10) = \frac{-10}{e} + 10 = 10\left(1 - \frac{1}{e}\right)$

3. Let $u = t$ and $dv = e^{t/3}\, dt$. Then $du = dt$ and $v = 3e^{t/3}$. Therefore,

$$\int_0^3 te^{t/3}\, dt = 3te^{t/3}\Big|_0^3 - 3\int_0^3 e^{t/3}\, dt = 9e - 9e^{t/3}\Big|_0^3 = 9e - 9e + 9 = 9.$$

5. Let $u = \ln x$. Then $du = \frac{1}{x}\, dx$. When $x = 1$, we have $u = \ln 1 = 0$, and when $x = e$, we have $u = \ln e = 1$. Therefore,

$$\int_1^e \frac{(\ln x)^2}{x}\, dx = \int_0^1 u^2\, du = \frac{1}{3}u^3\Big|_0^1 = \frac{1}{3}.$$

7. Let $u = x^3 - x$. Then $du = (3x^2 - 1)\, dx$, so that

$$\int (3x^2 - 1)(x^3 - x)^4\, dx = \int u^4\, du = \frac{u^5}{5} + c = \frac{(x^3 - x)^5}{5} + c.$$

Therefore, $\int_0^1 (3x^2 - 1)(x^3 - x)^4\, dx = \frac{(x^3 - x)^5}{5}\Big|_0^1 = 0.$

9. Let $u = t + 9$. Then $du = dt$. When $t = 0$, we have $u = 0 + 9 = 9$, and when $t = 7$, we have $u = 7 + 9 = 16$. Therefore,

$$\int_0^7 \frac{t}{\sqrt{t+9}}\, dt = \int_9^{16} \frac{u - 9}{\sqrt{u}}\, du = \int_9^{16} (u^{1/2} - 9u^{-1/2})\, du = \left(\frac{2}{3}u^{3/2} - 18u^{1/2}\right)\Big|_9^{16}$$

$$= \left(\frac{2}{3}\cdot 64 - 18\cdot 4\right) - (18 - 18\cdot 3) = \frac{20}{3}.$$

11. Let $u = 2x^2 + 4x + 1$. Then $du = (4x + 4)\, dx$ or $\frac{1}{4}\, du = (x + 1)\, dx$. Also $x = 0 \iff u = 1$ and $x = 3 \iff u = 31$, and

$$\int_0^3 \frac{x + 1}{2x^2 + 4x + 1}\, dx = \frac{1}{4}\int_1^{31} \frac{1}{u}\, du = \frac{1}{4}\ln|u|\,\Big|_1^{31} = \frac{1}{4}\ln 31.$$

13. Let $u = x^2 - 16$. Then $du = 2x\, dx$ or $x\, dx = \frac{1}{2}\, du$. When $x = 4$, we have $u = 0$, and when $x = 5$, we have $u = 9$. Therefore,

$$\int_4^5 x\sqrt{x^2 - 16}\, dx = \int_0^9 \frac{1}{2}\sqrt{u}\, du = \frac{1}{3}u^{3/2}\Big|_0^9 = \frac{1}{3}\cdot 9^{3/2} = 9.$$

15. Write $\frac{1}{t^2 - t} = \frac{1}{t(t-1)} = \frac{A}{t-1} + \frac{B}{t}$. Clearing denominators gives $1 = At + B(t - 1)$, so $A = 1$ and $B = -1$, and

$$\int_2^3 \frac{1}{t^2 - t}\, dt = \int_2^3 \left(\frac{1}{t-1} - \frac{1}{t}\right) dt = \ln\left|\frac{t-1}{t}\right|\,\Big|_2^3 = \ln\frac{2}{3} - \ln\frac{1}{2} = \ln\frac{4}{3}.$$

17. $\text{Area} = \displaystyle\int_{-1}^{2} [f(x) - g(x)] \, dx + \int_{2}^{4} [g(x) - f(x)] \, dx + \int_{4}^{6} [f(x) - g(x)] \, dx$

19. $f(x) - g(x) = (4 - x^2) - 3 = 1 - x^2 \geq 0$ for $0 \leq x \leq 1$. Therefore, $f(x) \geq g(x)$ for $0 \leq x \leq 1$, and

$$\text{Area} = \int_{0}^{1} [f(x) - g(x)] \, dx = \int_{0}^{1} (1 - x^2) \, dx = x - \frac{1}{3}x^3 \Big|_{0}^{1} = \frac{2}{3}.$$

21. $f(x) - g(x) = (3x^2 + 2) - (2x - 3) = 3x^2 - 2x + 5 \geq 0$, since the discriminant $b^2 - 4ac = (-2)^2 - 4 \cdot 3 \cdot 5 = -56 < 0$ and the leading coefficient is positive. Therefore, $f(x) \geq g(x)$.

$$\text{Area} = \int_{-1}^{2} [f(x) - g(x)] \, dx = \int_{-1}^{2} (3x^2 - 2x + 5) \, dx$$

$$= x^3 - x^2 + 5x \Big|_{-1}^{2} = (8 - 4 + 10) - (-1 - 1 - 5) = 21$$

23. $f(x) - g(x) = 2x^2 - (x^3 - 3x) = -(x^3 - 2x^2 - 3x) = -x(x - 3)(x + 1) \geq 0$ for $0 \leq x \leq 3$. Therefore, $f(x) \geq g(x)$ for $0 \leq x \leq 3$, and

$$\text{Area} = \int_{0}^{3} [f(x) - g(x)] \, dx = \int_{0}^{3} (2x^2 - x^3 + 3x) \, dx = \frac{2x^3}{3} - \frac{x^4}{4} + \frac{3x^2}{2} \Big|_{0}^{3}$$

$$= 18 - \frac{81}{4} + \frac{27}{2} = \frac{45}{4}.$$

25. To find the intersection points, we need to solve $\frac{2}{x} = x - 1$, which reduces to $x^2 - x - 2 = (x - 2)(x + 1) = 0$, whose solutions are $x = -1$ and $x = 2$. However, only $x = 2$ lies in $1 \leq x \leq 3$. The graphs intersect at $(2, 1)$, and

$$\text{Area} = \int_{1}^{2} \left(\frac{2}{x} - x + 1 \right) dx + \int_{2}^{3} \left(x - 1 - \frac{2}{x} \right) dx$$

$$= \left(2 \ln x - \frac{x^2}{2} + x \right) \Big|_{1}^{2} + \left(\frac{x^2}{2} - x - 2 \ln x \right) \Big|_{2}^{3}$$

$$= (2 \ln 2 - 2 + 2) - \left(-\frac{1}{2} + 1 \right) + \left(\frac{9}{2} - 3 - 2 \ln 3 \right) - (2 - 2 - 2 \ln 2)$$

$$= 2 \ln 2 - \frac{1}{2} + \frac{3}{2} - 2 \ln 3 + 2 \ln 2 = 4 \ln 2 - 2 \ln 3 + 1.$$

27. To find the intersection points, we need to solve $e^x = e^{-x}$ or $e^{2x} = 1$. The intersection point of these two graphs is $(0, 1)$. Since $f(x) - g(x) = e^x - e^{-x} = e^{-x}(e^{2x} - 1)$, we see that $f(x) - g(x) \leq 0$ for $-1 \leq x \leq 0$ and $f(x) - g(x) \geq 0$ for $0 \leq x \leq 1$. Therefore,

$$\text{Area} = \int_{-1}^{0} (e^{-x} - e^x) \, dx + \int_{0}^{1} (e^x - e^{-x}) \, dx = (-e^{-x} - e^x) \Big|_{-1}^{0} + (e^x + e^{-x}) \Big|_{0}^{1}$$

$$= (-1 - 1) - (-e - e^{-1}) + (e + e^{-1}) - (1 + 1) = 2e + 2e^{-1} - 4.$$

29. To find the intersection points, we need to solve $x^2 - 1 = -\frac{x}{2}$, which is equivalent to $2x^2 + x - 2 = 0$. There are two solutions $\frac{-1 \pm \sqrt{17}}{4}$, but only $\frac{\sqrt{17}-1}{4}$ lies in $[0, 1]$. $f(x) - g(x) = x^2 - 1 + \frac{x}{2} = x^2 + \frac{x}{2} - 1$, which is negative if $0 < x < \frac{\sqrt{17}-1}{4}$ and positive if $\frac{\sqrt{17}-1}{4} < x < 1$. Therefore,

$$\text{Area} = \int_0^{\frac{\sqrt{17}-1}{4}} [g(x) - f(x)]\, dx + \int_{\frac{\sqrt{17}-1}{4}}^1 [f(x) - g(x)]\, dx$$

$$= \int_0^{\frac{\sqrt{17}-1}{4}} \left(-x^2 - \frac{x}{2} + 1\right) dx + \int_{\frac{\sqrt{17}-1}{4}}^1 \left(x^2 + \frac{x}{2} - 1\right) dx$$

$$= \left(-\frac{x^3}{3} - \frac{x^2}{4} + x\right)\Big|_0^{\frac{\sqrt{17}-1}{4}} + \left(\frac{x^3}{3} + \frac{x^2}{4} - x\right)\Big|_{\frac{\sqrt{17}-1}{4}}^1$$

$$= \frac{17\sqrt{17} - 45}{48}.$$

31. $f(x) - g(x) = x^2 - x^3 = x^2(1 - x) \geq 0$ for $-1 \leq x \leq 1$. Therefore,

$$\text{Area} = \int_{-1}^1 [f(x) - g(x)]\, dx = \int_{-1}^1 (x^2 - x^3)\, dx$$

$$= \frac{1}{3}x^3 - \frac{1}{4}x^4\Big|_{-1}^1 = \frac{2}{3}.$$

33. To find the intersection points, we set $x^2 + x + 1 = -2x^2 + 4x + 7$, which reduces to $3(x - 2)(x + 1) = 0$, whose solutions are $x = -1$ and $x = 2$. Moreover, $(x^2 + x + 1) - (-2x^2 + 4x + 7) = 3(x - 2)(x + 1) \leq 0$ for $-1 \leq x \leq 2$. The area is

$$\int_{-1}^2 [(-2x^2 + 4x + 7) - (x^2 + x + 1)]\, dx = \int_{-1}^2 (-3x^2 + 3x + 6)\, dx$$

$$= \left(-x^3 + \frac{3x^2}{2} + 6x\right)\Big|_{-1}^2 = \frac{27}{2}.$$

35. $\dfrac{1}{3}\displaystyle\int_0^3 x^2\, dx = \dfrac{1}{9}x^3\Big|_0^3 = 3$

37. $\dfrac{1}{2 - \frac{1}{2}}\displaystyle\int_{1/2}^2 \dfrac{5}{t}\, dt = \dfrac{2}{3} \cdot 5 \int_{1/2}^2 \dfrac{1}{t}\, dt = \dfrac{10}{3}\ln|t|\Big|_{1/2}^2 = \dfrac{10}{3}\left(\ln 2 - \ln\dfrac{1}{2}\right)$

$$= \frac{10}{3}(\ln 2 + \ln 2) = \frac{20}{3}\ln 2$$

39. Using integration by parts,

$$\frac{1}{10 - 0}\int_0^{10} te^{0.1t}\, dt = \frac{1}{10}\left[10te^{0.1t}\Big|_0^{10} - 10\int_0^{10} e^{0.1t}\, dt\right]$$

$$= \frac{1}{10}\left[100e - 100e^{0.1t}\Big|_0^{10}\right] = \frac{1}{10} \cdot 100 = 10.$$

41. Write $\frac{1}{x^2-5x+4} = \frac{1}{(x-1)(x-4)} = \frac{A}{x-4} + \frac{B}{x-1}$. Clearing the denominators gives $1 = A(x-1) + B(x-4)$, so $A = \frac{1}{3}$ and $B = -\frac{1}{3}$.

$$\frac{1}{8-5} \int_5^8 \frac{1}{x^2 - 5x + 4}\,dx = \frac{1}{3}\int_5^8 \left(\frac{1/3}{x-4} - \frac{1/3}{x-1}\right)dx = \frac{1}{9}\ln\left|\frac{x-4}{x-1}\right|\Big|_5^8$$

$$= \frac{1}{9}\left(\ln\frac{4}{7} - \ln\frac{1}{4}\right) = \frac{1}{9}\ln\frac{16}{7}$$

43. $\dfrac{1}{4-(-1)}\displaystyle\int_{-1}^{4} f(x)\,dx = \dfrac{30}{5} = 6$

45. Since $s'(t) = v(t)$, we have

$$\frac{1}{b-a}\int_a^b v(t)\,dt = \frac{1}{b-a}s(t)\Big|_b^a = \frac{s(b) - s(a)}{b-a}.$$

The left-hand side is the average value of $v(t)$ over the interval $[a,b]$, and the right-hand side is the average velocity over $[a,b]$. Therefore, they are equal.

47. The slope is given by $y'(x) = 3x^2$, and its average over $[0,2]$ is

$$\frac{1}{2}\int_0^2 3x^2\,dx = \frac{1}{2}x^3\Big|_0^2 = 4.$$

On the other hand, the slope at $x = 0$ is 0, and the slope at $x = 2$ is 12. The average of these numbers is 6, which is not the same as the average slope computed above. Since $3x^2$ is not linear, we should not expect the averages to be the same.

49. The average amount in the patient's bloodstream is

$$\frac{1}{6}\int_0^6 (30 - 20e^{-0.4t})\,dt = \frac{1}{6}(30t + 50e^{-0.4t})\Big|_0^6$$

$$= 30 + \frac{50}{6}(e^{-2.4} - 1) = \frac{65}{3} + \frac{25}{3}e^{-2.4} \approx 22.4226.$$

51. $\dfrac{1}{24}\displaystyle\int_0^{24}(2{,}500e^{0.05t})\,dt = \dfrac{50{,}000}{24}e^{0.05t}\Big|_0^{24} = \dfrac{6{,}250}{3}(e^{1.2} - 1) \approx 4{,}833.58$

53. **(a)** Set $G'(y) = 10.736e^{-y}(1 - y) = 0$, which has a unique solution $y = 1$. Since $G''(y) = 10.736\,e^{-y}(y - 2)$ and $G''(1) = -10.736e^{-1} < 0$, the second derivative test and single critical point principle show that $G(y)$ has a maximum value at $y = 1$.

(b) As preparation, let us evaluate $\int e^{-y}(1 - y)\,dy$. Let $u = 1 - y$ and $dv = e^{-y}\,dy$. then $du = -dy$ and $v = -e^{-y}$. Therefore,

$$\int e^{-y}(1 - y)\,dy = -(1 - y)e^{-y} - \int e^{-y}\,dy = -(1 - y)e^{-y} + e^{-y} = ye^{-y} + c.$$

Since $G(0) = 0$, we see $G(y) = 10.736ye^{-y}$. The growth in $[0,1]$ equals $10.736e^{-1}$, and the growth in $[1,6]$ equals $64.416e^{-6} - 10.736e^{-1} \approx -3.79$.

(c) Averaging $G'(y)$ over $[0,1]$ gives

$$\int_0^1 G(y)\,dy = G(1) - G(0) = 10.736e^{-1} \approx 3.95,$$

and averaging it over $[1,6]$ gives

$$\frac{1}{5}\int_1^6 G'(y)\,dy = \frac{G(6) - G(1)}{5} = \frac{64.416e^{-6} - 10.736e^{-1}}{5} \approx -0.758.$$

(d) The population grows at an average rate of 3.95 hundred animals in the time interval $[0,1]$ and decreases at an average rate of 0.758 hundred animals in the time interval $[1,6]$.

55. Using the equation solver, we get $x \approx 0.567$ as the solution of $e^{-x} - x = 0$. Therefore, the area is approximately

$$\int_0^{0.567} (e^{-x} - x) = \left(-e^{-x} - \frac{x^2}{2}\right)\Big|_0^{0.567} = -e^{-0.567} - \frac{(0.567)^2}{2} + 1 \approx 0.272.$$

57. If $n = 8$, then $\Delta x = 0.25$ and the midpoints are $\{-0.875, -0.625,$ $\ldots, 0.625, 0.875\}$. Using the LIST editor, as shown, we get

$$\int_{-1}^1 \frac{1}{1+x^2}\,dx \approx 1.5734.$$

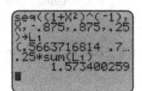

EXERCISES 5.7

1. (a) The integration interval is $[0,1]$, $\Delta x = \frac{1-0}{5} = 0.2$, and $f(x) = x^2$. Then $f''(x) = 2$ and the maximum value of $|f''(x)|$ on $[0,1]$ equals 2. The error using the midpoint rule satisfies $E_M \le \frac{2(1-0)^3}{24\cdot 5^2} = \frac{1}{300}$. The error using the trapezoidal rule satisfies $E_T \le \frac{2(1-0)^3}{12\cdot 5^2} = \frac{1}{150}$.

(b) The values of the function at the midpoints are as follows:

x	0.1	0.3	0.5	0.7	0.9
x^2	0.01	0.09	0.25	0.49	0.81

The midpoint rule gives $\int_0^1 x^2\,dx \approx 0.2(0.01 + 0.09 + 0.25 + 0.49 + 0.81) = 0.33$.
The values of the function at the endpoints are as follows:

x	0	0.2	0.4	0.6	0.8	1
x^2	0	0.04	0.16	0.36	0.64	1

The trapezoidal rule gives

$$\int_0^1 x^2\,dx \approx \frac{0.2}{2}(0 + 2\cdot 0.04 + 2\cdot 0.16 + 2\cdot 0.36 + 2\cdot 0.64 + 1) = 0.34.$$

Since $\int_0^1 x^2\,dx = \frac{1}{3}x^3\big|_0^1 = \frac{1}{3}$, the actual error for the midpoint rule is $|0.33 - \frac{1}{3}| = \frac{1}{300}$, and the actual error for the trapezoidal rule is $|0.34 - \frac{1}{3}| = \frac{1}{150}$.

3. **(a)** The integration interval is $[0, 2]$, and $\Delta x = \frac{2}{5}$. The function values at the midpoints are as follows:

x	1/5	3/5	5/5	7/5	9/5
$\dfrac{1}{x+1}$	5/6	5/8	5/10	5/12	5/14

The midpoint rule gives

$$\int_0^2 (x+1)^{-1}\, dx \approx \frac{2}{5}\left(\frac{5}{6} + \frac{5}{8} + \frac{5}{10} + \frac{5}{12} + \frac{5}{14}\right) = \frac{1}{3} + \frac{1}{4} + \frac{1}{5} + \frac{1}{6} + \frac{1}{7} = \frac{459}{420} \approx 1.0929.$$

To estimate the error, we compute $f''(x) = 2/(1+x)^3$. On the interval $[0, 2]$, $f''(x)$ is positive and achieves a maximum of 2 at $x = 0$. Therefore, the error satisfies $E_M \le \frac{2 \cdot 2^3}{24 \cdot 5^2} = \frac{2}{75} \approx$ 0.0267.

(b) Since $f(x) = 1/(x+1)$, we have $f''(x) = 2/(x+1)^3$ and the maximum value of $|f''(x)|$ on $[0, 2]$ is 2, achieved at $x = 0$. Therefore,

$$E_M \le \frac{2 \cdot (2-0)^3}{24 n^2} = \frac{2}{3 n^2}.$$

To make this less than 0.001, n must satisfy the inequality $n^2 \ge \frac{2{,}000}{3}$ or $n \ge \sqrt{2{,}000/3} \approx 25.8$. Thus, $n = 26$ will insure that $E_M < 0.001$.

If $n = 26$, then $\Delta x = 1/13$ and the midpoints are $\{\frac{1}{26}, \frac{3}{26}, \ldots, \frac{49}{26}, \frac{51}{26}\}$. The midpoint rule gives

$$\int_0^2 (x+1)^{-1}\, dx \approx 2 \cdot \left(\frac{1}{27} + \frac{1}{29} + \cdots + \frac{1}{75} + \frac{1}{77}\right) \approx 1.0984.$$

(c) With $n = 26$, the actual error is $|\ln 3 - 1.0984| \approx 0.0002$.

5. The area of the surface of the river from $x = 3$ to $x = 6$ equals

$$A = \int_3^6 \left(\frac{1}{(x-2)^{1/2}} + 1 - \frac{1}{x^{1/2}}\right) dx$$

The integration interval is $[3, 6]$, and $\Delta x = \frac{6-3}{6} = 0.5$. The function values at the midpoints are as follows:

x	3.25	3.75	4.25	4.75	5.25	5.75
$N(x) - S(x)$	1.3397	1.2395	1.1816	1.1442	1.1183	1.0994

By the midpoint rule, $A \approx 0.5 \cdot (1.3397 + 1.2395 + 1.1816 + 1.1442 + 1.1183 + 1.0994) = 3.5613$ hundred square feet.

7. If $n = 10$, then $\Delta x = 0.2$, and the endpoints are $\{-1, -0.8, -0.6, \ldots, 0.6, 0.8, 1\}$. Using the LIST editor to compute the trapezoidal rule, we get $\int_{-1}^1 e^{-x^2/2}\, dx \approx 1.7072$. Since $f''(x) = (x^2 - 1)e^{-x^2/2}$ we see that $|f''(x)| \le 1$ on $[-1, 1]$. (Note that both $|x^2 - 1|$ and $e^{-x^2/2}$ are less than or equal to 1 on $[-1, 1]$.) Therefore, $E_T \le 8/1{,}200 \approx 0.0067$.

9. If $n = 6$, then $\Delta x = 0.5$. The endpoints are $\{-1, -0.5, 0, 0.5, 1, 1.5, 2\}$, and the midpoints are $\{-0.75, -0.25, 0.25, 0.75, 1.25, 1.75\}$. The midpoint rule gives 1.8994, and the trapezoidal rule gives 1.8788.

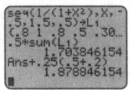

Midpoint   Trapezoidal

11. If $n = 10$, then $\Delta x = 0.1$. The endpoints are $\{-1, -0.9, -0.8, \ldots, -0.1, 0\}$, and the midpoints are $\{-0.95, -0.85, -0.75, \ldots, -0.15, -0.05\}$. The midpoint rule gives 1.4424, and the trapezoidal rule gives 1.4433.

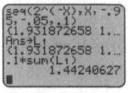
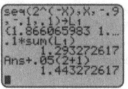

Midpoint   Trapezoidal

13. If $n = 10$, then $\Delta x = 0.1$. The endpoints are $\{0, 0.1, 0.2, \ldots, 0.9, 1\}$, and the midpoints are $\{0.05, 0.15, 0.25, \ldots, 0.85, 0.95\}$. The midpoint rule gives 0.4348, and the trapezoidal rule gives 0.3799.

Midpoint   Trapezoidal

15. If $n = 10$, then $\Delta x = 0.1$. The endpoints are $\{1, 1.1, 1.2, \ldots, 1.9, 2\}$, and the midpoints are $\{1.05, 1.15, 1.25, \ldots, 1.85, 1.95\}$. The midpoint rule gives 0.14739, and the trapezoidal rule gives 0.14688.

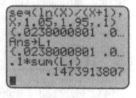
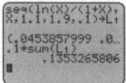

Midpoint   Trapezoidal

17. To estimate $\int_0^{10} v(t)\, dt$, we enter the function values at $t = 1, 2, \ldots 9$ as a sequence, say L_1. Then take the sum of the terms, and, finally, add $\frac{1}{2}(0 + 140) = 70$, as shown, to get 530.

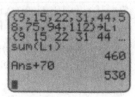

19. The graph shows that $n = 16$ and $\Delta x = 0.25$. The endpoints are $\{1, 1.25, 1.5, \ldots, 4.5, 4.75, 5\}$. We enter the values from 1.38 to 1.69 as a sequence, say L_1, take its sum and multiply by 0.25, as shown below. Then we add $0.125(1.0 + 1.75)$ to complete the trapezoidal approximation, which gives 8.09375 as the approximate area under the graph.

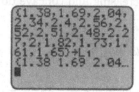   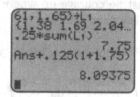

21. We need to approximate $\int_{-2}^{2} \frac{1{,}274}{h^2+1}\, dh = 1{,}274 \int_{-2}^{2} \frac{1}{h^2+1}\, dh$. If $n = 8$, then $\Delta h = 0.5$, and the values of the integrand at the endpoints are as follows:

h	-2	-1.5	-1	-0.5	0	0.5	1	1.5	2
$1/(h^2 + 1)$	0.2	0.3077	0.5	0.8	1	0.8	0.5	0.3077	0.2

The trapezoid rule gives

$$\int_{-2}^{2} \frac{1}{h^2 + 1}\, dh \approx (0.1 + 0.3077 + 0.5 + 0.8 + 1 + 0.8 + 0.5 + 0.3077 + 0.1) \cdot 0.5$$

$$= 2.2077.$$

The number of children is approximately $1{,}274 \cdot 2.2077 \approx 2{,}813$.

23. The fourth derivative, $f^{(4)} = 24x^{-5}$ is a decreasing function on $[1, 2]$ and takes a maximum value of 24 over that interval, achieved at $x = 1$. Therefore, the error in Simpson's rule satisfies $E_S \leq 24/180n^4 = 2/15n^4$. To insure that the error is less than 0.00001, we must have $n^4 > 200{,}000/15$, or $n > \sqrt[4]{200{,}000/15} \approx 10.75$. Since n must be divisible by 4, the first n that works is $n = 12$.

If $n = 12$, then $\Delta x = 1/12$, and the endpoints are $\{1, \frac{13}{12}, \frac{14}{12}, \ldots, \frac{23}{12}, 2\}$. To compute Simpson's rule, we use the calculator to compute and save the values $f(x_1), f(x_3), \ldots, f(x_{11})$ as a sequence, say L_1, and the values $f(x_2), f(x_4), \ldots, f(x_{10})$ as another sequence, say L_2. Then we compute $4 \times \text{sum}(L_1) + 2 \times \text{sum}(L_2)$ and divide by 36 (because $\Delta x/3 = 1/36$). Finally, we add $\frac{\Delta x}{3}[f(x_0) + f(x_{12})]$, which equals $(1 + 1/2)/36$, as shown below, to get 0.6931486622. The actual error is less than 0.0000015.

25. The integration interval is $[0, 14]$ and $\Delta t = 1$. By Simpson's rule, the total growth during this time equals

$$\int_{0}^{14} r(t)\, dt \approx \frac{1}{3}[0 + 4 \cdot 1.3 + 2 \cdot 2 + 4 \cdot 2.5 + 2 \cdot 3.5 + 4 \cdot 3.5 + 2 \cdot 3.6 + 4 \cdot 3.7 + 2 \cdot 3$$

$$+ 4 \cdot 2.6 + 2 \cdot 2.2 + 4 \cdot 3.5 + 2 \cdot 3.7 + 4 \cdot 4 + 3.2] = \frac{123.6}{3} \approx 41.2.$$

27.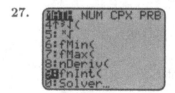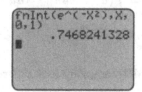

29.

CHAPTER 5 REVIEW EXERCISES

1. g is an antiderivative of f, but h is not. The antiderivative must be increasing where f is positive and decreasing where f is negative. The graphs show this behavior for g but not for h. Also, the antiderivative must be concave up where f is increasing and concave down where f is decreasing. Both g and h satisfy that criterion.

3. Let $u = e^{3x} + e^{-3x}$. Then $du = (3e^{3x} - 3e^{-3x})\,dx$ and

$$\int \frac{e^{3x} - e^{-3x}}{e^{3x} + e^{-3x}}\,dx = \frac{1}{3}\int \frac{du}{u} = \frac{1}{3}\ln|u| + c = \frac{1}{3}\ln(e^{3x} + e^{-3x}) + c.$$

So $\frac{1}{3}\ln(e^{3x} + e^{-3x})$ is an antiderivative of the given function.

5. Let $u = x^2 + 1$. Then $du = 2x\,dx$, or $x\,dx = \frac{1}{2}\,du$, so that

$$\int xe^{x^2+1}\,dx = \int e^u \cdot \frac{1}{2}\,du = \frac{1}{2}e^u + c = \frac{1}{2}e^{x^2+1} + c.$$

7. Let $u = t$ and $dv = e^{-0.25t}\,dt$. Then $du = dt$ and $v = -4e^{-0.25t}$. Integrating by parts gives

$$\int te^{-0.25t}\,dt = -4te^{-0.25t} + 4\int e^{-0.25t}\,dt = -4te^{-0.25t} - 16e^{-0.25t} + c.$$

9. Factoring gives $x^2 - 8x + 15 = (x - 3)(x - 5)$. Using the method of partial fractions, we write

$$\frac{1}{(x - 3)(x - 5)} = \frac{A}{x - 3} + \frac{B}{x - 5} \quad \text{or} \quad 1 = A(x - 5) + B(x - 3).$$

Setting $x = 3$ gives $A = -1/2$, and setting $x = 5$ gives $B = 1/2$. Therefore,

$$\int \frac{dx}{x^2 - 8x + 15} = \int \left(\frac{-1/2}{x - 3} + \frac{1/2}{x - 5}\right)dx$$

$$= -\frac{1}{2}\ln|x - 3| + \frac{1}{2}\ln|x - 5| + c = \frac{1}{2}\ln\left|\frac{x - 5}{x - 3}\right| + c.$$

11. Let $u = t$ and $dv = (t - 4)^{-1/2}\,dt$. Then $du = dt$ and $v = 2(t - 4)^{1/2}$. Integration by parts gives

$$\int \frac{t}{\sqrt{t - 4}}\,dt = 2t(t - 4)^{1/2} - \int 2(t - 4)^{1/2}\,dt = 2t(t - 4)^{1/2} - \frac{4}{3}(t - 4)^{3/2} + c.$$

A second method: by substitution with $u = t - 4$. Then $du = dt$ and $t = u + 4$, and

$$\int \frac{t}{\sqrt{t - 4}}\,dt = \int \frac{u + 4}{\sqrt{u}}\,du = \int (u^{1/2} + 4u^{-1/2})\,du$$

$$= \frac{2}{3}u^{3/2} + 8u^{1/2} + c = \frac{2}{3}(t - 4)^{3/2} + 8(t - 4)^{1/2} + c.$$

(Check that the two answers are the same although they look different.)

13. $y = \int (e^{-2x} - 4x^3 + 1)\,dx = -\frac{1}{2}e^{-2x} - x^4 + x + c$. Since, $y(0) = 12$ we have $12 = c - \frac{1}{2}$, so that $c = \frac{25}{2}$. Therefore, $y = -\frac{1}{2}e^{-2x} - x^4 + x + \frac{25}{2}$.

15. Since $y' = x^5 - \frac{1}{x}$, we obtain $y = \int (x^5 - \frac{1}{x})\, dx = \frac{1}{6}x^6 - \ln|x| + c$. The initial condition $y(1) = 3$ gives $\frac{1}{6} + c = 3$ or $c = 3 - \frac{1}{6} = \frac{17}{6}$. Therefore, $y(x) = \frac{1}{6}x^6 - \ln|x| + \frac{17}{6}$ and $y(e) = \frac{e^6}{6} - 1 + \frac{17}{6} = \frac{e^6 + 11}{6}$.

17. $C(x) = \int (\frac{1}{4}x + 5{,}000)\, dx = \frac{1}{8}x^2 + 5{,}000x + c$. Since $C(20) = 50{,}000$, we have $50 + 100{,}000 + c = 50{,}000$ or $c = -50{,}050$. Therefore,

$$C(x) = \frac{1}{8}x^2 + 5{,}000x - 50{,}050.$$

19. The integration interval is $[0, 2]$ and $\Delta x = \frac{2-0}{4} = 0.5$. The values of f at the midpoints are approximately as follows:

x	0.25	0.75	1.25	1.75
$f(x)$	1.9	1.5	0.85	0.4

Then $\int_0^2 f(x)\, dx \approx 0.5(1.9 + 1.5 + 0.85 + 0.4) = 2.325$.

21. $\Delta x = \frac{2-0}{8} = 0.25$ and the values at the left endpoints are

x	0	0.25	0.5	0.75	1	1.25	1.5	1.75
$f(x)$	2	1.9	1.75	1.5	1.2	0.85	0.6	0.4

$$\int_0^2 f(x)\, dx \approx 0.25(2 + 1.9 + 1.75 + 1.5 + 1.2 + 0.85 + 0.6 + 0.4) = 2.55.$$

23. Since $\int_{-1}^4 f(x)\, dx = \int_{-1}^2 f(x)\, dx + \int_2^4 f(x)\, dx$, we have

$$\int_2^4 f(x)\, dx = \int_{-1}^4 f(x)\, dx - \int_{-1}^2 f(x)\, dx = 3.6 - 4.8 = -1.2.$$

The area is $\int_{-1}^2 f(x)\, dx - \int_2^4 f(x)\, dx = 4.8 - (-1.2) = 6$.

25. $\int_1^3 f(x)\, dx \approx 0.5(0.5 - 1 - 2 - 1.5) = -2$

27. Let $u = x^3 - x$. Then $du = (3x^2 - 1)\, dx$. If $x = -3$, then $u = -24$, and if $x = -2$, then $u = -6$. Therefore,

$$\int_{-3}^{-2} \frac{6x^2 - 2}{x^3 - x}\, dx = 2 \int_{-24}^{-6} \frac{1}{u}\, du = 2\ln|u|\Big|_{-24}^{-6} = 2(\ln 6 - \ln 24) = 2\ln\frac{1}{4}.$$

29. Let $u = \sqrt{x}$. Then $du = \frac{1}{2\sqrt{x}}\, dx$ or $\frac{1}{\sqrt{x}}\, dx = 2\, du$. If $x = 1$, then $u = 1$, and if $x = 4$, then $u = 2$. Therefore,

$$\int_1^4 \frac{e^{\sqrt{x}}}{\sqrt{x}}\, dx = \int_1^2 2e^u\, du = 2e^u\Big|_1^2 = 2e^2 - 2e = 2e(e - 1).$$

31. Let $u = t^2$ and $dv = e^{-0.1t}\, dt$. Then $du = 2t\, dt$ and $v = -10e^{-0.1t}$, so that

$$\int_0^1 t^2 e^{-0.1t}\, dt = -10t^2 e^{-0.1t}\Big|_0^1 + \int_0^1 20t e^{-0.1t}\, dt$$

$$= -10e^{-0.1} + 20 \int_0^1 t e^{-0.1t}\, dt.$$

Now, let $u = t$ and $dv = e^{-0.1t}\,dt$. Then $du = dt$ and $v = -10e^{-0.1t}$, so that

$$\int_0^1 te^{-0.1t}\,dt = -10te^{-0.1t}\Big|_0^1 + \int_0^1 10e^{-0.1t}\,dt$$

$$= -10e^{-0.1} - 100e^{-0.1t}\Big|_0^1 = 100 - 110e^{-0.1}.$$

Putting these steps together, we get

$$\int_0^1 t^2 e^{-0.1t}\,dt = -10e^{-0.1} + 2{,}000 - 2{,}200e^{-0.1} = -2{,}210e^{-0.1} + 2{,}000.$$

33. To find the intersection points of these curves, we need to solve the equation $2x^2 - 4x + 6 = -x^2 - 2x + 1$, or $3x^2 - 2x + 5 = 0$. But this equation has no solution, since its discriminant $b^2 - 4ac = 4 - 60 = -56 < 0$. Therefore, two curves do not intersect. Since $(2x^2 - 4x + 6) - (-x^2 - 2x + 1) = 3x^2 - 2x + 5 > 0$, the area equals

$$\int_1^2 (3x^2 - 2x + 5)\,dx = (x^3 - x^2 + 5x)\Big|_1^2 = 9$$

35. The average value is $\frac{1}{3}\int_0^3 xe^{x/3}\,dx$. To compute it, we use integration by parts, with $u = x$ and $dv = e^{x/3}\,dx$. Then $du = dx$ and $v = 3e^{x/3}$, and

$$\frac{1}{3}\int_0^3 xe^{x/3}\,dx = xe^{x/3}\Big|_0^3 - \int_0^3 e^{x/3}\,dx = 3e - 3e^{x/3}\Big|_0^3$$

$$= 3e - (3e - 3) = 3.$$

37. Since $A(t) = \int_0^t \frac{1}{3+x^4}\,dx$, Theorem 5.5.2 gives $A'(t) = \frac{1}{3+t^4}$.

39. $A'(t) = \sqrt{t^2 - 2t + 8}$

41. $A'(t) = 3t^2 + e^t\sqrt{1+t^3}$, and $A'(2) = 3 \cdot 2^2 + e^2\sqrt{9} = 12 + 3e^2$.

43. In this case we take p as the independent variable, partition the interval $[1, 20]$ into 20 with equal subintervals $\Delta p = 1$, and use the right-hand endpoint of each subinterval to find the Riemann sum. The q-values at the endpoints, as determined from the picture are as follows:

p	1	2	3	4	5	6	7	8	9	10
q	0	50	100	150	200	250	300	350	400	350

p	11	12	13	14	15	16	17	18	19	20
q	300	250	220	180	150	110	80	50	25	0

Adding the q-values for $p = 2, 3, \ldots, 20$ gives an estimate for the area: 3,515.

45. If $n = 5$, $\Delta x = 0.2$.

(a) The midpoints are $\{0.1, 0.3, 0.5, 0.7, 0.9\}$, and the midpoint rule gives $\int_0^1 \sqrt{1 + x^3}\,dx \approx 1.10967$ (see below).

(b) The endpoints are $\{0, 0.2, 0.4, 0.6, 0.8, 1\}$. The trapezoidal rule gives $\int_0^1 \sqrt{1 + x^3}\, dx \approx 1.11499$ (see below).

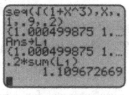

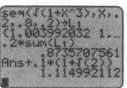

Midpoint   Trapezoidal

47. Write $R(t)$ for the revenue (in millions of dollars) of the company t years after 2000. Then $R(t) = \int r(t)\, dt = \int 10e^{0.1t}\, dt = 100e^{0.1t} + c$. The condition $R(0) = 500$ gives $100 + c = 500$ or $c = 400$. Thus, $R(t) = 100e^{0.1t} + 400$ and $R(10) = 100e + 400 \approx 671.83$ million.

49. Write $P(t)$ for the price of the car at the end of t years. Then

$$P(2) - P(0) = \int_0^2 1{,}000(t - 4)\, dt = 1{,}000 \left(\frac{1}{2}t^2 - 4t \right)\Big|_0^2 = -6{,}000.$$

The total depreciation of the car in the first 2 years is 6,000 dollars.

51. Write $v(t)$ for the velocity of the car at the time t. Then the distance travelled by the car in the first 40 seconds equals $\int_0^{40} v(t)\, dt$. Partition $[0, 40]$ in to 8 equal subintervals. Then the length of each subinterval is 5 seconds or $\frac{5}{3{,}600}$ of an hour. Taking $\Delta t = \frac{5}{3{,}600}$ and using the right-hand endpoints to calculate the Riemann sum, we find the distance (in miles) is approximately

$$\frac{5}{3{,}600} \cdot (20 + 34 + 44 + 51 + 56 + 59 + 60 + 60) \approx 0.533.$$

53. Using the symmetry of the stadium, we see that the area (in square yards) is approximately $4 \times (33 \times (50 - 5) + 4 \times (33 + 31 + 27)) = 7{,}396.$

CHAPTER 5 EXAM

1. The function $y = x^3 < 0$ in $(-1, 0)$ and $y = x^3 > 0$ in $(0, 2)$. The area of the region between the curve $y = x^3$ and the x-axis equals

$$-\int_{-1}^0 x^3\, dx + \int_0^2 x^3\, dx = -\left.\frac{1}{4}x^4\right|_{-1}^0 + \left.\frac{1}{4}x^4\right|_0^2 = \frac{1}{4} + 4 = \frac{17}{4}.$$

3. Using the substitution $u = x^2 - 5$ we obtain $du = 2x\, dx$ and the integral becomes $\int e^u\, du = e^u + c = e^{x^2 - 5} + c.$

5. We use integration by parts with $u = \ln t$ and $dv = t^3\, dt$. Then $du = \frac{1}{t}\, dt$ and $v = \frac{1}{4}t^4$, and

$$\int_1^e t^3 \ln t\, dt = \left.\frac{1}{4}t^4 \ln t\right|_1^e - \int_1^e \frac{1}{4}t^3\, dt = \frac{1}{4}e^4 - \left.\frac{1}{16}t^4\right|_1^e = \frac{1}{16}(3e^4 + 1).$$

7. Since $\int_{-1}^4 f(x)\, dx = \int_{-1}^2 f(x)\, dx + \int_2^4 f(x)\, dx$, we have

$$\int_2^4 f(x)\, dx = \int_{-1}^4 f(x)\, dx - \int_{-1}^2 f(x)\, dx = -4.2 + 6.3 = 2.1.$$

The area of the region is $-\int_{-1}^2 f(x)\, dx + \int_2^4 f(x)\, dx = 6.3 + 2.1 = 8.4.$

9. $\int_3^4 f(x)\,dx \approx (0.8 + 0.2 - 0.5 - 0.9 - 1.1) \cdot 0.2 = -0.3$

11. Using the substitution $u = x^4 + x^2 + 2$, we obtain $du = (4x^3 + 2x)\,dx$ or $(2x^3 + x)\,dx = \frac{1}{2}\,du$. Also, if $x = 1$, then $u = 4$, and if $x = 2$, then $u = 22$. The integral is equal to $\frac{1}{2}\int_4^{22} \frac{du}{u} = \frac{1}{2}\ln u\big|_4^{22} = \frac{1}{2}(\ln 22 - \ln 4) = \frac{1}{2}\ln\frac{11}{2}$.

13. Setting $x^2 = -x^2 + 8$ we find that the two curves intersect for $x = \pm 2$. Also, $-x^2 + 8 \geq x^2$ for $-2 \leq x \leq 2$. Thus, the area of the enclosed region is

$$\int_{-2}^2 (8 - x^2 - x^2)\,dx = \left(8x - \frac{2}{3}x^3\right)\Big|_{-2}^2 = \frac{64}{3}.$$

15. The revenue function $R(t)$ satisfies the differential equation $R'(t) = 20e^{0.2t}$ with initial condition $R(0) = 400$ ($t = 0$ is the year 2000). Integrating gives $R(t) = 100e^{0.2t} + c$, and using the initial condition gives $400 = 100 + c$. Therefore, $c = 300$, and $R(t) = 100e^{0.2t} + 300$.

17. We write $x^2 - 8x + 7 = (x - 1)(x - 7)$ and look for A, B such that

$$\frac{1}{(x - 1)(x - 7)} = \frac{A}{x - 1} + \frac{B}{x - 7}.$$

Clearing denominators gives $1 = A(x - 7) + B(x - 1)$. Setting $x = 1$ gives $A = -1/6$, and setting $x = 7$ gives $B = 1/6$. Thus, we obtain the partial fractions decomposition of $1/(x^2 - 8x + 7)$ and integrating we obtain

$$\int \frac{1}{x^2 - 8x + 7}\,dx = -\frac{1}{6}\int \frac{dx}{x - 1} + \frac{1}{6}\int \frac{dx}{x - 7} = \frac{1}{6}\ln\left|\frac{x - 7}{x - 1}\right| + c.$$

19. **(a)** Applying the trapezoid rule with $\Delta x = 0.2$ (corresponding to $n = 8$) we estimate the area (in square kilometers) of the upper region to be

$$[2(0.4) + 2(0.58) + 2(0.4) + 2(0.2) + 2(0.46) + 2(0.6) + 2(0.5)]\frac{0.2}{2} = 0.628.$$

Similarly, we estimate the area of the lower region by

$$[2(0.18) + 2(0.25) + 2(0.28) + 2(0.32) + 2(0.4) + 2(0.6) + 2(0.4)]\frac{0.2}{2} = 0.486.$$

Thus the area of the lake is $= 1.114$.

(b) Applying the midpoint rule for the upper region with $\Delta x = 0.4$ (corresponding to $n = 4$), we obtain the estimate $(0.4 + 0.4 + 0.46 + 0.5)(0.4) = 0.704$ for the area of the upper region. Similarly we estimate the area of the lower region by $(0.18 + 0.28 + 0.4 + 0.4)(0.4) = 0.504$. Thus the area of the lake is approximately 1.208 square kilometers.

21. Since $\frac{1}{7-2}\int_2^7 f(x)\,dx = 3$, we have $\int_2^7 f(x)\,dx = 15$; and since $\frac{1}{10-7}\int_7^{10} f(x)\,dx = -3$, we have $\int_7^{10} f(x)\,dx = -9$. Therefore,

$$\int_2^{10} f(x)\,dx = \int_2^7 f(x)\,dx + \int_7^{10} f(x)\,dx = 15 - 9 = 6,$$

so that the average value is $\frac{1}{10-2} \cdot 6 = \frac{3}{4}$.

CHAPTER 6
Further Applications and the Integral

EXERCISES 6.1

1. The graph gives $p_e = 3$ and $q_e = 2$. The consumer surplus, defined as $\int_0^{q_e} D(q)\,dq - p_e q_e$, is the area of a right triangle of base 2 and height 1 (formed by the demand line, the line $y = 3$, and the y-axis). Therefore, it equals $\frac{1}{2} \cdot 2 \cdot 1 = 1$.

    The producer surplus, defined by $p_e q_e - \int_0^{q_e} S(q)\,dq$ is the area of a right triangle of base 2 and height $= \frac{5}{2}$ (formed by the supply line, the line $y = 3$, and the y-axis). Therefore, it equals $\frac{1}{2} \cdot 2 \cdot \frac{5}{2} = \frac{5}{2} = 2.5$.

3. Solving $10 - 0.4q = 2 + 0.6q$ gives $q_e = 8$ and $p_e = 10 - 0.4 \cdot 8 = 6.8$.

    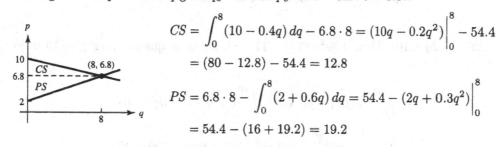

    $$CS = \int_0^8 (10 - 0.4q)\,dq - 6.8 \cdot 8 = (10q - 0.2q^2)\Big|_0^8 - 54.4$$
    $$= (80 - 12.8) - 54.4 = 12.8$$
    $$PS = 6.8 \cdot 8 - \int_0^8 (2 + 0.6q)\,dq = 54.4 - (2q + 0.3q^2)\Big|_0^8$$
    $$= 54.4 - (16 + 19.2) = 19.2$$

5. Rewrite the demand curve as $p = 100 - 0.02q$ and the supply curve as $p = 10 + 0.01q$. Solving $100 - 0.02q = 10 + 0.01q$ gives $q_e = 3{,}000$ and $p_e = 100 - 0.02 \cdot 3{,}000 = 40$.

    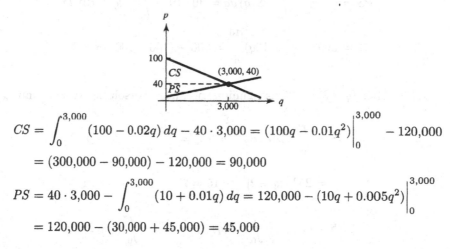

    $$CS = \int_0^{3{,}000} (100 - 0.02q)\,dq - 40 \cdot 3{,}000 = (100q - 0.01q^2)\Big|_0^{3{,}000} - 120{,}000$$
    $$= (300{,}000 - 90{,}000) - 120{,}000 = 90{,}000$$
    $$PS = 40 \cdot 3{,}000 - \int_0^{3{,}000} (10 + 0.01q)\,dq = 120{,}000 - (10q + 0.005q^2)\Big|_0^{3{,}000}$$
    $$= 120{,}000 - (30{,}000 + 45{,}000) = 45{,}000$$

7. The demand curve is $p = 30 - 0.6q$ and the supply curve is $p = 5 + 0.4q$. Solving $30 - 0.6q = 5 + 0.4q$ gives $q_e = 25$ and $p_e = 30 - 0.6 \cdot 25 = 15$.

    $$CS = \int_0^{25} (30 - 0.6q)\,dq - 15 \cdot 25 = (30q - 0.3q^2)\Big|_0^{25} - 375$$
    $$= (750 - 187.5) - 375 = 187.5$$
    $$PS = 15 \cdot 25 - \int_0^{25} (5 + 0.4q)\,dq = 375 - (5q + 0.2q^2)\Big|_0^{25}$$
    $$= 375 - (125 + 125) = 125$$

9. Set $-0.4q + 23 = 0.03q^2 + 3$. Then $0.03q^2 + 0.4q - 20 = 0$ has a unique non-negative solution $q_e = 20$, and $p_e = -0.4 \cdot 20 + 23 = 15$.

$$CS = \int_0^{q_e} D(q)\, dq - p_e \cdot q_e = \int_0^{20} (-0.4q + 23)\, dq - 15 \cdot 20$$

$$= (-0.2q^2 + 23q)\Big|_0^{20} - 300 = (-80 + 460) - 300 = 80$$

$$PS = p_e \cdot q_e - \int_0^{q_e} S(q)\, dq = 15 \cdot 20 - \int_0^{20} (0.03q^2 + 3)\, dq$$

$$= 300 - (0.01q^3 + 3q)\Big|_0^{20} = 300 - (80 + 60) = 160$$

11. Set $0.3(q - 20)^2 = 2q + 10$. Then $0.3q^2 - 14q + 110 = 0$ has a unique solution $q_e = 10$ in $[0, 20]$, and $p_e = 2 \cdot 10 + 10 = 30$.

$$CS = \int_0^{q_e} D(q)\, dq - p_e \cdot q_e = \int_0^{10} 0.3(q - 20)^2\, dq - 30 \cdot 10$$

$$= 0.1(q - 20)^3\Big|_0^{10} - 300 = (-100 + 800) - 300 = 400$$

$$PS = p_e \cdot q_e - \int_0^{q_e} S(q)\, dq = 30 \cdot 10 - \int_0^{10} (2q + 10)\, dq$$

$$= 300 - (q^2 + 10q)\Big|_0^{10} = 300 - (100 + 100) = 100$$

13. Set $\frac{25}{q+2} = q + 2$. Then $(q + 2)^2 = 25$ has a unique non-negative solution $q_e = 3$, and $p_e = 3 + 2 = 5$.

$$CS = \int_0^{q_e} D(q)\, dq - p_e \cdot q_e = \int_0^3 \frac{25}{q+2}\, dq - 5 \cdot 3$$

$$= 25 \ln|q + 2|\Big|_0^3 - 15 = 25 \ln \frac{5}{2} - 15 \approx 7.91$$

$$PS = p_e \cdot q_e - \int_0^{q_e} S(q)\, dq = 5 \cdot 3 - \int_0^3 (q + 2)\, dq$$

$$= 15 - \left(\frac{1}{2}q^2 + 2q\right)\Big|_0^3 = \frac{9}{2} = 4.50$$

15. Set $(q - 5)^2 = q^2 + q + 3$. Then $q_e = 2$ and $p_e = (2 - 5)^2 = 9$.

$$CS = \int_0^{q_e} D(d)\, dq - p_e q_e = \int_0^2 (q - 5)^2\, dq - 9 \cdot 2$$

$$= \frac{1}{3}(q - 5)^3\Big|_0^2 - 18 = \frac{1}{3}(-27 + 125) - 18 = \frac{44}{3} \approx 14.67$$

$$PS = p_e q_e - \int_0^{q_e} S(q)\, dq = 9 \cdot 2 - \int_0^2 (q^2 + q + 3)\, dq$$

$$= 18 - \left(\frac{1}{3}q^3 + \frac{1}{2}q^2 + 3q \right) \Big|_0^2 = \frac{22}{3} \approx 7.33$$

17. The producer surplus is the area of a right triangle with base 200 and height 8 (formed by the y-axis, the line $y = 10$, and the supply line.) Therefore, it equals $\frac{1}{2} \cdot 200 \cdot 8 = 800$.

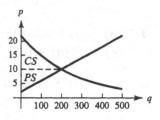

The consumer surplus, is the area of the region formed by the y-axis, the line $y = 10$, and the demand curve. To estimate it we count the number of rectangles fully contained in the region (they are 7) and estimate the proportion of each rectangle partially contained in the region (they are about 6 half rectangles). Since the area of each rectangle is $50 \cdot 2 = 100$, the producer surplus is approximately equal to 1,000.

19. (a) A supply function of the form $c = S(b)$, where b is the number of babies, and c is calories per baby.

(b) 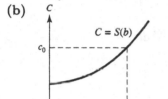   **(c)** $b_0 c_0$

21. (a) Since the equilibrium quantity is approximately 7, the equilibrium price is approximately $0.06 \cdot 7^2 + 5 = 7.94$. Then

$$PS \approx 7 \cdot (7.94) - \int_0^7 (0.06x^2 + 5)\, dx = 55.58 - (0.02x^3 + 5x) \Big|_0^7 = 13.72.$$

(b) $CS \approx \int_0^7 (0.2x^2 - 4.8x + 31.8)\, dx - 55.58$

$$= \left(\frac{0.2}{3}x^3 - 2.4x^2 + 31.8x \right) \Big|_0^7 - 55.58 = 72.29$$

23. $q_e = 429.6077, \quad p_e = 21.1747, \quad CS = 5{,}315.83, \quad PS = 5{,}655.10$

25. $q_e = 3.0172, \quad p_e = 4.5313, \quad CS = 6.1177, \quad PS = 4.6304$

EXERCISES 6.2

1. Given $PV = 5{,}000$, $t = 4$, $r = 0.06$, and $n = 12$,

$$FV = PV\left(1 + \frac{r}{n}\right)^{nt} = 5{,}000\left(1 + \frac{0.06}{12}\right)^{12\cdot 4} \approx 6{,}352.45 \text{ dollars.}$$

3. (a) $n = 1$, since the seeds sprout once a year. **(b)** $r = 0.23$ and $t = 5$.
 (c) $1{,}000(1 + 0.23)^5 \approx 2{,}815$

5. Given $PV = 4{,}000$, $t = 10$, and $r = 0.08$,

$$FV = PVe^{rt} = 4{,}000e^{0.08\cdot 10} = 4{,}000e^{0.8} \approx 8{,}902.16 \text{ dollars.}$$

7. We want to find the annual interest rate r so that $FV = 2PV$ at the end of 7 years. Set $2PV = PVe^{7r}$. Then $2 = e^{7r}$, or $\ln 2 = \ln e^{7r} = 7r$. Therefore, $r = \frac{1}{7}\ln 2 \approx 0.1$, or 10%.

9. We have a continuous income stream with $S(t) = 15{,}000$, $T = 5$, and $r = 0.075$. The present value is

$$PV = \int_0^T S(t)e^{-rt}\, dt = \int_0^5 15{,}000e^{-0.075t}\, dt$$

$$= -\frac{15{,}000}{0.075}e^{-0.075t}\Big|_0^5$$

$$= 200{,}000(1 - e^{-0.375}) \approx 62{,}542.14 \text{ dollars.}$$

11. The fish population in the lake can be thought of as a continuous income stream flowing at the rate of 1,000 fish per year, with an annual interest rate of 0.33 (the reproduction rate) compounded continuously. Therefore, its future size FV in six years is

$$\int_0^6 1{,}000e^{0.33(6-t)}\, dt = -1{,}000\int_6^0 e^{0.33u}\, du = -\frac{1{,}000}{0.33}e^{0.33u}\Big|_6^0 \approx 18{,}917.$$

13. If we compare present values, we see that the first plan pays 40,000,000 dollars, and the second plan pays

$$PV = 6{,}000{,}000 + \int_0^6 7{,}500{,}000e^{-0.08t}\, dt = 6{,}000{,}000 - \frac{7{,}500{,}000}{0.08}e^{-0.08t}\Big|_0^6$$

$$= 6{,}000{,}000 + 93{,}750{,}000(1 - e^{-0.48}) = 41{,}739{,}057.02 \text{ dollars.}$$

Thus, the second plan is better for the player.

15. With $S(t) = 5{,}000 + 500t$, $T = 18$, and $r = 0.055$, we obtain

$$FV = \int_0^T S(t)e^{r(T-t)}\, dt = \int_0^{18}(5{,}000 + 500t)e^{0.055(18-t)}\, dt$$

$$= 5{,}000\int_0^{18} e^{0.055(18-t)}\, dt + 500\int_0^{18} te^{0.055(18-t)}\, dt$$

$$= 5{,}000\int_0^{18} e^{0.055u}\, du + 500\int_0^{18}(18 - u)e^{0.055u}\, du \qquad \text{(setting } u = 18 - t\text{)}$$

$$= 5{,}000 \int_0^{18} e^{0.055u} \, du + 9{,}000 \int_0^{18} e^{0.055u} \, du - 500 \int_0^{18} u e^{0.055u} \, du$$

$$\approx 430{,}496.05 - 500 \int_0^u u e^{0.055u} \, du.$$

Integration by parts gives

$$\int_0^{18} u e^{0.055u} \, du = \frac{1}{0.055} \left[u e^{0.055u} \Big|_0^{18} - \int_0^{18} e^{0.055u} \, du \right]$$

$$= \frac{1}{0.055} \left[18 e^{0.99} - \frac{1}{0.055} e^{0.055u} \Big|_0^{18} \right] \approx 321.6819.$$

Thus, $FV \approx 430{,}496.05 - 500 \cdot 321.6819 = 269{,}655.10$ dollars.

17. This is a continuous income stream with $S(t) = 0.1t + 0.004$, $r = 0.02$, and $T = 10$. With an initial population of 4 million, its future value after 10 years is $4e^{0.02 \cdot 10} + \int_0^{10} (0.1t + 0.004) e^{0.02(10-t)} \, dt$, or, equivalently,

$$FV = 4e^{0.2} + 0.1e^{0.2} \int_0^{10} t e^{-0.02t} \, dt + 0.004 e^{0.2} \int_0^{10} e^{-0.02t} \, dt$$

$$= 4e^{0.2} + 0.1e^{0.2} \int_0^{10} t e^{-0.02t} \, dt - \frac{0.004}{0.02} e^{0.2} \left(e^{-0.2} - 1 \right)$$

$$= 4.2e^{0.2} + 0.1e^{0.2} \int_0^{10} t e^{-0.02t} \, dt - 0.2.$$

Using integration by parts with $u = t$ and $v' = e^{-0.02t}$ gives

$$\int_0^{10} t e^{-0.02t} \, dt = -50 t e^{-0.02t} \Big|_0^{10} + 50 \int_0^{10} e^{-0.02t} \, dt$$

$$= -500 e^{-0.2} - 2{,}500 e^{-0.02t} \Big|_0^{10} = 2{,}500 - 3{,}000 e^{-0.2}.$$

Putting it all together, we find that after 10 years the population's size is $254.2e^{0.2} - 300.2 \approx 10.2806$ million.

19. Denote by S the investment rate. The income at the end of 40 years is given by

$$FV = \int_0^{40} S e^{0.07(40-t)} \, dt = \frac{S}{0.07} (e^{2.8} - 1).$$

This amount must be the present value of an income stream flowing at the rate of $60,000 per year for 20 years, that is,

$$\int_0^{20} 60{,}000 e^{-0.07t} \, dt = \frac{60{,}000}{0.07} (1 - e^{-1.4}).$$

Thus, to find S we solve the equation $\frac{S}{0.07}(e^{2.8} - 1) = \frac{60{,}000}{0.07}(1 - e^{-1.4})$ which gives

$$S = 60{,}000 \frac{(1 - e^{-1.4})}{e^{2.8} - 1} \approx 2{,}926.85 \text{ dollars.}$$

21. Denote by T the number of years this account will last. Then

$$\int_0^T 5{,}000e^{-0.045t}\,dt = 60{,}000,$$

so that

$$12 = \int_0^T e^{-0.045t}\,dt = -\frac{1}{0.045}e^{-0.045t}\Big|_0^T = \frac{1}{0.045}(1 - e^{-0.045T}),$$

which reduces to $e^{-0.045T} = 0.46$. Thus, $T = \frac{1}{0.045}\ln 0.46 \approx 17.26$ dollars.

23. **(a)** Since the initial deposit is $\frac{S}{n}$, we have $A(1) = (1 + \frac{r}{n})\frac{S}{n}$. At the end of the $(k-1)$st period, which is also the beginning of the kth period, we have $A(k-1)$ and add $\frac{S}{n}$ to get $A(k-1) + \frac{S}{n}$. At the end of the kth period we get that amount plus $\frac{r}{n}$ times that amount (the interest), so that

$$A(k) = \left[A(k-1) + \frac{S}{n}\right] + \frac{r}{n}\left[A(k-1) + \frac{S}{n}\right] = \left(1 + \frac{r}{n}\right)\left[A(k-1) + \frac{S}{n}\right].$$

(b) We have $\frac{r}{n} = \frac{0.06}{4} = 0.015$ and $\frac{S}{n} = \frac{1{,}000}{4} = 250$. Therefore,

$$A(k) = 1.015[A(k-1) + 250] = 1.015A(k-1) + 253.75.$$

With $A(0) = 0$, repeated computations give us

$$A(1) = 253.75, \quad A(2) = 511.30625, \quad A(3) = 772.7258438,$$
$$A(4) = 1{,}038.066731, \quad A(5) = 1{,}307.387732, \quad A(6) = 1{,}580.748548,$$
$$A(7) = 1{,}858.209777, \quad A(8) = 2{,}139.832923.$$

Thus, (i) the amount at the end of 1 year is $A(4) \approx 1{,}038.07$ dollars, and (ii) the amount at the end of 2 years is $A(8) \approx 2{,}139.83$ dollars.

25.

`=1.01*C2`

6.2.25.wb

Quarter	1	2	3	4	5	6	7	8	9	10	11	12	13	14	15
Begin	1500.00	3015.00	4545.15	6090.60	7651.51	9228.02	10820.30	12428.51	14052.79	15693.32	17350.25	19023.75	20713.99	22421.13	24145.34
End	1515.00	3045.15	4590.60	6151.51	7728.02	9320.30	10928.51	12552.79	14193.32	15850.25	17523.75	19213.99	20921.13	22645.34	24386.80

Quarter	16	17	18	19	20	21	22	23	24	25	26	27	28	29	30
Begin	25886.80	27645.66	29422.12	31216.34	33028.51	34858.79	36707.38	38574.45	40460.20	42364.80	44289.45	46231.33	48193.65	50175.58	52177.34
End	26145.66	27922.12	29716.34	31528.51	33358.79	35207.38	37074.45	38960.20	40864.80	42788.45	44731.33	46693.65	48675.58	50677.34	52699.11

Quarter	31	32	33	34	35	36	37	38	39	40
Begin	54199.11	56241.10	58303.51	60386.55	62490.41	64615.32	66761.47	68929.09	71118.38	73329.56
End	54741.10	56803.51	58886.55	60990.41	63115.32	65261.47	67429.09	69618.38	71829.56	74062.86

27.

D3 `=(1+0.075/52)*D2`

6.2.27.wb

Week	1	2	3	4	5	6	7	8	9	10	11	12	13
Begin	96.15	192.45	288.88	385.45	482.16	579.01	676.00	773.13	870.39	967.80	1065.35	1163.04	1260.87
End	96.29	192.72	289.29	386.00	482.85	579.84	676.97	774.24	871.65	969.20	1066.89	1164.72	1262.69

Week	14	15	16	17	18	19	20	21	22	23	24	25	26
Begin	1358.85	1456.96	1555.22	1653.61	1752.15	1850.83	1949.66	2048.62	2147.73	2246.98	2346.38	2445.91	2545.60
End	1360.81	1459.06	1557.46	1656.00	1754.68	1853.50	1952.47	2051.56	2150.83	2250.22	2349.76	2449.44	2549.27

29. (a)

$$A(2) = \left(1 + \frac{r}{n}\right)\left[A(1) + \frac{S}{n}\right] = \left(1 + \frac{r}{n}\right)\left[\left(1 + \frac{r}{n}\right)\frac{S}{n} + \frac{S}{n}\right]$$

$$= \left(1 + \frac{r}{n}\right)^2 \frac{S}{n} + \left(1 + \frac{r}{n}\right)\frac{S}{n} = \frac{S}{n}\left[\left(1 + \frac{r}{n}\right)^2 + \left(1 + \frac{r}{n}\right)\right],$$

$$A(3) = \left(1 + \frac{r}{n}\right)\left[A(2) + \frac{S}{n}\right] = \left(1 + \frac{r}{n}\right)\left[\left(1 + \frac{r}{n}\right)^2 \frac{S}{n} + \left(1 + \frac{r}{n}\right)\frac{S}{n} + \frac{S}{n}\right]$$

$$= \left(1 + \frac{r}{n}\right)^3 \frac{S}{n} + \left(1 + \frac{r}{n}\right)^2 \frac{S}{n} + \left(1 + \frac{r}{n}\right)\frac{S}{n}$$

$$= \frac{S}{n}\left[\left(1 + \frac{r}{n}\right)^3 + \left(1 + \frac{r}{n}\right)^2 + \left(1 + \frac{r}{n}\right)\right]$$

If we continue in this way up to the formula

$$A(k-1) = \frac{S}{n}\left[\left(1 + \frac{r}{n}\right)^{k-1} + \left(1 + \frac{r}{n}\right)^{k-2} + \cdots + \left(1 + \frac{r}{n}\right)\right],$$

then

$$A(k) = \left(1 + \frac{r}{n}\right)\left[A(k-1) + \frac{S}{n}\right]$$

$$= \left(1 + \frac{r}{n}\right)\left[\left(1 + \frac{r}{n}\right)^{k-1}\frac{S}{n} + \left(1 + \frac{r}{n}\right)^{k-2}\frac{S}{n} + \cdots + \left(1 + \frac{r}{n}\right)\frac{S}{n} + \frac{S}{n}\right]$$

$$= \frac{S}{n}\left[\left(1 + \frac{r}{n}\right)^k + \left(1 + \frac{r}{n}\right)^{k-1} + \cdots + \left(1 + \frac{r}{n}\right)\right].$$

(b) Letting $t = 1 + \frac{r}{n}$, we obtain

$$A(k) = \frac{S}{n}\left[\left(1 + \frac{r}{n}\right)^k + \left(1 + \frac{r}{n}\right)^{k-1} + \cdots + \left(1 + \frac{r}{n}\right)\right]$$

$$= \frac{S}{n}(t^k + t^{k-1} + \cdots + t) = t\frac{S}{n}(t^{k-1} + t^{k-2} + \cdots + 1)$$

$$= t\frac{S}{n}\left(\frac{t^k - 1}{t - 1}\right) = \frac{S}{n}\left(\frac{t^{k+1} - t}{t - 1}\right)$$

$$= \frac{S}{n}\frac{\left(1 + \frac{r}{n}\right)^{k+1} - \left(1 + \frac{r}{n}\right)}{r/n} = \frac{S}{r}\left[\left(1 + \frac{r}{n}\right)^{k+1} - \left(1 + \frac{r}{n}\right)\right].$$

(c) $A(3,650) = \dfrac{1,200}{0.05}\left[\left(1 + \dfrac{0.05}{365}\right)^{3,651} - \left(1 + \dfrac{0.05}{365}\right)\right] \approx 15{,}570.09$

If the interest is compounded continuously, the amount at the end of 10 years equals

$$\int_0^{10} 1{,}200e^{0.05(10-t)}\,dt = 24{,}000(e^{0.5} - 1) \approx 15{,}569.31.$$

EXERCISES 6.3

1. By the product and chain rules, $\frac{dy}{dt} = C(e^{t^2/2} + t^2 e^{t^2/2}) = (1 + t^2)Ce^{t^2/2}$. Therefore, $t\frac{dy}{dt} = t(1 + t^2)Ce^{t^2/2} = (1 + t^2)Cte^{t^2/2} = (1 + t^2)y.$

3. By the chain rule, $\frac{dz}{dx} = \frac{1}{C+x^3} \cdot \frac{d}{dx}(C+x^3) = 3x^2 \frac{1}{C+x^3}$. Also,

$$z = \ln(C+x^3) \iff e^z = (C+x^3) \iff \frac{1}{C+x^3} = e^{-z}.$$

Therefore, $\frac{dz}{dx} = 3x^2 e^{-z}$.

5. Write $dy/dt = -y$ in the form $dy/y = -dt$. Taking integrals, we get

$$\ln|y| = \int \frac{dy}{y} = \int -dt = -t + c,$$

and exponentiating gives $|y| = e^c e^{-t}$, or $y = \pm e^c e^{-t}$. Therefore, $y = Ce^{-t}$, where $C = \pm e^c$.

7. Writing the equation in the form $\frac{dM}{0.1M+10} = dt$, we obtain

$$\int \frac{dM}{0.1M+10} = \int dt \quad \text{or} \quad \frac{1}{0.1}\ln|0.1M+10| = t + c.$$

Therefore, $\ln|0.1M+10| = 0.1t + 0.1c$, and exponentiating gives

$$0.1M + 10 = \pm e^{0.1c}e^{0.1t} \quad \text{or} \quad M = \pm 10 e^{0.1c}e^{0.1t} - 100.$$

Thus, $M = Ce^{0.1t} - 100$, where $C = \pm 10 e^{0.1c}$.

9. Since $\frac{dy}{dt} = ye^t - e^t = (y-1)e^t$, we have $\frac{dy}{y-1} = e^t\, dt$. Integrating, we get

$$\ln|y-1| = e^t + c \quad \text{or} \quad y - 1 = \pm e^c \cdot e^{e^t}.$$

Therefore, $y = Ce^{e^t} + 1$, where $C = \pm e^c$.

11. Write the equation in the form $\frac{dy}{(y-1)(y-2)} = dt$. Then the method of partial fractions gives

$$\int \left(\frac{1}{y-2} - \frac{1}{y-1} \right) dy = \int dt \quad \text{or} \quad \ln\left| \frac{y-2}{y-1} \right| = t + c.$$

Therefore, $\frac{y-2}{y-1} = \pm e^c \cdot e^t = Ce^t$, where $C = \pm e^c$, $y - 2 = Ce^t y - Ce^t$. Solving for y gives $y = \frac{Ce^t - 2}{Ce^t - 1}$.

13. Write $\frac{dy}{dt} = r(y-a)$ in the form $\frac{dy}{y-a} = r\, dt$. Integrating gives

$$\ln|y-a| = rt + c \quad \text{or} \quad y - a = \pm e^c e^{rt}.$$

Therefore, $y = Ce^{rt} + a$, where $C = \pm e^c$.

15. Integrating $\frac{dy}{y^2} = dt$ gives $-\frac{1}{y} = t + c$, or $y = -\frac{1}{t+c}$. The initial condition $y(0) = 4$ gives $-\frac{1}{c} = 4$, or $c = -\frac{1}{4}$. Therefore, $y = -\frac{1}{t-(1/4)} = \frac{4}{1-4t}$.

17. Given $\frac{dy}{y-9} = -dt$, we integrate to obtain $\int \frac{dy}{y-9} = \int -dt$. Then

$$\ln|y-9| = -t + c \quad \text{or} \quad y - 9 = \pm e^c e^{-t}.$$

Therefore, $y = Ce^{-t} + 9$, where $C = \pm e^c$. The initial condition $y(0) = 12$ gives $C + 9 = 12$, or $C = 3$. Therefore, $y = 3e^{-t} + 9$.

19. Write $dM/dt = 0.1M + 100 = 0.1(M + 1,000)$ as $dM/(M + 1,000) = 0.1\,dt$. Integrate to get $\ln|M + 1,000| = 0.1t + c$ or $M + 1,000 = \pm e^c e^{0.1t}$. Therefore, $M = Ce^{0.1t} - 1,000$, where $C = \pm e^c$. The initial condition $M(0) = 50$ gives $C - 1,000 = 50$, or $C = 1,050$. Therefore, $M = 1,050e^{0.1t} - 1,000$.

21. Write $\frac{dy}{dx} = xy^2 - x = x(y^2 - 1)$ as $\frac{dy}{y^2-1} = x\,dx$. By partial fractions,

$$\int\left(\frac{1/2}{y-1} - \frac{1/2}{y+1}\right)dy = \int x\,dx \quad \text{and} \quad \frac{1}{2}\ln\left|\frac{y-1}{y+1}\right| = \frac{1}{2}x^2 + c.$$

Multiplying by 2 and exponentiating, we get $\frac{y-1}{y+1} = \pm e^{2c}e^{x^2} = Ce^{x^2}$, and solving for y gives $y = \frac{1+Ce^{x^2}}{1-Ce^{x^2}}$. From the initial condition $y(0) = 2$ we get $\frac{1+C}{1-C} = 2$ or $C = \frac{1}{3}$. Therefore, $y = \frac{1+\frac{1}{3}e^{x^2}}{1-\frac{1}{3}e^{x^2}} = \frac{3+e^{x^2}}{3-e^{x^2}}$.

23. From $dy/dx = y/x$ we obtain $dy/y = dx/x$, and integrating gives $\ln|y| = \ln|x| + c$. Therefore, $|y| = e^c|x|$, or $y = Cx$, where $C = \pm e^c$. The initial values correspond to $C = -3, -2, -1, 0, 1, 2, 3$.

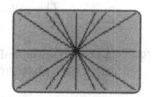

25. Write $dy/dt = e^{-y}$ as $e^y\,dy = dt$. Integrating gives $e^y = t + c$. Therefore, $y = \ln(t + c)$. The initial values correspond to $c = 1, e^{0.5}, e, e^{1.5}, e^2, e^{2.5}, e^3$.

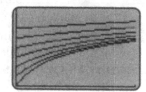

27. Write $e^{x^2}y\frac{dy}{dx} = -2x$ in the form $y\,dy = -2xe^{-x^2}\,dx$. Integrating gives

$$\frac{1}{2}y^2 = e^{-x^2} + c \quad \text{or} \quad y^2 = 2e^{-x^2} + 2c = 2e^{-x^2} + C,$$

where $C = 2c$.

(a) The initial condition $y(0) = 1$ gives $\sqrt{2 + C} = 1$, or $C = -1$, and we take the positive square root. Thus, $y = \sqrt{2e^{-x^2} - 1}$.

(b) The initial condition $y(0) = -2$ gives $\sqrt{2 + C} = -2$, or $C = 2$, and we take the negative square root. Thus, $y = -\sqrt{2e^{-x^2} + 2}$.

29. Let $A(t)$ be the amount of money in the account at the time t, and let r be the rate at which the withdrawals should be made each year. We have to solve the initial value problem.

$$\frac{dA}{dt} = 0.09A - r, \quad A(0) = 1,800,000$$

Rewriting the differential equation in the form $\frac{dA}{0.09A-r} = dt$ and integrating, we get

$$\frac{1}{0.09}\ln|0.09A - r| = t + c \quad \text{or} \quad 0.09A - r = \pm e^{0.09c}e^{0.09t}.$$

Therefore, $A = Ce^{0.09t} + \frac{r}{0.09}$, where $C = \pm\frac{1}{0.09}e^{0.09c}$. The initial condition $A(0) = 1,800,000$ gives $C + \frac{r}{0.09} = 1,800,000$, or $C = 1,800,000 - \frac{r}{0.09}$. Therefore, $A(t) = (1,800,000 - \frac{r}{0.09})e^{0.09t} + \frac{r}{0.09}$.
Since $A(30) = 500,000$, we have $(1,800,000 - \frac{r}{0.09})e^{2.7} + \frac{r}{0.09} = 500,000$, and solving for r gives $r = \frac{0.09(1,800,000e^{2.7}-500,000)}{e^{2.7}-1} \approx 170,429.56$ dollars.

31. Let $M(t)$ be the amount of money in the account at time t. The rate at which the investment grows due the interest is $0.1M$ (an annual rate of 10%, compounded continuously), and it decreases at a rate of k dollars. Therefore, M satisfies the differential equation $dM/dt = 0.1M - k$, with initial condition $M(0) = 33,000,000$. However, if $M(t)$ is constant, its derivative is zero, which means $0 = dM/dt = 0.1M - k$. Therefore, $0.1M = k$, and since $M = 33,000,000$ (the same amount for all t), $k = 3,300,000$. Thus, he can withdraw \$3,300,000 per year.

33. (a) $\dfrac{dy}{dt} = 100 - 0.12y$

(b) We rewrite the differential equation as $\dfrac{dy}{0.12y-100} = -dt$ and integrate to get

$$\frac{1}{0.12}\ln|0.12y - 100| = -t + c \quad \text{or} \quad 0.12y - 100 = \pm e^{0.12c}e^{-0.12t}.$$

Therefore, $y = Ce^{-0.12t} + \frac{2,500}{3}$, where $C = \pm\frac{1}{0.12}e^{0.12c}$.
The initial condition $y(0) = 50$ gives $C + \frac{2,500}{3} = 50$, or $C = -\frac{2,350}{3}$. Therefore, $y = -\frac{2,350}{3}e^{-0.12t} + \frac{2,500}{3}$, and $y(10) = -\frac{2,350}{3}e^{-1.2} + \frac{2,500}{3} \approx 597.40$ mg.

35. (a) The initial value problem is $dV/dt = 200 - 0.01V$, $V(0) = 50$.
(b) Writing $dV/(0.01V - 200) = -dt$, we integrate to get

$$\frac{1}{0.01}\ln|0.01V - 200| = -t + c \quad \text{or} \quad \ln|0.01V - 200| = -0.01t + 0.01c.$$

Thus, $0.01V - 200 = \pm e^{0.01c}e^{-0.01t}$, and we have $V = Ce^{-0.01t} + 20,000$, where $C = \pm 100 e^{0.01c}$. The initial condition $V(0) = 50$ gives $C + 20,000 = 50$, or $C = -19,950$. Therefore, $V = -19,950e^{-0.01t} + 20,000$.
(c) $V(3) = -19,950e^{-0.03} + 20,000 \approx 640$ words

37. (a) The pollution is modeled by the initial value problem

$$\frac{dP}{dt} = 1,000 - 0.02P, \quad P(0) = 40,000,$$

where t is the time in years and $P(t)$ amount of chemical polution (in pounds) in the lake at time t. To find the solution, we write the differential equation in the form $dP/(0.02P - 1,000) = -dt$ and integrate to get

$$\frac{1}{0.02}\ln|0.02P - 1,000| = -t + c \quad \text{or} \quad \ln|0.02P - 1,000| = -0.02t + 0.02c.$$

Therefore, $0.02P - 1,000 = \pm e^{0.02c}e^{-0.02t}$, and $P = Ce^{-0.02t} + 50,000$, where $C = \pm 50e^{0.02c}$.
The initial condition $P(0) = 40,000$ gives $C + 50,000 = 40,000$, so that $C = -10,000$. Therefore, $P = -10,000e^{-0.02t} + 50,000$.

(b)

t	0	1	2	3
$P(t)$	40,000	40,198	40,392	40,582

The clean-up effort does not appear to be successful because the amount of pollution is increasing.

39. (a) Let M be the balance of the fund at time t. Its rate of increase (from earning interest) is $0.09M$, and its rate of decrease (from paying out scholarships is $20 \cdot 25,000$, which equals $500,000$. Therefore, M satisfies the initial value problem

$$\frac{dM}{dt} = 0.09M - 500,000, \quad M(0) = M_0,$$

where $M_0 = M(0)$, the initial amount put in the fund.

(b) We write the differential equation in the form $\frac{dM}{0.09M-500,000} = dt$ and integrate to get

$$\frac{1}{0.09} \ln |0.09M - 500,000| = t + c \quad \text{or} \quad \ln |0.09M - 500,000| = 0.09t + 0.09c.$$

Then $0.09M - 500,000 = \pm e^{0.09c}e^{0.09t}$, and $M = Ce^{0.09t} + \frac{50,000,000}{9}$, where $C = \pm\frac{1}{0.09}e^{0.09c}$. The initial condition $M(0) = M_0$ gives $M_0 = C + \frac{50,000,000}{9}$, or $C = M_0 - \frac{50,000,000}{9}$. Therefore, $M = (M_0 - \frac{50,000,000}{9})e^{0.09t} + \frac{50,000,000}{9}$. To keep the fund paying solvent forever, we need $M(t) > 0$ for all t, which means that M_0 should be at least $\frac{50,000,000}{9} \approx 5,555,555.56$ dollars.

41. (a) The corresponding initial value problem is

$$\frac{dM}{dt} = 0.12M - 12A, \quad M(0) = 10,000.$$

Writing the differential equation in the form $\frac{dM}{0.12M-12A} = dt$ and integrating, we get

$$\frac{1}{0.12} \ln |0.12M - 12A| = t + c \quad \text{or} \quad 0.12M - 12A = \pm e^{0.12c}e^{0.12t}.$$

Therefore, $M = Ce^{0.12t} + 100A$, where $C = \pm\frac{1}{0.12}e^{0.12c}$. From the initial condition $M(0) = 10,000$ we get $C + 100A = 10,000$, or $C = 10,000 - 100A$. Therefore, $M = (10,000 - 100A)e^{0.12t} + 100A$. Since the loan is paid off in 5 years, $M(5) = (10,000 - 100A)e^{0.6} + 100A = 0$, and solving for A gives $A = \frac{100e^{0.6}}{e^{0.6}-1} \approx 221.64$ dollars.

(b) Following the procedure of the last exercise, we have

$$M_n = (1.01)^n \cdot 10,000 - A\frac{(1.01)^n - 1}{0.01}.$$

Since the balance reduces to zero at the end of 60 months,

$$M_{60} = (1.01)^{60} \cdot 10,000 - A\frac{(1.01)^{60} - 1}{0.01} = 0,$$

and solving for A gives $A = \frac{100(1.01)^{60}}{(1.01)^{60}-1} \approx 222.444$ dollars.

43. Let $w(t)$ be the wind energy generating t years after 2000.

$$\frac{dw}{dt} = 0.12w, \quad w(0) = 2,500$$

45. Letting t be the time (in years) and $P(t)$ the population of the region (in millions), we have

$$\frac{dP}{dt} = 0.025P + 0.05, \quad P(0) = 30.$$

EXERCISES 6.4

1. $p_0 = 327$, $r = 0.015$, and $k = \frac{r}{K} = \frac{0.015}{K}$.

3. (a) The population is modeled by a logistic equation with initial value,

$$\frac{dP}{dt} = 0.7P - 0.01P^2, \quad P(0) = 35.$$

In this case, $r = 0.7$, $k = 0.01$, $p_0 = 35$, and $K = r/k = 70$, and the solution is

$$P(t) = \frac{p_0 K}{p_0 + (K - p_0)e^{-rt}} = \frac{35 \cdot 70}{35 + 35e^{-0.7t}} = \frac{70}{1 + e^{-0.7t}}.$$

(b) The carrying capacity is 70, which is $\lim_{t \to \infty} P(t)$.

(c) The predicted population after 16 weeks equals $P(8) = \frac{70}{1+e^{-5.6}} \approx 70$. (Recall that the time unit is 2 weeks.)

5. (a) Under the given assumptions, the U.S. population can be modelled by the initial value problem

$$\frac{dP}{dt} = rP, \quad P(0) = 5,$$

where t is the number of years after 1800, P is the population in millions, and r is an unknown constant. The solution is $P(t) = 5e^{rt}$. To find r, we substitute $t = 200$ (for the year 2000 and $P(200) = 280$ to get $280 = 5e^{200r}$. Then $r = \frac{1}{200} \ln 56$, and $P(t) = 5e^{t(\ln 56)/200}$. Therefore, the population in 1900 is given by $P(100) = 5e^{(\ln 56)/2} = 5\sqrt{56} = 10\sqrt{14} \approx 37.42$ million.

(b) The actual population in 1900 was much larger than the population obtained from the model. Note that the model takes no account of immigration, which was a major factor in population growth during that period.

7. The intrinsic growth rate is $r = 0.04$ and the damping factor is $k = 0.001$. The equilibrium solutions are $p = 0$ and $p = 40$, obtained by solving $0.04p - 0.001p^2 = 0.001p(40 - p) = 0$.

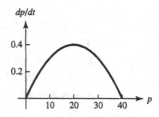

9. The intrinsic growth rate is 0.25 and the damping factor is 0.05. The equilibrium solutions are $p = 0$ and $p = 5$, obtained by solving $0.25p - 0.05p^2 = 0.05p(5 - p) = 0$.

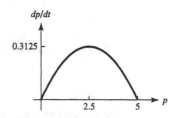

11. Since $r = 0.03$ and $K = 9$, we have $k = \frac{r}{K} = \frac{1}{300}$. The logistic equation is $\frac{dp}{dt} = 0.03p(1 - \frac{p}{9})$, or $\frac{dp}{dt} = 0.03p - \frac{1}{300}p^2$.

13. (a) To solve the differential equation $\frac{dy}{dt} = ay(3{,}000 - y)$, we write it in the form $\frac{dy}{y(3{,}000-y)} = a\,dt$. Integrating by partial fractions gives

$$\frac{1}{3{,}000} \ln \left| \frac{y}{3{,}000 - y} \right| = at + c \quad \text{or} \quad \ln \left| \frac{y}{3{,}000 - y} \right| = 3{,}000at + 3{,}000c.$$

Therefore, $\frac{y}{3{,}000-y} = Ce^{3{,}000at}$, where $C = \pm e^{3{,}000c}$, and solving for y gives $y = 3{,}000Ce^{3{,}000at}/(1 + Ce^{3{,}000at})$.

(b) Given $y(0) = 20$ and $y(3) = 314$, we use the first condition to obtain $3{,}000C/(1 + C) = 20$, so that $C = 1/149$ and $y = 3{,}000e^{3{,}000at}/(149 + e^{3{,}000at})$. The second condition gives $3{,}000e^{9{,}000a}/(149 + e^{9{,}000a}) = 314$. By clearing denominators and collecting terms, we get $2{,}686e^{9{,}000a} = 46{,}786$, and solving for a gives $a = \frac{1}{9{,}000} \ln \frac{23{,}393}{1{,}343} \approx 0.000318$.

(c) Using $a \approx 0.000318$, we find $y(5) \approx \frac{3{,}000e^{15{,}000 \cdot (0.000318)}}{149 + e^{15{,}000 \cdot (0.000318)}} \approx 1{,}325$.

15. In this case, $r = 0.03$ and $p_0 = 76$ (million). The solution of the logistic equation is

$$p(t) = \frac{p_0 K}{p_0 + (K - p_0)e^{-rt}} = \frac{76K}{76 + (K - 76)e^{-0.03t}},$$

where t is the number of years after 1900 and K is the carrying capacity. To find K, we use the information that $p(10) = 92$ to get $\frac{76K}{76+(K-76)e^{-0.3}} = 92$, whose solution is $K = \frac{6,992(1-e^{-0.3})}{76-92e^{-0.3}} \approx 231$ million. Therefore,

$$p(t) = \frac{76 \cdot 231}{76 + (231 - 76)e^{-0.03t}} = \frac{17,556}{76 + 155e^{-0.03t}}.$$

The tables for 1900 to 2000 (t from 0 to 100) in steps of 4 are as follows:

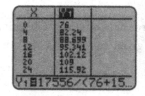
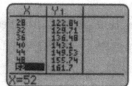
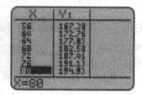
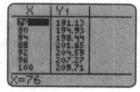

17. If $dp/dt = 0.05p(1 - p) = 0.05p - 0.05p^2$, then $r = 0.05$, $k = 0.05$, and $K = r/k = 1$. Since $p_0 = 0.8$, the solution is

$$p(t) = \frac{p_0 K}{p_0 + (K - p_0)e^{-rt}} = \frac{0.8}{0.8 + 0.2e^{-0.05t}} = \frac{4}{4 + e^{-0.05t}}.$$

19. If $dp/dt = 0.03p(1 - \frac{1}{5}p)$, then $r = 0.03$ and $K = 5$. Since $p_0 = 4$, the solution is

$$p(t) = \frac{p_0 K}{p_0 + (K - p_0)e^{-rt}} = \frac{20}{4 + e^{-0.03t}}.$$

21. (a) From the given data, we know $M = 36,000$ and $y_0 = 50$. Therefore, the differential equation has the form $dy/dt = 36,000ay - ay^2$, which is a logistic equation with $r = 36,000a$ and $k = a$. Thus, $K = \frac{r}{k} = 36,000$, and the solution is

$$y(t) = \frac{y_0 K}{y_0 + (K - y_0)e^{-rt}} = \frac{1,800,000}{50 + 35,950e^{-36,000at}}.$$

To find a, we use the condition $y(2) = 250$ to get $\frac{1,800,000}{50+35,950e^{-72,000a}} = 250$. Solving for a, we obtain $a = \frac{1}{72,000} \ln \frac{719}{143} \approx 0.00002243$. Therefore,

$$y(t) \approx \frac{1,800,000}{50 + 35,950e^{-0.81t}}.$$

(b) $y(10) \approx \dfrac{1,800,000}{50 + 35,950e^{-8.1}} \approx 29,551$

23. The population (in hundreds of deer) is modeled by the differential equation $p' = 0.42p - 0.02p^2$, where t is the time in years. In order to have a population that does not change over time, we must have $p'(t) = 0$ for all t, which means that $0.42p - 0.02p^2 = 0$. The only positive solution is $p = \frac{0.42}{0.02} = 21$, and we conclude that an initial population size of 2,100 deer will not change.

25. (a) Using the chain rule and the fundamental theorem of calculus, we obtain

$$\frac{dp}{dt} = -\frac{Kp_0}{[p_0 + (K - p_0)e^{-\int_0^t r(s)\,ds}]^2}(K - p_0)e^{-\int_0^t r(s)\,ds} \cdot (-r(t))$$

$$= \frac{Kp_0(K - p_0)e^{-\int_0^t r(s)\,ds}r(t)}{[p_0 + (K - p_0)e^{-\int_0^t r(s)\,ds}]^2}.$$

By algebra, we also obtain

$$r(t)p\left(1 - \frac{p}{K}\right) = r(t) \cdot \frac{Kp_0}{p_0 + (K - p_0)e^{-\int_0^t r(s)\,ds}} \cdot \left(1 - \frac{p_0}{p_0 + (K - p_0)e^{-\int_0^t r(s)\,ds}}\right)$$

$$= \frac{Kp_0(K - p_0)e^{-\int_0^t r(s)\,ds}r(t)}{[p_0 + (K - p_0)e^{-\int_0^t r(s)\,ds}]^2}.$$

Thus, $\frac{dp}{dt} = r(t)p(1 - \frac{p}{K})$, so that the function satisfies the differential equation. Checking the initial condition gives $p(0) = \frac{Kp_0}{p_0+(K-p_0)} = p_0$. Therefore, the given function is the solution to the initial value problem.

(b) We have

$$\int_0^t r(s)\,ds = \int_0^t \frac{0.2s}{s+1}\,ds = 0.2\int_0^t \left(1 - \frac{1}{s+1}\right)\,ds = 0.2[s - \ln(s+1)]_0^t = 0.2[t - \ln(t+1)].$$

Thus, $e^{-\int_0^t r(s)\,ds} = e^{-0.2[t-\ln(t+1)]} = (t+1)^{0.2}e^{-0.2t}$, and the solution is

$$p(t) = \frac{10 \cdot 6}{6 + 4(t+1)^{0.2}e^{-0.2t}} = \frac{30}{3 + 2(t+1)^{0.2}e^{-0.2t}}.$$

(c) $p(5) = \dfrac{30}{3 + 2 \cdot 6^{0.2}e^{-1}} \approx 7.4022$ and $\lim\limits_{t\to\infty} p(t) = \dfrac{30}{3} = 10$.

27. (a) Since p must be positive, the only solution of $rp\ln\frac{K}{p} = 0$ is $\frac{K}{p} = 1$. Thus, the only equilibrium solution is $p = K$.

(b) Solving $rp\frac{K-p}{K+ap} = 0$ gives the equilibrium solutions $p = 0$ and $p = K$.

(c) Solving $rp[1 - (\frac{p}{K})^\theta] = 0$ gives the equilibrium solutions $p = 0$ and $p = K$.

(d) From $p(re^{1-p/K} - d) = 0$ we obtain $p = 0$ or $re^{1-p/K} - d = 0$. Solving the last equation gives $e^{1-p/K} = \frac{d}{r}$, or $1 - \frac{p}{K} = \ln\frac{d}{r}$, or $\frac{p}{K} = 1 - \ln\frac{d}{r}$, or $p = K(1 - \ln\frac{d}{r})$. The equilibrium solutions are $p = 0$ and $p = K(1 - \ln\frac{d}{r})$.

EXERCISES 6.5

1. $\lim\limits_{b\to\infty} \displaystyle\int_2^b \frac{dx}{x^5} = \lim\limits_{b\to\infty} -\frac{x^{-4}}{4}\bigg|_2^b = \lim\limits_{b\to\infty}\left(\frac{2^{-4}}{4} - \frac{b^{-4}}{4}\right) = \frac{1}{64} - \lim\limits_{b\to\infty}\frac{b^{-4}}{4} = \frac{1}{64}$

3. $\lim\limits_{a\to-\infty} \displaystyle\int_a^0 e^{0.1t}\,dt = \lim\limits_{a\to-\infty} 10e^{0.1t}\bigg|_a^0 = \lim\limits_{a\to-\infty}[10 - 10e^{0.1a}] = 10$

5. $\lim\limits_{b\to\infty} \displaystyle\int_0^b e^{-0.05t}\,dt = \lim\limits_{b\to\infty} -20e^{-0.05t}\bigg|_0^b = \lim\limits_{b\to\infty}[20 - 20e^{-0.05b}] = 20$

7. $\lim\limits_{b\to\infty} \displaystyle\int_0^b \frac{x^2}{(x^3+1)^2}\,dx = \lim\limits_{b\to\infty} \frac{1}{3}\int_1^{b^3+1} u^{-2}\,du$ (substitution $u = x^3 + 1$)

$$= \lim\limits_{b\to\infty} \frac{-1}{3u}\bigg|_1^{b^3+1} = \lim\limits_{b\to\infty} \frac{-1}{3(b^3+1)} + \frac{1}{3} = \frac{1}{3}$$

9. By integration by parts,

$$\lim_{b\to\infty} \int_0^b te^{-t}\, dt = \lim_{b\to\infty} -(t+1)e^{-t}\Big|_0^b = \lim_{b\to\infty} [1-(b+1)e^{-b}] = 1.$$

11. First, $f(x) = \lambda e^{-\lambda x} \geq 0$ if $\lambda > 0$. Second,

$$\int_0^\infty \lambda e^{-\lambda x}\, dx = \lim_{b\to\infty} \int_0^b \lambda e^{-\lambda x}\, dx = \lim_{b\to\infty} (-e^{-\lambda x})\Big|_0^b = \lim_{b\to\infty} [1-e^{-\lambda b}] = 1.$$

13. We must have $c > 0$ and $1 = \int_0^\infty cx^2 e^{-x}\, dx = c\lim_{b\to\infty} \int_0^b x^2 e^{-x}\, dx$. By integration by parts, $\int x^2 e^{-x}\, dx = -(x^2 + 2x + 2)e^{-x}$, so that

$$\int_0^b x^2 e^{-x}\, dx = -(x^2 + 2x + 2)e^{-x}\Big|_0^b = 2 - (b^2 + 2b + 2)e^{-b}.$$

Therefore, $1 = c \cdot \lim_{b\to\infty} (2 - (b^2 + 2b + 2)e^{-b}) = 2c$, and $c = 1/2$.

15. Using integration by parts with $u = \ln x$ and $dv = dx$, we obtain

$$\int_0^1 \ln x\, dx = \lim_{a\to 0^+} \int_a^1 \ln x\, dx = \lim_{a\to 0^+} (x\ln x - x)\Big|_a^1 = \lim_{a\to 0^+} (-1 - a\ln a + a) = -1.$$

17. The area equals

$$\int_0^1 x^{-1/3}\, dx = \lim_{a\to 0^+} \int_a^1 x^{-1/3}\, dx = \lim_{a\to 0^+} \frac{3}{2}x^{2/3}\Big|_a^1 = \lim_{a\to 0^+} \left(\frac{3}{2} - \frac{3}{2}a^{2/3}\right) = \frac{3}{2}.$$

19. $\int_0^1 \dfrac{dx}{x^2}$ diverges since

$$\int_0^1 \frac{dx}{x^2} = \lim_{a\to 0^+} \int_a^1 \frac{dx}{x^2} = \lim_{a\to 0^+} \left(-\frac{1}{x}\right)\Big|_a^1 = \lim_{a\to 0^+} \left(\frac{1}{a} - 1\right) = \infty.$$

21. Using integration by parts with $u = \ln x$ and $dv = x\, dx$, we have

$$\lim_{a\to 0^+} \int_a^1 x\ln x\, dx = \lim_{a\to 0^+} \left(\frac{x^2}{2}\ln x - \frac{x^2}{4}\right)\Big|_a^1 = \lim_{a\to 0^+} \left(-\frac{a^2}{2}\ln a + \frac{a^2}{4} - \frac{1}{4}\right) = -\frac{1}{4}.$$

Therefore, $\int_0^1 x\ln x\, dx = -\dfrac{1}{4}$.

23. By substitution with $u = x^2 + 1$,

$$\lim_{b\to\infty} \int_0^b \frac{x}{x^2+1}\, dx = \lim_{b\to\infty} \frac{1}{2}\ln(x^2+1)\Big|_0^b = \lim_{b\to\infty} \frac{1}{2}\ln(b^2+1) = \infty.$$

Therefore, $\int_0^\infty \dfrac{x}{x^2+1}\, dx$ diverges.

25. By the partial fraction method,

$$\int_a^2 \frac{dx}{x^2-1} = \frac{1}{2}\int_a^2\left(\frac{1}{x-1}-\frac{1}{x+1}\right)dx = \frac{1}{2}\ln\left|\frac{x-1}{x+1}\right|\Big|_a^2 = \frac{1}{2}\ln\frac{1}{3}-\frac{1}{2}\ln\left|\frac{a-1}{a+1}\right|.$$

Therefore,

$$\int_1^2 \frac{dx}{x^2-1} = \lim_{a\to1+}\int_a^2\frac{dx}{x^2-1} = \lim_{a\to1+}\left[\frac{1}{2}\ln\frac{1}{3}-\frac{1}{2}\ln\left|\frac{a-1}{a+1}\right|\right] = \infty,$$

so that $\int_1^2 \dfrac{dx}{x^2-1}$ diverges.

27. By the partial fraction method,

$$\int_0^b \frac{1}{(x+1)(x+2)}dx = \int_0^b\left(\frac{1}{x+1}-\frac{1}{x+2}\right)dx = \ln\left|\frac{x+1}{x+2}\right|\Big|_0^b = \ln\left|\frac{b+1}{b+2}\right| - \ln\frac{1}{2}.$$

Therefore,

$$\int_0^\infty \frac{1}{(x+1)(x+2)}dx = \lim_{b\to\infty}\left[\ln\left|\frac{b+1}{b+2}\right|-\ln\frac{1}{2}\right] = \ln 1 - \ln\frac{1}{2} = -\ln\frac{1}{2} = \ln 2.$$

29. Using the substitution $u = \ln x$, we obtain

$$\lim_{a\to1+}\int_a^2 \frac{1}{x\ln x}dx = \lim_{a\to1+}\ln(\ln x)\Big|_a^2 = \lim_{a\to1+}[\ln(\ln 2)-\ln(\ln a)] = \infty.$$

Therefore, $\int_1^2 \dfrac{1}{x\ln x}dx$ diverges.

31. For any $p \neq 1$,

$$\int_0^1 \frac{1}{x^p}dx = \lim_{a\to0+}\int_a^1 x^{-p}dx = \lim_{a\to0+}\frac{1}{1-p}x^{1-p}\Big|_a^1 = \lim_{a\to0+}\frac{1}{1-p}(1-a^{1-p}).$$

If $p < 1$, then $\lim\limits_{a\to0+}\dfrac{1}{1-p}(1-a^{1-p}) = \dfrac{1}{1-p}$. If $p > 1$, then $\lim\limits_{a\to0+}\dfrac{1}{1-p}(1-a^{1-p}) = \infty$.

33.
$$\int_0^\infty 12{,}000e^{-0.09t}\,dt = 12{,}000\lim_{b\to\infty}\int_0^b e^{-0.09t}\,dt = 12{,}000\lim_{b\to\infty}\left(\frac{1}{0.09}-\frac{1}{0.09}e^{-0.09b}\right)$$

$$= \frac{12{,}000}{0.09} \approx 133{,}333.33$$

35.
$$\int_0^\infty 100{,}000e^{-0.1t}\,dt = 100{,}000\lim_{b\to\infty}\int_0^b e^{-0.1t}\,dt$$

$$= 100{,}000\lim_{b\to\infty}(10-10e^{-0.1b}) = 1{,}000{,}000$$

37. The total revenue (in dollars) the book is expected to make equals

$$\int_0^\infty 1{,}000{,}000e^{-0.5t}\,dt = 1{,}000{,}000\lim_{b\to\infty}\int_0^b e^{-0.5t}\,dt = 1{,}000{,}000\lim_{b\to\infty}(2-2e^{-0.5b}) = 2{,}000{,}000.$$

39. (a) $\displaystyle\int_0^h A'(t)\,dt = \int_0^h \frac{180}{(t+1)^3}\,dt$

(b) The area is bounded since the total growth (in square millimeters) is

$$\int_0^\infty \frac{180}{(t+1)^3}\,dt = \lim_{h\to\infty}\left(-\frac{180}{2(t+1)^2}\right)\Big|_0^h = \lim_{h\to\infty}\left(90 - \frac{180}{2(h+1)^2}\right) = 90.$$

41. The number of animals the manager needs to establish is

$$\int_0^\infty 75e^{-0.03t}\,dt = 75\lim_{b\to\infty}\int_0^b e^{-0.03t}\,dt = 75\lim_{b\to\infty}\left(\frac{1}{0.03} - \frac{1}{0.03}e^{-0.03b}\right) = 2{,}500.$$

43. Since $\dfrac{1}{\sqrt{x^3+1}} \le \dfrac{1}{\sqrt{x^3}} = \dfrac{1}{x^{3/2}}$ for $1 \le x < \infty$ and $\displaystyle\int_1^\infty \frac{1}{x^{3/2}}\,dx$ converges,

$$\int_0^\infty \frac{1}{\sqrt{x^3+1}}\,dx = \int_0^1 \frac{1}{\sqrt{x^3+1}}\,dx + \int_1^\infty \frac{1}{\sqrt{x^3+1}}\,dx \quad \text{converges.}$$

45. $\displaystyle\int_0^1 \frac{1}{\sqrt{x^3(1-x)}}\,dx$ diverges since

$$\frac{1}{\sqrt{x^3(1-x)}} > \frac{1}{\sqrt{x^3}} = \frac{1}{x^{3/2}} \quad \text{for } 0 < x < 1 \quad \text{and} \quad \int_0^1 \frac{1}{x^{3/2}}\,dx \quad \text{diverges.}$$

47. $\displaystyle\int_0^1 \frac{1}{x\ln(x+1)}\,dx$ diverges since

$$\frac{1}{x\ln(x+1)} \ge \frac{1}{x} \quad \text{for } 0 < x \le 1 \quad \text{and} \quad \int_0^1 \frac{1}{x}\,dx \quad \text{diverges.}$$

49. By the previous exercise, $\int_1^\infty e^{-x^2}\,dx$ converges, and, by symmetry, $\int_{-\infty}^{-1} e^{-x^2}\,dx$ also converges. Since

$$\int_{-\infty}^\infty e^{-x^2}\,dx = \int_{-\infty}^{-1} e^{-x^2}\,dx + \int_{-1}^1 e^{-x^2}\,dx + \int_1^\infty e^{-x^2}\,dx,$$

we conclude that $\int_{-\infty}^\infty e^{-x^2}\,dx$ converges.

51. Observe that $(x^2-1)-(x-1)^2 = 2x-2$. If $x \ge 1$, then $2x-2 \ge 0$, which means that $x^2-1 \ge (x-1)^2$. Thus, if $x \ge 2$, then $\frac{1}{x^2-1} \le \frac{1}{(x-1)^2}$. Since

$$\int_2^\infty \frac{1}{(x-1)^2}\,dx = \lim_{b\to\infty}\int_2^b \frac{1}{(x-1)^2}\,dx = \lim_{b\to\infty}\left(1 - \frac{1}{b-1}\right) = 1,$$

we conclude that $\int_2^\infty \frac{1}{x^2-1}\,dx$ converges.

53.
```
fnInt(1/(1+X²),X
,0,1000)
        1.569796327
■
```

55. It appears that $\int_{-4}^{4} e^{-x^2/2}\, dx \approx \sqrt{2\pi}$, and we guess that $\int_{-\infty}^{\infty} e^{-x^2/2}\, dx = \sqrt{2\pi}$.

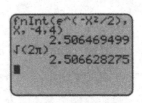

57. The error equals $\int_{0}^{0.0001} \frac{1}{\sqrt{x^2+x}}\, dx$. Since $\frac{1}{\sqrt{x^2+x}} \le \frac{1}{\sqrt{x}}$, we have

$$\int_{0}^{0.0001} \frac{1}{\sqrt{x^2+x}}\, dx \le \int_{0}^{0.0001} \frac{1}{\sqrt{x}} = \lim_{a \to 0^+} 2\sqrt{x}\,\Big|_{a}^{0.0001} = 0.02.$$

59. **(a)** $f_3(0) = e^0 = 1 > 0$, and $f_3'(x) = e^x - \frac{x^2}{2} > 0$ for $x \ge 0$ (from step c of the previous exercise). By the first derivative test, $f_3(x)$ is increasing in $(0, \infty)$, and it follows that $f_3(x) \ge f_3(0) > 0$, and $e^x > \frac{x^3}{3 \cdot 2}$ for $x \ge 0$.

(b) If $x > 0$, dividing both sides of the previous inequality by x^2 gives

$$\frac{e^x}{x^2} \ge \left(\frac{x^3}{3 \cdot 2} \right) \Big/ x^2 = \frac{x}{3 \cdot 2}.$$

Since $\lim_{x \to \infty} \frac{x}{3 \cdot 2} = \infty$, we have $\lim_{x \to \infty} \frac{e^x}{x^2} = \infty$. Therefore,

$$\lim_{x \to \infty} x^2 e^{-x} = \lim_{x \to \infty} \frac{x^2}{e^x} = \frac{1}{e^x/x^2} = 0.$$

61. Set $u = -\ln x$. Then $x = e^{-u}$, and $x \to 0^+ \iff u \to \infty$. Therefore,

$$\lim_{x \to 0^+} x(\ln x)^m = \lim_{u \to \infty} e^{-u} \cdot (-1)^m u^m = 0 \quad \text{for} \quad m = 1, 2, 3, 4.$$

CHAPTER 6: REVIEW EXERCISES

1. The diagram shows that the graphs intersect at $(q_e, p_e) = (400, 9)$. The approximate height of the demand curve above the q-axis is given by the following table.

q	0	50	100	150	200	250	300	350	400
p	20	18	16.3	14.9	13.4	12.1	11	10	9

Using the trapezoidal rule with $\Delta q = 50$ to estimate $\int_0^{400} D(q)\, dq$, we obtain

$$\int_0^{400} D(q)\, dq = \frac{\Delta q}{2}[D(0) + 2D(50) + \cdots + 2D(350) + D(400)] \approx 5{,}510.$$

Therefore,

$$CS = \int_0^{q_e} D(q)\, dq - p_e q_e = \int_0^{400} D(q)\, dq - 3{,}600 \approx 5{,}510 - 3{,}600 = 1{,}910.$$

3. To find the equilibrium point we solve $(q - 5)^2 = q^2 + q + 3$, which reduces to $-10q + 25 = q + 3$, or $11q = 22$. Hence, the equilibrium quantity is $q_e = 2$ and the equilibrium price is $p_e = D(q_e) = (2 - 5)^2 = 9$.

$$CS = \int_0^2 (q - 5)^2\, dq - 18 = \frac{1}{3}(q - 5)^3\,\Big|_0^2 - 18 = \frac{1}{3}(-27 + 125) - 18 = \frac{98}{3} - 18 = \frac{44}{3}$$

5. (a) To find the equilibrium point, we solve $-0.1q + 40 = 0.1q + 20$, or $0.2q = 20$. The equilibrium quantity is $q_e = 100$ and the equilibrium price is $p_e = D(q_e) = -0.1 \cdot 100 + 40 = 30$.

$$CS = \int_0^{100} (-0.1q + 40)\, dq - 3{,}000 = (-0.05q^2 + 40q)\Big|_0^{100} - 3{,}000 = 3{,}500 - 3{,}000 = 500$$

$$PS = 3{,}000 - \int_0^{100} (0.1q + 20)\, dq = 3{,}000 - (0.05q^2 + 20q)\Big|_0^{100} = 3{,}000 - 2{,}500 = 500$$

(b) The new supply curve is $p = 0.1q + 24$, and the demand curve stays the same. Solving $-0.1q + 40 = 0.1q + 24$ gives $q_e = 80$, and $p_e = D(q_e) = 32$.

$$CS = \int_0^{80} (-0.1q + 40)\, dq - 32 \cdot 80 = (-0.05q^2 + 40q)\Big|_0^{80} - 2{,}560 = 2{,}880 - 2{,}560 = 320$$

$$PS = 32 \cdot 80 - \int_0^{80} (0.1q + 24)\, dq = 2{,}560 - (0.05q^2 + 24q)\Big|_0^{80} = 2{,}560 - 2{,}240 = 320$$

7. $\int_0^8 3{,}600 e^{0.06(8-t)}\, dt = -60{,}000 e^{0.06(8-t)}\Big|_0^8 = 60{,}000(e^{0.48} - 1) \approx 36{,}964.46$

9. (a) The present value after 20 years of \$50,000 at 10% interest, compounded continuously, is

$$\int_0^{20} 50{,}000 e^{-0.1t}\, dt = -500{,}000 e^{-0.1t}\Big|_0^{20} = 500{,}000(1 - e^{-2}) \approx 432{,}332.$$

Therefore, option (2), which has a present value of \$500,000 is better.

(b) The present value after 20 years of \$50,000 at 5% interest, compounded continuously, is

$$\int_0^{20} 50{,}000 e^{-0.05t}\, dt = -1{,}000{,}000 e^{-0.05t}\Big|_0^{20} = 1{,}000{,}000(1 - e^{-1}) \approx 632{,}121,$$

which is greater than 500,000. Therefore, Option (1) is better.

11. (a) $PV = \int_0^5 1{,}000 \dfrac{5-t}{t+1} e^{-0.04t}\, dt$

(b) We apply the trapezoidal rule wih $f(t) = \frac{5-t}{t+1} e^{-0.04t}$ and $\Delta t = 1$. The (approximate) values at the endpoints are as follows:

t	0	1	2	3	4	5
$f(t)$	5	1.9216	0.9231	0.4435	0.1704	0

Then $PV \approx 1{,}000(\frac{5}{2} + 1.9216 + 0.9231 + 0.4435 + 0.1704) = 5{,}958.60$.

13. Taking the derivative gives $y' = 10x$, so that

$$xy' - 2y = x \cdot 10x - 2(5x^2 - k) = 2k.$$

Therefore, $xy' - 2y = 8 \iff 2k = 8 \iff k = 4$.

15. Since $dy/dx = xy + x = x(y + 1)$, separating variables gives $dy/(y + 1) = x\, dx$. By integrating, we get $\ln|y + 1| = \frac{1}{2}x^2 + c$, so that $y + 1 = \pm e^c e^{x^2/2} = Ce^{x^2/2}$, where $C = \pm e^c$. Thus, $y = Ce^{x^2/2} - 1$. The initial condition $y(0) = 2$ gives $C - 1 = 2$, or $C = 3$. Therefore, $y = 3e^{x^2/2} - 1$.

17. By separating variables and integrating, we get $\int e^{-y}\,dy = \int x\,dx$. Therefore, $-e^{-y} = \frac{1}{2}x^2 + c$, or $e^{-y} = -\frac{x^2}{2} - c$. Solving for y gives $y = -\ln(-\frac{x^2}{2} - c)$. The initial condition $y(0) = \ln 2$ gives $-\ln(-c) = \ln 2$, or $\ln(-c) = \ln\frac{1}{2}$, or $c = -\frac{1}{2}$. Therefore, $y = -\ln(\frac{1}{2} - \frac{x^2}{2})$.

19. The initial value problem is $\frac{dM}{dt} = 0.09M + 9{,}000$, $M(0) = 2{,}000$. Writing $\frac{dM}{dt} = 0.09(M + 100{,}000)$, we separate variables and integrate to get $\int \frac{dM}{M+100{,}000} = \int 0.09\,dt$, or $\ln|M + 100{,}000| = 0.09t + c$. Therefore, $M + 100{,}000 = Ce^{0.09t}$, where $C = \pm e^c$. Thus, $M = Ce^{0.09t} - 100{,}000$. The initial condition $M(0) = 2{,}000$ gives $C - 100{,}000 = 2{,}000$, or $C = 102{,}000$. Therefore, $M(t) = 102{,}000e^{0.09t} - 100{,}000$ and $M(25) = 102{,}000e^{2.25} - 100{,}000 \approx 867{,}749.06$ dollars.

21. **(a)** Twelve monthly payments of \$900 total \$10,800, considered as a continuous income stream. Therefore, the mortgage balance M satisfies the equation $dM/dt = 0.08M - 10{,}800$. The initial condition $M(0)$ is the original size of the loan, which we must determine, given the condition $M(30) = 0$ (the balance becomes zero in 30 years). By using separation of variables (as in the previous exercise), we obtain the solution $M(t) = Ce^{0.08t} + 135{,}000$, and setting $M(30) = 0$ gives $C = -135{,}000e^{-2.4} \approx -12{,}247$. Therefore, $M(t) = 135{,}000 - 12{,}247e^{0.08t}$, and $M(0) = 135{,}000 - 12{,}247 = 122{,}753$.

   (b) The buyer paid \$10,800 per year for 30 years for a total \$324,000 against a loan of \$122,753. The total interest paid is \$201,247, the difference between those two amounts.

23. Method 1: Since $p_0 = 5$, $r = 1$, $k = 1$, and $K = r/k = 1$, the logistic equation has the solution

$$p(t) = \frac{p_0 K}{p_0 + (K - p_0)e^{-rt}} = \frac{5}{5 - 4e^{-t}}.$$

Method 2: Separate variables and integrate to obtain $\frac{p-1}{p} = Ce^{-t}$ and solve for p to get $p = \frac{1}{1-Ce^{-t}}$. Then use the initial condition, which gives $C = \frac{4}{5}$, and simplify to get $p = \frac{5}{5-4e^{-t}}$.

   Using either method, we have $\lim_{t\to\infty} p(t) = \frac{5}{5} = 1$.

25. The intrinsic growth rate $r = 0.015$ and the damping factor $k = 0.0003$. Therefore, the enviromental carrying capacity is $K = \frac{r}{k} = \frac{0.015}{0.0003} = 50$.

   As an alternate method, rewrite the differential equation in the form $\frac{dp}{dt} = 0.015p(1 - \frac{p}{50}) = rp(1 - \frac{p}{K})$, where K is the carrying capacity. Therefore, $K = 50$.

27. We find the equilibrium solutions $y = 0$ and $y = 5$ by solving $5y - y^2 = 0$.

29. By using the substitution $u = x^3 + 1$, we obtain

$$\int \frac{x^2}{(x^3+1)^4}\,dx = \frac{1}{3}\int u^{-4}\,du = \frac{1}{3}\left(-\frac{1}{3}u^{-3}\right) + c = -\frac{1}{9(x^3+1)^3} + c.$$

Therefore,

$$\int_0^\infty \frac{x^2}{(x^3+1)^4}\,dx = \lim_{b\to\infty}\int_0^b \frac{x^2}{(x^3+1)^4}\,dx = \lim_{b\to\infty}\left(\frac{1}{9} - \frac{1}{9(b^3+1)^3}\right) = \frac{1}{9}.$$

31. We first observe that $f(x) \geq 0$. Then we compute

$$\int_0^\infty 0.5e^{-0.5x}\,dx = \lim_{b\to\infty}\int_0^b 0.5e^{-0.5x}\,dx = \lim_{b\to\infty}\left(-e^{-0.5x}\right)\Big|_0^b = \lim_{b\to\infty}\left(1 - e^{-0.5b}\right) = 1.$$

33. $\displaystyle\int_0^\infty 10{,}000e^{-0.1t}\,dt = 10{,}000 \lim_{b\to\infty}\int_0^b e^{-0.1t}\,dt = 10{,}000 \lim_{b\to\infty}\left(-10e^{-0.1t}\right)\Big|_0^b$

$$= 10{,}000 \lim_{b\to\infty}(10 - 10e^{-0.1b}) = 100{,}000$$

35. Using the substitution $u = e^t - 1$, we get

$$\int \frac{e^t}{\sqrt{e^t - 1}}\,dt = \int \frac{1}{\sqrt{u}}\,du = 2\sqrt{u} = 2\sqrt{e^t - 1} + c.$$

Therefore,

$$\int_0^1 \frac{e^t}{\sqrt{e^t - 1}}\,dt = \lim_{a\to 0^+}\int_a^1 \frac{e^t}{\sqrt{e^t - 1}}\,dt = \lim_{a\to 0^+} 2\sqrt{e^t - 1}\,\Big|_a^1$$

$$= \lim_{a\to 0^+}\left(2\sqrt{e-1} - 2\sqrt{e^a - 1}\right) = 2\sqrt{e-1}.$$

37. Using the substitution $u = e^x$, we get

$$\int \frac{e^x}{e^x + e^{-x}}\,dx = \int \frac{1}{u + u^{-1}}\,du = \int \frac{u}{u^2 + 1}\,du$$

$$= \frac{1}{2}\ln(u^2 + 1) + c \qquad \text{(using the substitution } w = u^2 + 1\text{)}$$

$$= \frac{1}{2}\ln(e^{2x} + 1) + c.$$

Therefore,

$$\int_0^\infty \frac{e^x}{e^x + e^{-x}}\,dx = \lim_{b\to\infty}\int_0^b \frac{e^x}{e^x + e^{-x}}\,dx = \lim_{b\to\infty}\left(\frac{1}{2}\ln(e^{2b} + 1) - \frac{1}{2}\ln 2\right) = \infty.$$

We conclude that $\int_0^\infty \frac{e^x}{e^x + e^{-x}}\,dx$ diverges.

39. Using integration by parts three times, we have

$$\int_0^1 (\ln x)^3\,dx = \lim_{a\to 0^+}\int_a^1 (\ln x)^3\,dx = \lim_{a\to 0^+}\left[x(\ln x)^3 - 3x(\ln x)^2 + 6x\ln x - 6x\right]\Big|_a^1$$

$$= \lim_{a\to 0^+}\left(-6 - a(\ln a)^3 + 3a(\ln a)^2 - 6a\ln a + 6a\right) = -6.$$

41. Since $\frac{1}{x^3+1} < \frac{1}{x^3}$ for $1 \le x < \infty$ and $\int_1^\infty \frac{1}{x^3}\,dx$ converges, we conclude that $\int_1^\infty \frac{1}{x^3+1}\,dx$ converges.

43. If $x \ge 2$, then $\frac{x^3}{2} \ge 4 > 1$. Therefore, $x^3 - 1 \ge \frac{x^3}{2}$ and $\frac{1}{\sqrt{x^3-1}} \le \frac{\sqrt{2}}{x^{3/2}}$ for $2 \le x < \infty$. Since $\int_1^\infty \frac{\sqrt{2}}{x^{3/2}}$ converges, we conclude that $\int_1^\infty \frac{1}{\sqrt{x^3-1}}\,dx$ converges.

45. We first observe that $\ln(x^2 + 1) < x^2$ if $x > 0$. (To check that, let $f(x) = x^2 - \ln(x^2 + 1)$. Then $f(0) = 0$ and $f'(x) = 2x - \frac{2x}{x^2+1} = 2x(\frac{x^2}{x^2+1}) > 0$ if $x > 0$.) Therefore, $\frac{1}{\ln(x^2+1)} > \frac{1}{x^2}$. Since $\int_0^1 \frac{1}{x^2}\,dx$ diverges, we conclude that $\int_0^1 \frac{1}{\ln(x^2+1)}\,dx$ diverges.

47. Step 1: Separate variables and integrate, so that

$$\int \frac{dk}{(A\sqrt{k}+1)(B\sqrt{k}+1)} = \int s\, dt = st + c.$$

Step 2: Substitute $u = \sqrt{k}$. Then $du = \frac{1}{2\sqrt{k}}\, dk$, or $dk = 2\sqrt{k}\, du = 2u\, du$, and the equation becomes

$$\int \frac{2u\, du}{(Au+1)(Bu+1)} = st + c.$$

Step 3: Use partial fractions, $\frac{2u}{(Au+1)(Bu+1)} = \frac{a}{Au+1} + \frac{b}{Bu+1}$.

Clearing denominators, we get

$$2u = a(Bu+1) + b(Au+1) = (aB + bA)u + (a+b),$$

from which we see that $b = -a$ and $a(B - A) = 2$. Therefore, $a = 2/(B - A)$ and $b = -2/(B - A)$, and the equation becomes

$$\frac{2}{B - A} \int \left(\frac{1}{Au+1} - \frac{1}{Bu+1} \right) du = st + c.$$

ntegrating, multiplying by $(B - A)/2$ (which equals $\sqrt{b/s}$), and substituting $u = \sqrt{k}$ gives

$$\frac{1}{A} \ln(A\sqrt{k}+1) - \frac{1}{B} \ln(B\sqrt{k}+1) = \frac{B - A}{2} st + C = (\sqrt{bs})\, t + C.$$

Step 4: Use the initial condition $k(0) = k_0$, to get

$$C = \frac{1}{A} \ln(A\sqrt{k_0}+1) - \frac{1}{B} \ln(B\sqrt{k_0}+1).$$

Step 5: Collecting terms, exponentiating, and observing that

$$\frac{1}{A} \ln(A\sqrt{k}+1) - \frac{1}{B} \ln(B\sqrt{k}+1) = \ln\left[(A\sqrt{k}+1)^{1/A}(B\sqrt{k}+1)^{-1/B} \right]$$

gives the desired solution.

CHAPTER 6: PRACTICE EXAM

1. The producer surplus region is slightly bigger than the triangle with vertices $(0,2)$, $(0,5)$ and $(5,3)$. Thus, the producer surplus is slightly bigger than $\frac{1}{2} \cdot 3 \cdot 3 = \frac{9}{2} = 4.5$. Therefore, 5 is the best approximation, and (d) is the correct answer.

3. To find the equilibrium point, we solve $q + 2 = 16/(q+2)$, or $(q+2)^2 = 16$. Thus, $q + 2 = \pm 4$, and the only positive solution is $q_e = 2$. The equilibrium price is $S(2) = 2 + 2 = 4$. The consumer surplus is equal to

$$\int_0^2 \frac{16}{q+2}\, dq - 2 \cdot 4 = 16 \ln(q+2)\Big|_0^2 - 8 = 16\ln 2 - 8 \approx 3.1.$$

5. The future value of this investment equals $50{,}000e^{(0.08) \cdot 20} = 50{,}000e^{1.6} \approx 247{,}651.62$ dollars.

7. (a) The present value of this 5-year continuous income stream is

$$\int_0^5 360{,}000 e^{-0.09t}\, dt = \left.\frac{360{,}000}{-0.09}e^{-0.09t}\right|_0^5$$

$$= 4{,}000{,}000(1 - e^{-0.45}) \approx 1{,}449{,}487.39 \text{ dollars.}$$

   (b) The lump sum payment of 2,000,000 is better since the other option has a present value of $500{,}000 + 1{,}449{,}487.39 = 1{,}949{,}487.39$.

9. The rate at which the balance $M(t)$ changes is equal to the rate of growth due to interest minus the rate of withdrawal. That is, $dM/dt = 0.07M - 36{,}000$. Her 100th birthday will be 35 years later at, which time her money will run out. That gives the initial condition $M(35) = 0$.

11. Separating variables, we obtain $y\, dy = (x+1)\, dx$, and integrating this equation, we obtain $\frac{1}{2}y^2 = \frac{1}{2}x^2 + x + c$. From the initial condition $y(2) = 4$, we obtain $\frac{1}{2} \cdot 4^2 = \frac{1}{2} \cdot 2^2 + 2 + c$, or $c = 4$. Using this c and solving for y, we obtain the solution $y(x) = \sqrt{x^2 + 2x + 8}$. (Note: we choose the positive square root because $y(2) > 0$.)

13. The intrinsic growth rate is $r = 0.06$ and the damping factor is $k = 0.003$. Thus, the environmental carrying capacity is $K = r/k = 0.06/0.003 = 20$.

15. $$\int_0^\infty 80{,}000 e^{-0.1t}\, dt = \lim_{b \to \infty}\left.\left(-\frac{80{,}000}{0.1}e^{-0.1t}\right)\right|_0^b$$

$$= \lim_{b \to \infty} 800{,}000(1 - e^{-0.1b}) = 800{,}000$$

17. Let $u = x - 2$. Then $du = dx$ and $x = a \iff u = a - 2$, and

$$\int_2^3 (x-2)^{-1/4}\, dx = \lim_{a \to 2+}\int_a^3 (x-2)^{-1/4}\, dx = \lim_{a \to 2+}\int_{a-2}^1 u^{-1/4}\, du$$

$$= \lim_{a \to 2+}\left.\frac{4}{3}u^{3/4}\right|_{a-2}^1 = \lim_{a \to 2+}\left(\frac{4}{3} - \frac{4}{3}(a-2)^{3/4}\right) = \frac{4}{3}.$$

CHAPTER 7

Functions of Several Variables and Applications

EXERCISES 7.1

1.

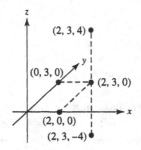

3.

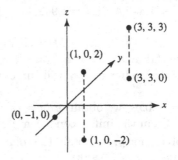

• **5.** Let (x, y, z) be the coordinates of the final position of that point. Then $x = 5 - 2 = 3$, $y = 4 - 3 = 1$, and $z = 8 - 7 + 1 = 2$. Hence the coordinates of the new point are $(3, 1, 2)$.

7.

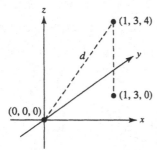

$$d = \sqrt{(0-1)^2 + (0-3)^2 + (0-4)^2} = \sqrt{26}$$

9.

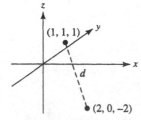

$$d = \sqrt{(1-2)^2 + (1-0)^2 + (1-(-2))^2} = \sqrt{11}$$

11. The distances between the three pairs of points are:

$$d((0,0,0),(3,2,0)) = \sqrt{(3-0)^2 + (2-0)^2 + (0-0)^2} = \sqrt{13}$$
$$d((3,2,0),(1,5,1)) = \sqrt{(1-3)^2 + (5-2)^2 + (1-0)^2} = \sqrt{14}$$
$$d((1,5,1),(0,0,0)) = \sqrt{(0-1)^2 + (0-5)^2 + (0-1)^2} = \sqrt{27}.$$

Since $(\sqrt{13})^2 + (\sqrt{14})^2 = (\sqrt{27})^2$, the three points determine a right triangle.

13. The distances between the three pairs of points are:

$$d((0,3,-5),(4,-1,5)) = \sqrt{(0-4)^2 + (3+1)^2 + (-5-5)^2} = \sqrt{132}$$

$$d((4,-1,5),(10,-4,2)) = \sqrt{(10-4)^2 + (-4+1)^2 + (2-5)^2} = \sqrt{54}$$

$$d((10,-4,2),(0,3,-5)) = \sqrt{(0-10)^2 + (3+4)^2 + (-5-2)^2} = \sqrt{198}.$$

Since $\sqrt{132} + \sqrt{54} \neq \sqrt{198}$, the points do not determine a right triangle.

15. The distances (in hundreds of meters) are as follows.

Day 1: $\sqrt{0.6^2 + (-3)^2 + (-0.02)^2} = \sqrt{9.3604} \approx 3.06$

Day 2: $\sqrt{(-1.3)^2 + (-6)^2 + (-0.3)^2} = \sqrt{37.78} \approx 6.15$

Day 3: $\sqrt{(-10.6)^2 + (-4)^2 + 0^2} = \sqrt{128.36} \approx 11.33$

Day 4: $\sqrt{(-21)^2 + 2^2 + (-0.4)^2} = \sqrt{445.16} \approx 21.10$

Day 5: $\sqrt{(-39)^2 + 7^2 + (-0.09)^2} = \sqrt{1570.0081} \approx 39.62$

The z-axis represents the change in depth from the original position.

17. $f(1,0) = 8 - 2 \cdot 1 + 4 \cdot 0 = 6, \quad f(0,1) = 8 - 2 \cdot 0 + 4 \cdot 1 = 12,$
    $f(-3,2) = 8 - 2 \cdot (-3) + 4 \cdot 2 = 22, \quad f(-\frac{1}{2},-\frac{1}{4}) = 8 - 2 \cdot (-\frac{1}{2}) + 4 \cdot (-\frac{1}{4}) = 8$

19. $f(0,0) = 2 \cdot 0^2 - 3 \cdot 0 \cdot 0 + 5 \cdot 0^2 - 0 = 0, \quad f(0,1) = 2 \cdot 0^2 - 3 \cdot 0 \cdot 1 + 5 \cdot 1^2 - 0 = 5,$
    $f(-2,3) = 2 \cdot (-2)^2 - 3 \cdot (-2) \cdot 3 + 5 \cdot 3^2 - (-2) = 73,$
    $f(2,-3) = 2 \cdot 2^2 - 3 \cdot 2 \cdot (-3) + 5 \cdot (-3)^2 - 2 = 69$

21. $r(1,0) = \dfrac{2 \cdot 1 \cdot 0}{1^2 + 0^2} = 0, \quad r(-3,4) = \dfrac{2 \cdot (-3) \cdot 4}{(-3)^2 + 4^2} = -\dfrac{24}{25},$
    $r(\sqrt{5},\sqrt{5}) = \dfrac{2 \cdot \sqrt{5} \cdot \sqrt{5}}{(\sqrt{5})^2 + (\sqrt{5})^2} = 1, \quad r(8,8) = \dfrac{2 \cdot 8 \cdot 8}{8^2 + 8^2} = 1, \quad r(a,a) = \dfrac{2 \cdot a \cdot a}{a^2 + a^2} = 1$

23. $l(1,0) = \ln\sqrt{1^2 + 0^2} = 0, \quad l(0,-1) = \ln\sqrt{0 + (-1)^2} = 0,$
    $l(-e,0) = \ln\sqrt{(-e)^2 + 0^2} = \ln e = 1, \quad l(-2,-3) = \ln\sqrt{(-2)^2 + (-3)^2} = \ln\sqrt{13} = \dfrac{1}{2}\ln 13$

25. $g(x,y) = (y-x)^{-1/2}$ is defined for (x,y) with $y - x > 0$. The natural domain of $g(x,y)$ is set of all (x,y) with $x < y$.

27. $f(x,y) = \frac{1}{1-x-y}$ is defined for (x,y) with $1 - x - y \neq 0$. The natural domain of $f(x,y)$ is the set of all (x,y) with $x + y \neq 1$.

29. $f(x,y) = \frac{1}{\sqrt{xy-x}}$ is defined for (x,y) with $xy - x = x(y-1) > 0$. The natural domain of $f(x,y)$ is the set of all (x,y) with either $x < 0$ and $y < 1$ or $x > 0$ and $y > 1$.

31. Solving the equation $3 + 2x + 3y = z_0$ for y, we see that the level curves are parallel lines of the form $y = -\frac{2}{3}x - 1 + \frac{z_0}{3}$.

The x-section at $x = 1$ is the line $z = 5 + 3y$; the y-section at $y = 0$ is the line $z = 3 + 2x$.

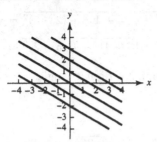
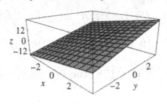

33. Solving the equation $x - y^2 = z_0$ for x, we see that the level curves are parabolas (with horizontal axes of symmetry) of the form $x = y^2 + z_0$.

The x-section at $x = 0$ is the parabola $z = -y^2$; the y-section at $y = -2$ is the line $z = x - 4$.

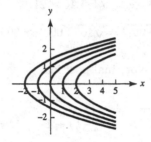
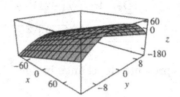

35. Writing the equation $1 + x^2 + y^2 = z_0$ in the form $x^2 + y^2 = z_0 - 1$, we see that the level curves are circles (centered at the origin), if $z_0 > 1$. If $z_0 = 1$, then the level curve is the point $(0, 0)$. If $z_0 < 1$, then the level curve is the empty set.

The x-section at $x = 2$ is the parabola $z = y^2 + 5$; the y-section at $y = -3$ is the parabola $z = x^2 + 10$.

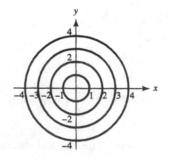
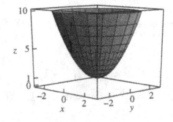

37. For $z_0 \neq 0$, the level curves are called *hyperbolas* and are of the form $xy = z_0$. If $z_0 = 0$ the level curve consist of the two coordinate axes.

The x-section at $x = 1$ is the line $z = y$; the y-section at $y = -1$ is the line $z = -x$.

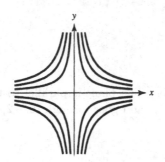

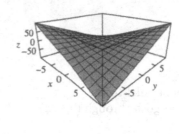

39. (a) When $L = L_0$, we have $Q = Q(r, L_0) = \pi r^4 P/8 L_0 v = Kr^4$, where $K = \pi P/8 L_0 v$ (a positive constant).

(b) When $r = r_0$, we have $Q = Q(r_0, L) = \pi r_0^4 P/8 L v = \frac{K}{L}$, where $K = \pi r_0^4 P/8v$ (a positive constant).

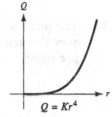

$Q = Kr^4$

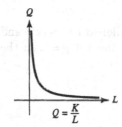

$Q = \frac{K}{L}$

41. (a) The revenue (in dollars) is given by the formula $R(x, y) = 2{,}000x + 3{,}000y$, where x is the number of units of product X sold and y is the number of units of product Y.

(b) If we fix R at 60,000 we get the equation $2{,}000x + 3{,}000y = 60{,}000$, or $y = -\frac{2}{3}x + 20$.

When fix R at 120,000, we get the equation $2{,}000x + 3{,}000y = 120{,}000$, or $y = -\frac{2}{3}x + 40$.

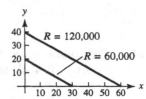

43. If we fix M at 5,000 dollars, we obtain the equation $1{,}000e^{rt} = 5{,}000$. Solving for t in terms of r, we get $t = \frac{\ln 5}{r}$. If $r = 0.08$, then the needed time is $t = \frac{\ln 5}{r} = \frac{\ln 5}{0.08} \approx 20.12$ years.

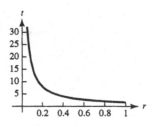

45. The equation $x = 3$ puts no restrictions on y and z. So its graph is
the plane of all points $(3, y, z)$, which is the plane through $(3, 0, 0)$
and parallel to the y, z-plane.

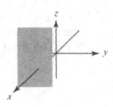

47. The plane is parallel to the
x, y-plane and goes through $(0, 0, 4)$.

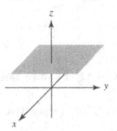

49. The plane is parallel to the x, z-plane
and goes through $(0, -2, 0)$.

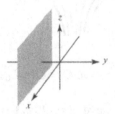

51. The plane is parallel to the z-axis and
goes through the line $x + y = 2$ in the
x, y-plane.

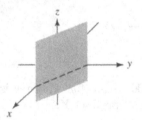

53. The plane is parallel to the x-axis and
goes through the line $y + z = 1$ in the
y, z-plane.

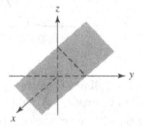

55. The plane goes through the points
$(6, 0, 0)$, $(0, 4, 0)$, and $(0, 0, 2)$.

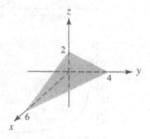

57. The plane goes through the points
$(3, 0, 0)$, $(0, 2, 0)$, and $(0, 0, -6)$.

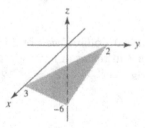

59. In this case, $(x_0, y_0, z_0) = (4, -3, 3)$, $a = 2$, and $b = 5$. Therefore, the equation of the plane is
$z - 3 = 2(x - 4) + 5(y + 3)$, or $z = 2x + 5y + 10$.

61. Since $(x_0, y_0, z_0) = (-4, -2, \frac{9}{2})$ and $a = -\frac{5}{2}$, $b = \frac{7}{2}$, the equation of the plane is $z - \frac{9}{2} = -\frac{5}{2}(x + 4) + \frac{7}{2}(y + 2)$, or $z = -\frac{5}{2}x + \frac{7}{2}y + \frac{3}{2}$.

63. Let $(x_0, y_0, z_0) = (1, 0, -1)$. Then $z - (-1) = a(x - 1) + b(y - 0)$, or $z + 1 = a(x - 1) + by$. To find a and b, we substitute

$$(x, y, z) = (0, 1, 3) \qquad \text{to get} \quad 3 + 1 = a(0 - 1) + b, \quad \text{or} \quad 4 = -a + b,$$
$$(x, y, z) = (1, -1, -2) \quad \text{to get} \quad -2 + 1 = a(1 - 1) + b(-1), \quad \text{or} \quad -1 = -b.$$

Thus, $a = -3$ and $b = 1$. The equation of the plane is $z + 1 = -3(x - 1) + y$, or $z = -3x + y + 2$.

65. Let $(x_0, y_0, z_0) = (2, -3, -1)$. Then $z + 1 = a(x - 2) + b(y + 3)$. To find a and b, we substitute

$$(x, y, z) = (-2, -4, 4) \quad \text{to get} \quad 5 = -4a - b, \text{ or } 4a + b = -5,$$
$$(x, y, z) = (1, \tfrac{1}{2}, 4) \qquad \text{to get} \quad 5 = -a + \tfrac{7}{2}b, \text{ or } a - \tfrac{7}{2}b = -5.$$

Solving for a and b, we get $a = -\frac{3}{2}$ and $b = 1$. The equation of the plane is $z + 1 = -\frac{3}{2}(x - 2) + (y + 3)$, or $z = -\frac{3}{2}x + y + 5$.

67. (a) Since the enrollment decreases as the tuition increases, the sign of the slope in the tuition direction is negative.

(b) Since the enrollment increases as the reputation increases, the sign of the slope in the reputation direction is positive.

(c) The school would enroll 2,000 students if both the reputation level and the tuition were zero. Therefore, the plane passes through the point $(0, 0, 2,000)$, which gives the coordinates of the enrollment intercept.

69. To find the intersection set of these three planes, we must solve the equations

$$x - 2y + 2z = 4$$
$$x - y + z = 3$$
$$2x - 3y + 3z = 0.$$

Subtracting the first equation from the second gives $y - z = -1$. Multiplying the first equation by 2 and then subtracting from the third gives $y - z = -8$. Since these two equations are incompatible (i.e., have no common solution), we conclude that the three original equations have no common solution, and therefore the planes do not intersect.

71. To find the intersection set of these planes, we must solve the equations

$$x + y - z = 4$$
$$x + 3y + z = 10$$
$$x + 2y + z = 8$$
$$x + z = 4.$$

Subtracting the fourth equation from the third gives $2y = 4$, or $y = 2$. Substituting that into each of the first three equation reduces the system to the set of equations

$$x - z = 2$$
$$x + z = 4,$$

whose solution is $x = 3$, $z = 1$. Therefore, the four planes intersect at the point $(3, 2, 1)$.

73. Let $C(x, y)$ be the cost of producing x units of the first item and y units of the second. Write $C(x, y) = ax + by + c$ for some constants a, b, and c. To find them, we substitute the data given in the table, as follows:

$$800 = C(200, 100) = 200a + 100b + c$$
$$900 = C(200, 150) = 200a + 150b + c$$
$$850 = C(300, 100) = 300a + 100b + c.$$

The solution of this system of equations is $a = \frac{1}{2}$, $b = 2$, and $c = 500$. All other data satisfy $C(x, y) = \frac{1}{2}x + 2y + 500$. Therefore, the cost function is $C(x, y) = \frac{1}{2}x + 2y + 500$. (Note: You can pick any three points from the table for the calculation of a, b and c.)

75. The following pictures were created with Mathematica, using the commands as shown.

(a) The command for the graph:

```
In[4]:= Plot3D[{y^2 - x^2, GrayLevel[0.75]}, {x, -2, 2}, {y, -2, 2},
          ViewPoint ->{1.2, -1.2, .5},
        AxesLabel → "x", "y", "z"}, Ticks → {{-1, 1}, {-1, 1}, {-2, 2}}]
```

The command for the level curves (contour plot):

```
In[43]:= ContourPlot[y^2 - x^2, {x, -2, 2}, {y, -2, 2}, ContourShading → False, Contours → 20]
```

The graph and level curves:

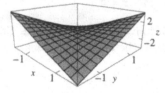
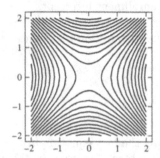

(b) The command for the graph:

```
In[29]:= Plot3D[{x*y, GrayLevel[0.75]}, {x, -2, 2}, {y, -2, 2},
          ViewPoint ->{1.2, -1.2, .5},
        AxesLabel → { "x", "y", "z"}, Ticks → {{-1, 1}, {-1, 1}, {-2, 2}}]
```

The command for the level curves (contour plot):

```
In[41]:= ContourPlot[x*y, {x, -2, 2}, {y, -2, 2}, ContourShading → False, Contours → 30]
```

The graph and level curves:

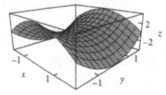
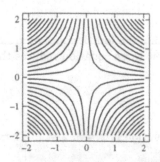

(c) The command for the graph:

```
In[1]:= Plot3D[{x^3 + 3 x * y^2 - 3 x^2 - 3 y^2, GrayLevel[0.75]}, {x, -5, 5}, {y, -5, 5},
            ViewPoint ->{1.2, -1.2, .5}, AxesLabel →} "x", "y", "z"},
        Ticks → {{4, 0, 4}, {-4, 0, 4}, {-400, 0, 200}}]
```

The command for the level curves (contour plot):

```
In[2]:= ContourPlot[x^3 + 3x*y^2 - 3x^2 - 3y^2,
        {x, -5, 5}, {y, -5, 5}, ContourShading → False, Contours → 45]
```

The graph and level curves:

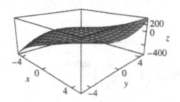 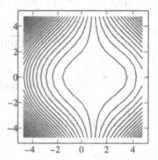

(d) The command for the graph:

```
In[67]:= Plot3D[{Exp[-x^2] + Exp[-9 y^2], GrayLevel[0.75]}, {x, -2, 2}, {y, -2, 2},
            ViewPoint ->{1, -1, .5}, AxesLabel → {"x", "y", "z"},
        PlotRange → {-.5, 2}, Ticks → {-1, 0, 1, 2}, {-1, 0, 1, 2}, {0, 1, 2}}]
```

The command for the level curves (contour plot):

```
In[69]:= ContourPlot[Exp[-x^2] + Exp[-9y^2], {x, -2, 2},
        {y, -2, 2}, ContourShading → False, Contours → 15]
```

The graph and level curves:

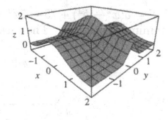 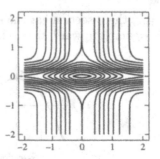

EXERCISES 7.2

1. $\dfrac{\partial f}{\partial x} = -8, \quad \dfrac{\partial f}{\partial y} = 5$

3. $\dfrac{\partial f}{\partial x} = 3x^2y^2 - 4xy, \quad \dfrac{\partial f}{\partial y} = 2x^3y - 2x^2 + 1$

5. $\dfrac{\partial M}{\partial r} = 1{,}000te^{rt}, \quad \dfrac{\partial M}{\partial t} = 1{,}000re^{rt}$

7. $\dfrac{\partial R}{\partial s} = -\dfrac{5t^2}{s^2}, \quad \dfrac{\partial R}{\partial t} = \dfrac{10t}{s}$

9. $\dfrac{\partial l}{\partial x} = \dfrac{1}{x^2 - 5y + 3} \dfrac{\partial}{\partial x}(x^2 - 5y + 3) = \dfrac{2x}{x^2 - 5y + 3}$,

$\dfrac{\partial l}{\partial y} = \dfrac{1}{x^2 - 5y + 3} \dfrac{\partial}{\partial y}(x^2 - 5y + 3) = -\dfrac{5}{x^2 - 5y + 3}$

11. $\dfrac{\partial Q}{\partial K} = 4K^{-0.6}L^{0.6}$, $\quad \dfrac{\partial Q}{\partial L} = 6K^{0.4}L^{-0.4}$

13. $\dfrac{\partial M}{\partial t} = 1{,}000\left(1 + \dfrac{r}{360}\right)^{360t}\left[\ln\left(1 + \dfrac{r}{360}\right)\right] \cdot 360$

$\qquad = 360{,}000\left(1 + \dfrac{r}{360}\right)^{360t}\ln\left(1 + \dfrac{r}{360}\right)$

$\dfrac{\partial M}{\partial r} = 1{,}000 \cdot 360t\left(1 + \dfrac{r}{360}\right)^{360t-1}\left(\dfrac{1}{360}\right) = 1{,}000t\left(1 + \dfrac{r}{360}\right)^{360t-1}$

15. $\dfrac{\partial f}{\partial x} = 2x + y$, $\quad \dfrac{\partial f}{\partial x}(2,1) = 2 \cdot 2 + 1 = 5$

17. $\dfrac{\partial f}{\partial x} = 50x^{-1/2}y^{1/2}$, $\quad \dfrac{\partial f}{\partial x}(25,16) = 50 \cdot 25^{-1/2} \cdot 16^{1/2} = 50 \cdot \dfrac{1}{5} \cdot 4 = 40$

19. Since the flavor decreases as time increases, we should have $\partial F/\partial t < 0$. However, for $F(t,q)$ as given, $\partial F/\partial t = 2t + 2q \geq 0$ (since t and q are both non-negative). We conclude that the formula is not an accurate model.

21. **(a)** $\dfrac{\partial T}{\partial x} = 10x$, $\quad \dfrac{\partial T}{\partial y} = -10y$

(b) If $x = a$, the x-section has the equation $T = 5(a^2 - y^2) + 75 = -5y^2 + (75 + 5a^2)$. It is a parabola, opening downward, that achieves a maximum at $y = 0$. The case of $a = 0$ is shown below.

(c) If $y = b$, the y-section has the equation $T = 5(x^2 - b^2) + 75 = 5x^2 + (75 - 5b^2)$. It is a parabola, opening upward, that achieves a minimum at $x = 0$. The case of $b = 5$ is shown below.

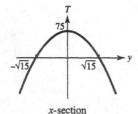

x-section

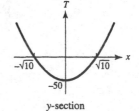

y-section

(d) If there were a local minimum (or maximum), both the x- and y-sections through the point would have a local minimum (or maximum), which cannot be the case.

23. The equation of the tangent plane has the form

$$z = f(x_0, y_0) + \dfrac{\partial f}{\partial x}(x_0, y_0)(x - x_0) + \dfrac{\partial f}{\partial y}(x_0, y_0)(y - y_0).$$

Since $\frac{\partial f}{\partial x} = 6x$ and $\frac{\partial f}{\partial y} = 10y$, we have $\frac{\partial f}{\partial x}(1,2) = 6$ and $\frac{\partial f}{\partial y}(1,2) = 20$. Also, $f(1,2) = 3 \cdot 1 + 5 \cdot 4 = 23$. The equation of the tangent plane is given by $z = 23 + 6(x - 1) + 20(y - 2) = -23 + 6x + 20y$, and the linear approximation is $f(x,y) \approx -23 + 6x + 20y$.

25. Since $\frac{\partial f}{\partial x} = 2xe^{x^2-y^4}$ and $\frac{\partial f}{\partial y} = -4y^3e^{x^2-y^4}$, we have $\frac{\partial f}{\partial x}(1,-1) = 2$ and $\frac{\partial f}{\partial y}(1,-1) = 4$. Also, $f(1,-1) = e^0 = 1$. The equation of the tangent plane is given by $z = 1 + 2(x-1) + 4(y+1) = 3 + 2x + 4y$, and the linear approximation is $f(x,y) \approx 3 + 2x + 4y$.

27. Since $\frac{\partial f}{\partial x} = 4x^{-1/2}y^2$ and $\frac{\partial f}{\partial y} = 16x^{1/2}y$, we have $\frac{\partial f}{\partial x}(4,1) = 2$ and $\frac{\partial f}{\partial y}(4,1) = 32$. Also, $f(4,1) = 16$. The equation of the tangent plane is given by $z = 16 + 2(x-4) + 32(y-1) = -24 + 2x + 32y$, and the linear approximation is $f(x,y) \approx -24 + 2x + 32y$.

29. The equation of the tangent plane has the form

$$z = f(x_0, y_0) + \frac{\partial f}{\partial x}(x_0, y_0)(x - x_0) + \frac{\partial f}{\partial y}(x_0, y_0)(y - y_0).$$

In this case, we have $z = 8 + 5(x-0) + (-3)(y-0)$, which simplifies to $z = 8 + 5x - 3y$. The linear approximation is $f(x,y) \approx 5x - 3y + 8$.

31. $z = -2 + 2(x+3) - \frac{8}{7}(y-5) = 2x - \frac{8}{7}y + \frac{68}{7}$

The linear approximation is $f(x,y) \approx 2x - \frac{8}{7}y + \frac{68}{7}$.

33. The volume of a rectangular box whose height is x inches and whose base is a square with each side y inches long equals $V(x,y) = xy^2$.

(a) $V(6,4) = 6 \cdot 4^2 = 96$ cubic inches.

(b) Since $\frac{\partial V}{\partial x} = y^2$ and $\frac{\partial V}{\partial y} = 2xy$, we obtain $\frac{\partial V}{\partial x}(6,4) = 16$ and $\frac{\partial V}{\partial y}(6,4) = 48$. The linear approximation formula gives

$$V(6.02,\ 4.01) - V(6,4) \approx \frac{\partial V}{\partial x}(6,4) \cdot (6.02 - 6) + \frac{\partial V}{\partial y}(6,4) \cdot (4.01 - 4)$$

$$= 16 \cdot (0.02) + 48 \cdot (0.01) = 0.8.$$

35. Since $\frac{\partial V}{\partial r} = 2\pi rh$ and $\frac{\partial V}{\partial h} = \pi r^2$, linear approximation gives

$$V(r,h) - V(r_0, h_0) \approx \frac{\partial V}{\partial r}(r_0, h_0)(r - r_0) + \frac{\partial V}{\partial h}(r_0, h_0)(h - h_0)$$

$$= 2\pi r_0 h_0(r - r_0) + \pi r_0^2(h - h_0).$$

If $r - r_0 = 1$ and $h = h_0$ (as given in the exercise), then the change in volume is approximately $2\pi r_0 h_0$. The actual change is

$$V(r_0 + 1, h_0) - V(r_0, h_0) = \pi(r_0 + 1)^2 h_0 - \pi r_0^2 h_0 = \pi h_0(2r_0 + 1) = 2\pi r_0 h_0 + \pi h_0.$$

37. (a) Given $V(P,T) = \frac{0.5T}{P}$, we have $\frac{\partial V}{\partial P} = -\frac{0.5T}{P^2}$ and $\frac{\partial V}{\partial T} = \frac{0.5}{P}$.

(b) By linear approximation,

$$V(22,396) - V(20,400) \approx \frac{\partial V}{\partial P}(20,400)(22 - 20) + \frac{\partial V}{\partial T}(20,400)(396 - 400)$$

$$= -\left(\frac{0.5 \cdot 400}{20^2}\right) \cdot 2 - \left(\frac{0.5}{20}\right) \cdot 4 = -1.1.$$

39. Since $\frac{\partial P}{\partial K} = 6K^{-0.7}L^{0.7}$ and $\frac{\partial P}{\partial L} = 14K^{0.3}L^{-0.3}$,

$$MPK = \frac{\partial P}{\partial K}(1{,}000,\ 4{,}000) = 6 \cdot (1{,}000)^{-0.7} \cdot (4{,}000)^{0.7} = 6 \cdot \left(\frac{4{,}000}{1{,}000}\right)^{0.7} \approx 15.83$$

$$MPL = \frac{\partial P}{\partial L}(1{,}000,\ 4{,}000) = 14 \cdot (1{,}000)^{0.3} \cdot (4{,}000)^{-0.3} = 14 \cdot \left(\frac{1{,}000}{4{,}000}\right)^{0.3} \approx 9.24.$$

41. $\dfrac{\partial f}{\partial x} = \dfrac{2x}{y}, \qquad \dfrac{\partial f}{\partial y} = -\dfrac{x^2}{y^2}, \qquad \dfrac{\partial^2 f}{\partial x^2} = \dfrac{2}{y}, \qquad \dfrac{\partial^2 f}{\partial y^2} = \dfrac{2x^2}{y^3},$

$\dfrac{\partial^2 f}{\partial y \partial x} = \dfrac{\partial}{\partial y}\left(\dfrac{2x}{y}\right) = -\dfrac{2x}{y^2}, \qquad \dfrac{\partial^2 f}{\partial x \partial y} = \dfrac{\partial}{\partial x}\left(-\dfrac{x^2}{y^2}\right) = -\dfrac{2x}{y^2}$

43. $\dfrac{\partial f}{\partial x} = \ln y, \qquad \dfrac{\partial f}{\partial y} = \dfrac{x}{y}, \qquad \dfrac{\partial^2 f}{\partial x^2} = 0, \qquad \dfrac{\partial^2 f}{\partial y^2} = -\dfrac{x}{y^2}$

$\dfrac{\partial^2 f}{\partial y \partial x} = \dfrac{\partial}{\partial y}\ln y = \dfrac{1}{y}, \qquad \dfrac{\partial^2 f}{\partial x \partial y} = \dfrac{\partial}{\partial x}\left(\dfrac{x}{y}\right) = \dfrac{1}{y}$

45. $\dfrac{\partial f}{\partial x}(1, 2) \approx \dfrac{f(1.5, 2) - f(1, 2)}{0.5} = \dfrac{10 - 8}{0.5} = 4,$

$\dfrac{\partial f}{\partial y}(1, 2) \approx \dfrac{f(1, 2.6) - f(1, 2)}{0.6} = \dfrac{8.3 - 8}{0.6} = 0.5$

47. $\dfrac{\partial f}{\partial x}(5, 6) \approx \dfrac{f(4.5, 6) - f(5, 6)}{-0.5} = \dfrac{20 - 21}{-0.5} = 2,$

$\dfrac{\partial f}{\partial y}(5, 6) \approx \dfrac{f(5, 5.7) - f(5, 6)}{-0.3} = \dfrac{23 - 21}{-0.3} = -\dfrac{20}{3}$

49. $\dfrac{\partial f}{\partial x}(0, 0) \approx \dfrac{f(0.05, 0) - f(-0.05, 0)}{0.1} = \dfrac{7.8 - 8}{0.1} = -2,$

$\dfrac{\partial f}{\partial y}(0, 0) \approx \dfrac{f(0, 0.05) - f(0, -0.05)}{0.1} = \dfrac{0 - (-1)}{0.1} = 10$

51. Let $T(x, y)$ for the rising time (in minutes), where x is the sugar content (in teaspoons) and y is the temperature (in degrees Fahrenheit).

(a) $\dfrac{\partial T}{\partial y}(2, 89) \approx \dfrac{T(2, 91) - T(2, 87)}{4} = \dfrac{56 - 59}{4} = -0.75$

(b) $\dfrac{\partial T}{\partial x}(2, 89) \approx \dfrac{T(2.25, 89) - T(1.75, 89)}{0.5} = \dfrac{51 - 63}{0.5} = -24$

(c) $T\left(2\dfrac{1}{8}, 90\right) \approx T(2, 89) + \dfrac{\partial T}{\partial x}(2, 89)\left(2\dfrac{1}{8} - 2\right) + \dfrac{\partial T}{\partial y}(2, 89)(90 - 89)$

$$= 57 - 24 \cdot \dfrac{1}{8} - 0.75 = 53.25 \text{ minutes}$$

53. $\dfrac{\partial f}{\partial x} = -\dfrac{z}{(x+y^2)^2}\dfrac{\partial}{\partial x}(x+y^2) = -\dfrac{z}{(x+y^2)^2},$

$\dfrac{\partial f}{\partial y} = -\dfrac{z}{(x+y^2)^2}\dfrac{\partial}{\partial y}(x+y^2) = -\dfrac{2yz}{(x+y^2)^2}, \qquad \dfrac{\partial f}{\partial z} = \dfrac{1}{x+y^2}$

55. $\dfrac{\partial f}{\partial x} = y^2 \ln z, \qquad \dfrac{\partial f}{\partial y} = 2xy \ln z, \qquad \dfrac{\partial f}{\partial z} = \dfrac{xy^2}{z}$

57. $\dfrac{\partial f}{\partial x} = \dfrac{x}{\sqrt{x^2+y^2+z^2}}, \qquad \dfrac{\partial f}{\partial y} = \dfrac{y}{\sqrt{x^2+y^2+z^2}}, \qquad \dfrac{\partial f}{\partial z} = \dfrac{z}{\sqrt{x^2+y^2+z^2}}$

59. The volume of a rectangular box of length x, width y, and height z is given by $V(x,y,z) = xyz$. Then $V(10,12,5) = 600$, and the partial derivatives are

$$\frac{\partial V}{\partial x} = yz, \qquad \frac{\partial V}{\partial y} = xz, \qquad \frac{\partial V}{\partial z} = xy,$$

so that

$$\frac{\partial V}{\partial x}(10,12,5) = 60, \qquad \frac{\partial V}{\partial y}(10,12,5) = 50, \qquad \frac{\partial V}{\partial z}(10,12,5) = 120.$$

(a) $V(11,12,5) \approx V(10,12,5) + \dfrac{\partial V}{\partial x}(10,12,5)(11-10) = 600 + 60 = 660$

(b) $V(9,13,5.25) \approx V(10,12,5) + \dfrac{\partial V}{\partial x}(10,12,5)(9-10) + \dfrac{\partial V}{\partial y}(10,12,5)(13-12)$

$$+ \frac{\partial V}{\partial z}(10,12,5)(5.25-5) = 600 + 60\cdot(-1) + 50\cdot 1 + 120\cdot 0.25 = 620$$

61. The first-order partial derivatives are

$$\frac{\partial u}{\partial x} = \frac{1}{2}[f'(x+ct) + f'(x-ct)] \quad \text{and} \quad \frac{\partial u}{\partial t} = \frac{1}{2}[cf'(x+ct) - cf'(x-ct)],$$

so that

$$\frac{\partial^2 u}{\partial x^2} = \frac{1}{2}[f''(x+ct) + f''(x-ct)] \quad \text{and} \quad \frac{\partial^2 u}{\partial t^2} = \frac{1}{2}[c^2 f''(x+ct) + c^2 f''(x-ct)].$$

Comparing them, we see that $\frac{\partial^2 u}{\partial t^2} = c^2 \frac{\partial^2 u}{\partial x^2}$. Checking the initial conditions gives

$$u(x,0) = \frac{1}{2}[f(x) + f(x)] = f(x) \quad \text{and} \quad \frac{\partial u}{\partial t}(x,0) = \frac{1}{2}[cf'(x) - cf'(x)] = 0.$$

EXERCISES 7.3

1. We must solve the system of equations

$$\frac{\partial f}{\partial x} = 2x - 2 = 0$$

$$\frac{\partial f}{\partial y} = 2y + 4 = 0.$$

From the first equation we get $x = 1$, and from the second we get $y = -2$. Thus, $(1,-2)$ is the only critical point.

3. Setting $\frac{\partial f}{\partial x} = -5y - 4x = 0$ and $\frac{\partial f}{\partial y} = 2y - 5x = 0$, we obtain the system of equations

$$4x + 5y = 0$$
$$5x - 2y = 0,$$

whose unique solution is $x = 0$, $y = 0$. Thus, $(0,0)$ is the only critical point.

5. Setting $\frac{\partial f}{\partial x} = 3y - 3x^2 = 0$ and $\frac{\partial f}{\partial y} = 3x - 3y^2 = 0$, we obtain the system of equations

$$y = x^2 \quad \text{and} \quad x = y^2,$$

which leads to $y = y^4$, whose solutions are $y = 0$ and $y = 1$. If $y = 0$, then $x = 0$, and if $y = 1$, then $x = 1$. Thus, $(0,0)$ and $(1,1)$ are the critical points.

7. Setting $\frac{\partial f}{\partial x} = y - e^x = 0$ and $\frac{\partial f}{\partial y} = x = 0$, we see that $x = 0$ and $y = e^0 = 1$. Thus, $(0,1)$ is the only critical point.

9. Set $\frac{\partial f}{\partial x} = 2x = 0$ and $\frac{\partial f}{\partial y} = 2ye^{y^2} = 0$. The only solution is $x = 0$ and $y = 0$. Thus, $(0,0)$ is the only critical point.

11. Setting $\frac{\partial g}{\partial x} = 2x - y = 0$ and $\frac{\partial g}{\partial y} = -x + 2y + 3 = 0$, we obtain the system of equations $2x - y = 0$ and $x - 2y = 3$. This system has the unique solution $x = -1$ and $y = -2$. Thus, $(-1, -2)$ is the only critical point. The second-order partial derivatives are $\frac{\partial^2 g}{\partial x^2} = 2$, $\frac{\partial^2 g}{\partial x \partial y} = -1$, $\frac{\partial^2 g}{\partial y^2} = 2$, and $D(x,y) = \frac{\partial^2 g}{\partial x^2} \cdot \frac{\partial^2 g}{\partial y^2} - (\frac{\partial^2 g}{\partial x \partial y})^2 = 2 \cdot 2 - (-1)^2 = 3 > 0$. Since $D(x,y) > 0$ and $\frac{\partial^2 g}{\partial x^2} = 2 > 0$, g has a local minimum at $(-1, -2)$.

13. Set $\frac{\partial f}{\partial x} = 2x - y + 1 = 0$ and $\frac{\partial f}{\partial y} = -x + 2 = 0$. From the second equation we see that $x = 2$, and substituting that into the first equation gives solution $5 - y = 0$, or $y = 5$. Thus, $(2,5)$ is the only critical point. Moreover, $\frac{\partial^2 f}{\partial x^2} = 2$, $\frac{\partial^2 f}{\partial x \partial y} = -1$, $\frac{\partial^2 f}{\partial y^2} = 0$, so that $D(x,y) = 2 \cdot 0 - (-1)^2 = -1 < 0$. Therefore, $(2,5)$ is a saddle point.

15. Setting $\frac{\partial f}{\partial x} = 3x^2 - 3y = 0$ and $\frac{\partial f}{\partial y} = 3y^2 - 3x = 0$, we obtain the system of equations $y = x^2$ and $x = y^2$, whose solutions are $x = 0$, $y = 0$ and $x = 1$, $y = 1$ (see exercise 5). Thus, $(0,0)$ and $(1,1)$ are the critical points. In addition, $\frac{\partial^2 f}{\partial x^2} = 6x$, $\frac{\partial^2 f}{\partial x \partial y} = -3$, $\frac{\partial^2 f}{\partial y^2} = 6y$, and $D(x,y) = 36xy - 9$. Thus, $D(1,1) = 27 > 0$ and $\frac{\partial^2 f}{\partial x^2}(1,1) = 6 > 0$, so that f has a local minimum at $(1,1)$; and $D(0,0) = -9 < 0$, so that f has a saddle point at $(0,0)$.

17. Set $\frac{\partial f}{\partial x} = 4 - 2x = 0$ and $\frac{\partial f}{\partial y} = 2 - 2y = 0$. From the first equation we get $x = 2$, and from the second we get $y = 1$. Thus, $(2,1)$ is the only critical point. In addition, $\frac{\partial^2 f}{\partial x^2} = -2$, $\frac{\partial^2 f}{\partial x \partial y} = 0$, and $\frac{\partial^2 f}{\partial y^2} = -2$. Since $D(x,y) = 4 > 0$ and $\frac{\partial^2 f}{\partial x^2} = -2 < 0$, we conclude that f has a local maximum at $(2,1)$.

19. Set $\frac{\partial f}{\partial x} = 4x^3 = 0$ and $\frac{\partial f}{\partial y} = 4y^3 = 0$. The unique solution is $x = 0$ and $y = 0$, so that $(0,0)$ is the only critical point. In addition,

$$\frac{\partial^2 f}{\partial x^2} = 12x^2, \quad \frac{\partial^2 f}{\partial x \partial y} = 0, \quad \frac{\partial^2 f}{\partial y^2} = 12y^2, \quad \text{and} \quad D(x,y) = 144x^2y^2.$$

Since $D(0,0) = 0$, the second derivative is inconclusive. Note, however, that $f(x,y) > 0$ if $(x,y) \neq (0,0)$ and $f(0,0) = 0$. Therefore, f has a local (and global) minimum at $(0,0)$.

21. Set $\frac{\partial f}{\partial x} = -4x^3 = 0$ and $\frac{\partial f}{\partial y} = -4y^3 = 0$. The unique solution is $x = 0$ and $y = 0$, so that $(0,0)$ is the only critical point. In addition,

$$\frac{\partial^2 f}{\partial x^2} = -12x^2, \quad \frac{\partial^2 f}{\partial x \partial y} = 0, \quad \frac{\partial^2 f}{\partial y^2} = -12y^2 \quad \text{and} \quad D(x,y) = 144x^2 y^2.$$

Since $D(0,0) = 0$, the second derivative is inconclusive. Note, however, that $f(x,y) < 1$ if $(x,y) \neq (0,0)$, and $f(0,0) = 1$. Therefore, f has a local (and global) maximum at $(0,0)$.

23. Set $\frac{\partial f}{\partial x} = \ln y = 0$ and $\frac{\partial f}{\partial y} = \frac{x}{y} = 0$. From the first equation, we see that $y = 1$, and from the second we see that $x = 0$. Thus, $(0,1)$ is the only critical point. Taking the second-order derivatives, we find that

$$\frac{\partial^2 f}{\partial x^2} - 0, \quad \frac{\partial^2 f}{\partial x \partial y} = \frac{1}{y}, \quad \frac{\partial^2 f}{\partial y^2} = -\frac{x}{y^2}, \quad \text{and} \quad D(x,y) = -\frac{1}{y^2}.$$

Therefore, $D(0,1) = -1 < 0$ which means that $(0,1)$ is a saddle point.

25. Set $\frac{\partial f}{\partial x} = 2x = 0$ and $\frac{\partial f}{\partial y} = 2ye^{y^2} = 0$. The only solution is $x = 0$ and $y = 0$, which means that $(0,0)$ is the only critical point. In addition,

$$\frac{\partial^2 f}{\partial x^2} = 2, \quad \frac{\partial^2 f}{\partial x \partial y} = 0, \quad \frac{\partial^2 f}{\partial y^2} = 4y^2 e^{y^2} + 2e^{y^2} = 2e^{y^2}(2y^2 + 1).$$

Then $D(x,y) = 4e^{y^2}(2y^2 + 1)$ and $D(0,0) = 4 > 0$. Since $\frac{\partial^2 f}{\partial x^2} = 2 > 0$, we conclude that f has a local minimum at $(0,0)$.

27. Setting $\frac{\partial f}{\partial x} = 3x + 6y = 0$ and $\frac{\partial f}{\partial y} = 6x + \frac{4}{3}y = 0$. The only solution is $x = y = 0$. (You can verify that by subtracting one-half the second equation from the first, which gives $\frac{16}{3}y = 0$. Thus, $y = 0$, and it follows that $x = 0$.) Taking the second-order derivatives, we find that

$$\frac{\partial^2 f}{\partial x^2} = 3, \quad \frac{\partial^2 f}{\partial x \partial y} = 6, \quad \frac{\partial^2 f}{\partial y^2} = \frac{4}{3}, \quad \text{and} \quad D(x,y) = 3 \cdot \frac{4}{3} - 6^2 = -32 < 0.$$

Thus, $(0,0)$ is the only critical point of f and it is a saddle point and not a local maximum or minimum.

29. Set $\frac{\partial f}{\partial x} = 4x - 12 = 0$ and $\frac{\partial f}{\partial y} = 6y - 6 = 0$. From the first equation, we get $x = 3$, and from the second we get $y = 1$. In addition, $\frac{\partial^2 f}{\partial x^2} = 4$, $\frac{\partial^2 f}{\partial x \partial y} = 0$, $\frac{\partial^2 f}{\partial y^2} = 6$, and $D(x,y) = 24$, so that $D(x,y) > 0$ and $\frac{\partial^2 f}{\partial x^2} > 0$. Therefore, f has a local minimum at $(3,1)$, and since f is a quadratic function the local minimum is also a global one.

31. Set $\frac{\partial f}{\partial x} = 6 - 6x = 0$ and $\frac{\partial f}{\partial y} = 4 - 8y = 0$. The only solution is $x = 1$ and $y = 1/2$. Thus, $(1, \frac{1}{2})$ is the only critical point. In addition, $\frac{\partial^2 f}{\partial x^2} = -6$, $\frac{\partial^2 f}{\partial x \partial y} = 0$, $\frac{\partial^2 f}{\partial y^2} = -8$, and $D(x,y) = 48$, so that $D(x,y) > 0$ and $\frac{\partial^2 f}{\partial x^2} < 0$. Therefore, f has a local maximum at $(1, \frac{1}{2})$, and since f is a quadratic function the local maximum is also a global one.

33. Setting $\frac{\partial f}{\partial x} = -2 - 6x - 10y = 0$ and $\frac{\partial f}{\partial y} = -6 - 10x - 18y = 0$, we obtain the system of equations $3x + 5y = -1$ and $5x + 9y = -3$, whose solution is $x = 3$ and $y = -2$. In addition, $\frac{\partial^2 f}{\partial x^2} = -6 < 0$, $\frac{\partial^2 f}{\partial x \partial y} = -10$, $\frac{\partial^2 f}{\partial y^2} = -18$, and $D(x,y) = 8 > 0$. Thus, f has a local maximum at $(3, -2)$, and since f is a quadratic function the local maximum is also a global one.

35. Setting $\frac{\partial P}{\partial x} = 8 - 0.02x - 0.01y = 0$ and $\frac{\partial P}{\partial y} = 11 - 0.01x - 0.04y = 0$, we obtain the system of equations $2x + y = 800$ and $x + 4y = 1,100$. The system has the solution $x = 300$ and $y = 200$. In addition, $\frac{\partial^2 P}{\partial x^2} = -0.02 < 0$, $\frac{\partial^2 P}{\partial x \partial y} = -0.01$, $\frac{\partial^2 P}{\partial y^2} = -0.04$, and $D(x, y) = (-0.02) \cdot (-0.04) - (-0.01)^2 = 0.0007 > 0$. Thus, the profit is maximized if $x = 300$ and $y = 200$, with the maximum profit being

$$P(300, 200) = 2,400 + 2,200 - 900 - 0.01(300^2 + 300 \cdot 200 + 2 \cdot 200^2) = 1,400.$$

37. Setting $\frac{\partial P}{\partial x} = 90 - 0.04x - 0.01y = 0$ and $\frac{\partial P}{\partial y} = 80 - 0.01x - 0.06y = 0$, we obtain a system of equations $4x + y = 9,000$ and $x + 6y = 8,000$, whose solution is $x = 2,000$ and $y = 1,000$. In addition, $\frac{\partial^2 P}{\partial x^2} = -0.04 < 0$, $\frac{\partial^2 P}{\partial x \partial y} = -0.01$, $\frac{\partial^2 P}{\partial y^2} = -0.06$, and $D(x, y) = (-0.04)(-0.06) - (-0.01)^2 = 0.0023 > 0$. Thus, the maximum profit of $P(2,000, 1,000) = 120,000$ is achieved if $x = 2,000$ and $y = 1,000$.

39. (a) The weekly revenue of the toy store is

$$R(x, y) = x(50 - 3x + 2y) + y(6 + \frac{1}{2}x - y)$$

$$= 50x + 6y + \frac{5}{2}xy - 3x^2 - y^2.$$

(b) Setting $\frac{\partial R}{\partial x} = 50 + \frac{5}{2}y - 6x = 0$ and $\frac{\partial R}{\partial y} = 6 + \frac{5}{2}x - 2y = 0$, we obtain the system of equations $12x - 5y = 100$ and $5x - 4y = -12$, whose solution is $x = 20$ and $y = 28$. In addition, $\frac{\partial^2 R}{\partial x^2} = -6$, $\frac{\partial^2 R}{\partial x \partial y} = \frac{5}{2}$, and $\frac{\partial^2 R}{\partial y^2} = -2$. Since $D(x, y) = (-6)(-2) - (\frac{5}{2})^2 = \frac{23}{4} > 0$ and $\frac{\partial^2 R}{\partial x^2} = -6 < 0$, the revenue function has a local maximum at $x = 20$ and $y = 28$. Because the function is quadratic, it also has a global maximum at that point.

(c) The weekly profit of the toy store is

$$P(x, y) = R(x, y) - 9(50 - 3x + 2y) - 14(6 + \frac{1}{2}x - y)$$

$$= 70x + 2y + \frac{5}{2}xy - 3x^2 - y^2 - 534.$$

Setting $\frac{\partial P}{\partial x} = 70 + \frac{5}{2}y - 6x = 0$ and $\frac{\partial P}{\partial y} = 2 + \frac{5}{2}x - 2y = 0$, we obtain the system of equations $12x - 5y = 140$ and $5x - 4y = -4$, whose solution is $x = \frac{580}{23} \approx 25.22$ and $y = \frac{748}{23} \approx 32.52$. Applying again the second derivative test, we see that those are the prices the store should charge for the standard and deluxe games, respectively, to maximize its profit.

41. (a) If the stove is placed at the point (x, y), the total distance (as shown in Figure A) is $r(x, y) = \sqrt{x^2 + y^2} + \sqrt{(x - 3)^2 + (y - 5)^2}$. Setting $\frac{\partial r}{\partial x} = 0$ and $\frac{\partial r}{\partial y} = 0$ leads to the system of equations

$$\frac{x}{\sqrt{x^2 + y^2}} = \frac{3 - x}{\sqrt{(x - 3)^2 + (y - 5)^2}} \quad \text{and} \quad \frac{y}{\sqrt{x^2 + y^2}} = \frac{5 - y}{\sqrt{(x - 3)^2 + (y - 5)^2}}.$$

Assuming that $y \neq 0$ and $y \neq 5$ (which must be true for the second equation to be valid), we can divide the first equation by the second to get

$$\frac{x}{y} = \frac{3 - x}{5 - y},$$

which reduces to $x(5-y) = y(3-x)$ and further simplifies to $5x = 3y$, or $y = \frac{5}{3}x$. Conversely, if $y = \frac{5}{3}x$ and $0 < x < 3$, then

$$\frac{\partial r}{\partial x} = \frac{x}{\sqrt{x^2 + y^2}} + \frac{x - 3}{\sqrt{(x-3)^2 + (y-5)^2}}$$

$$= \frac{3x}{\sqrt{34x^2}} + \frac{3(x-3)}{\sqrt{34(x-3)^2}} = \frac{3}{\sqrt{34}} - \frac{3}{\sqrt{34}} = 0,$$

and similarly $\frac{\partial r}{\partial y} = 0$.

(b) We have seen that any pair (x, y) satisfying $y = \frac{5}{3}x$ and $0 < x < 3$ is a critical point. These are the points of the line segment between $(0,0)$ and $(3,5)$, as shown in Figure B. A geometric argument shows that choosing any such point minimizes the total distance. In fact, this exercise is an analytic verification of the well-known geometric principle that the shortest distance between two points is along the straight line bewteen them.

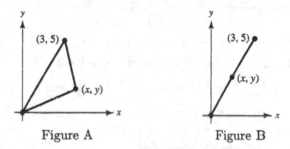

Figure A   Figure B

43. $D(2,3) = (-4)(-7) - 5^2 = 3 > 0$ and $\frac{\partial^2 f}{\partial x^2}(2,3) = -4 < 0$. Therefore, f has a local maximum at $(2,3)$.

45. $D(2,3) = 4 \cdot 7 - 8^2 = -36 < 0$. Therefore, $(2,3)$ is a saddle point.

47. Setting

$$\frac{\partial E}{\partial a} = 2(a + b - 2) + 2(2a + b - 2) \cdot 2 + 2(3a + b - 4) \cdot 3 = 28a + 12b - 36 = 0$$

$$\frac{\partial E}{\partial b} = 2(a + b - 2) + 2(2a + b - 2) + 2(3a + b - 4) = 12a + 6b - 16 = 0,$$

we obtain the system of equations $7a + 3b = 9$ and $6a + 3b = 8$. This system has the solution $a = 1$ and $b = \frac{2}{3}$. In addition,

$$\frac{\partial^2 E}{\partial a^2} = 28, \quad \frac{\partial^2 E}{\partial a \partial b} = 12, \quad \text{and} \quad \frac{\partial^2 E}{\partial b^2} = 6.$$

Therefore, $D(a,b) = 28 \cdot 6 - 12^2 = 24 > 0$ and $\frac{\partial^2 E}{\partial a^2} = 28 > 0$. Since E is a quadratic function, it has a global minimum at $(1, \frac{2}{3})$.

EXERCISES 7.4

1. We have $n = 3$ and $\begin{array}{c||c|c|c} x_i & 1 & 2 & 3 \\ \hline y_i & 1 & 2 & 4 \end{array}$. Then

   $x_1 y_1 + x_2 y_2 + x_3 y_3 = 1 \cdot 1 + 2 \cdot 2 + 3 \cdot 4 = 17$, $x_1 + x_2 + x_3 = 1 + 2 + 3 = 6$,

   $y_1 + y_2 + y_3 = 1 + 2 + 4 = 7$, and $x_1^2 + x_2^2 + x_3^2 = 1 + 4 + 9 = 14$.

   Therefore, $a = \frac{3 \cdot 17 - 6 \cdot 7}{3 \cdot 14 - 6^2} = \frac{3}{2}$, $b = \frac{7 - (3/2) \cdot 6}{3} = -\frac{2}{3}$, and $y = \frac{3}{2} x - \frac{2}{3}$.

3. We have $n = 3$ and $\begin{array}{c||c|c|c} x_i & 0 & 1 & 2 \\ \hline y_i & 1 & 2 & 4 \end{array}$. Then

   $x_1 y_1 + x_2 y_2 + x_3 y_3 = 0 \cdot 1 + 1 \cdot 2 + 2 \cdot 4 = 10$, $x_1 + x_2 + x_3 = 0 + 1 + 2 = 3$,

   $y_1 + y_2 + y_3 = 1 + 2 + 4 = 7$, $x_1^2 + x_2^2 + x_3^2 = 0 + 1 + 4 = 5$.

   Therefore, $a = \frac{3 \cdot 10 - 3 \cdot 7}{3 \cdot 5 - 3^2} = \frac{3}{2}$, $b = \frac{7 - (3/2) \cdot 3}{3} = \frac{5}{6}$, and $y = \frac{3}{2} x + \frac{5}{6}$.

5. We have $n = 3$ and $\begin{array}{c||c|c|c} x_i & 0 & 1 & 2 \\ \hline y_i & 1 & 2 & 5/2 \end{array}$. Then

   $x_1 y_1 + x_2 y_2 + x_3 y_3 = 0 \cdot 1 + 1 \cdot 2 + 2 \cdot \frac{5}{2} = 7$, $x_1 + x_2 + x_3 = 0 + 1 + 2 = 3$,

   $y_1 + y_2 + y_3 = 1 + 2 + \frac{5}{2} = \frac{11}{2}$, $x_1^2 + x_2^2 + x_3^2 = 0 + 1 + 4 = 5$.

   Therefore, $a = \frac{3 \cdot 7 - 3 \cdot (11/2)}{3 \cdot 5 - 3^2} = \frac{3}{4}$, $b = \frac{11/2 - (3/4) \cdot 3}{3} = \frac{13}{12}$, and $y = \frac{3}{4} x + \frac{13}{12}$.

7. We have $n = 4$ and $\begin{array}{c||c|c|c|c} x_i & 0 & 1 & 2 & 3 \\ \hline y_i & 1 & 2 & 1 & 2 \end{array}$. Then

   $x_1 y_1 + x_2 y_2 + x_3 y_3 + x_4 y_4 = 0 \cdot 1 + 1 \cdot 2 + 2 \cdot 1 + 3 \cdot 2 = 10$,

   $x_1 + x_2 + x_3 + x_4 = 0 + 1 + 2 + 3 = 6$,

   $y_1 + y_2 + y_3 + y_4 = 1 + 2 + 1 + 2 = 6$, $x_1^2 + x_2^2 + x_3^2 + x_4^2 = 0 + 1 + 4 + 9 = 14$.

   Therefore, $a = \frac{4 \cdot 10 - 6 \cdot 6}{4 \cdot 14 - 6^2} = \frac{1}{5}$, $b = \frac{6 - (1/5) \cdot 6}{4} = \frac{6}{5}$, and $y = \frac{1}{5} x + \frac{6}{5}$.

9. Letting x denote the year and y denote the profit, we have

   $$x_1 y_1 + x_2 y_2 + x_3 y_3 + x_4 y_4 + x_5 y_5 = 1 \cdot (2.3) + 2 \cdot (2.7) + 3 \cdot (2.8) + 4 \cdot (3.0) + 5 \cdot (3.5) = 45.6$$

   $$x_1 + x_2 + x_3 + x_4 + x_5 = 1 + 2 + 3 + 4 + 5 = 15$$

   $$y_1 + y_2 + y_3 + y_4 + y_5 = 2.3 + 2.7 + 2.8 + 3.0 + 3.5 = 14.3$$

   $$x_1^2 + x_2^2 + x_3^2 + x_4^2 + x_5^2 = 1 + 4 + 9 + 16 + 25 = 55.$$

   Therefore, $a = \frac{5 \cdot (45.6) - 15 \cdot (14.3)}{5 \cdot 55 - 15^2} = 0.27$, $b = \frac{14.3 - (0.27) \cdot 15}{5} = 2.05$, and $y = 0.27 x + 2.05$. For the sixth year $x = 6$, and $y = (0.27) \cdot 6 + 2.05 = 3.67$ is the predicted profit.

11. Letting x be the number of years after 1960 and y the GNP in billions of dollars,

    $$x_1 y_1 + x_2 y_2 + x_3 y_3 + x_4 y_4 + x_5 y_5 = 0 \cdot (530.6) + 10 \cdot (1{,}046.08) + 20 \cdot (2{,}830.83)$$
    $$+ 30 \cdot (5{,}832.23) + 40 \cdot (9{,}860.8) = 636{,}476.3$$

    $$x_1 + x_2 + x_3 + x_4 + x_5 = 0 + 10 + 20 + 30 + 40 = 100$$

    $$y_1 + y_2 + y_3 + y_4 + y_5 = 530.6 + 1{,}046.08 + 2{,}830.83 + 5{,}832.23$$
    $$+ 9{,}860.8 = 20{,}100.54$$

    $$x_1^2 + x_2^2 + x_3^2 + x_4^2 + x_5^2 = 0 + 100 + 400 + 900 + 1{,}600 = 3{,}000.$$

Therefore, $a = \frac{5 \cdot (636,476.3) - 100 \cdot (20,100.54)}{5 \cdot (3,000) - 100^2} = 234.4655$, $b = \frac{20,100.54 - (234.4655) \cdot 100}{5} = -669.202$, and $y = 234.4655\,x - 669.202$. If $x = 50$ (corresponding to the year 2010), then $y = (234.4655) \cdot 50 - 669.202 = 11,054.073$ is the predicted GNP.

13. Letting x be the number of years after 1900 and y population in millions,

$$x_1 y_1 + x_2 y_2 + \cdots + x_{10} y_{10} + x_{11} y_{11} = 0 \cdot 76 + 10 \cdot 92 + \cdots + 90 \cdot 249 + 100 \cdot 281 = 113,180$$

$$x_1 + x_2 + \cdots + x_{10} + x_{11} = 0 + 10 + \cdots + 90 + 100 = 550$$

$$y_1 + y_2 + \cdots + y_{10} + y_{11} = 76 + 92 + \cdots + 249 + 281 = 1,819$$

$$x_1^2 + x_2^2 + \cdots + x_4^2 + x_5^2 = 0 + 100 + \cdots + 90^2 + 100^2 = 38,500.$$

Therefore, $a = \frac{11 \cdot (113,180) - 550 \cdot (1,819)}{11 \cdot (38,500) - 550^2} \approx 2.021$, $b = \frac{1,819 - a \cdot 550}{11} \approx 64.318$, and $y \approx 2.021x + 64.318$. If $x = 110$, then $y \approx 2.021 \cdot 110 + 64.318 = 286.628$, the predicted population in 2010.

15. (a) All of the data points except one are below the line and it seems as if lowering the line would give a better fit. In addition, a scatter diagram of the points (shown in Figure A) suggests that the slope of the regression line should be much closer to zero. Figure B shows the points together with the correct and incorrect lines. To show the lines more clearly, the horizontal axis in Figure B is placed at $y = 210$.

(b) The correct regression line, found by using formulas (16) and (17), is $y = -0.914x + 236.6$.

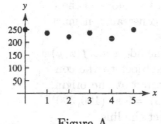

Figure A

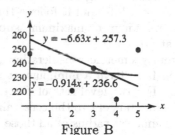

Figure B

19. (a) By the procedure of exercise 18, with a $[1, 12] \times [120, 220]$ window:

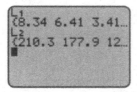

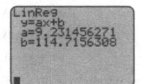

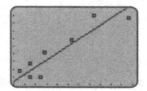

(b) $y(x + 1) - y(x) \approx 9.23(x + 1) + 114.7 - [9.23x + 114.7] = 9.23$. Note that this is the (approximate) slope of the regression line, and recall that the slope of a line is the rate of increase of y per unit increase of x.

(c) $y(0) \approx 114.7$

21. The scatter diagram does not look good for linear regression. Quadratic regression gives a reasonably good fit.

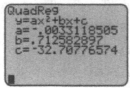

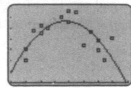

23. The scatter diagram suggests an exponential curve. Using exponential regression with a $[0, 8] \times [0, 120]$ window, we get

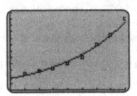

EXERCISES 7.5

1. Given $f(x, y) = x^2 + y^2$ with constraint $g(x, y) = 2x + 6y - 2{,}000 = 0$, we set $\frac{\partial f}{\partial x} = \lambda \frac{\partial g}{\partial x}$ and $\frac{\partial f}{\partial y} = \lambda \frac{\partial g}{\partial y}$ leading to the system of equations

$$2x = 2\lambda$$
$$2y = 6\lambda$$
$$2x + 6y = 2{,}000.$$

From the first two equations we obtain $x = \lambda$ and $y = 3\lambda$, so that $y = 3x$. Substituting that into the third equation gives $2x + 18x = 2{,}000$, or $x = 100$, and it follows that $y = 300$. Therefore, if there is local minimum or maximum, subject to the constraint, it must occur at $(100, 300)$.

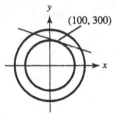

From geometric considerations, we can conclude that $f(x, y)$ does, in fact, have a minimum at $(100, 300)$, subject to the constraint. For the level sets of f are circles centered at the origin, and the one tangent to the constraint line $g(x, y) = 0$ at $(100, 300)$ has the smallest radius of all those that intersect the line.

3. Given $f(x, y) = x^2 + 3xy + y^2$ with constraint $g(x, y) = x + y - 1{,}000 = 0$, we set $\frac{\partial f}{\partial x} = \lambda \frac{\partial g}{\partial x}$ and $\frac{\partial f}{\partial y} = \lambda \frac{\partial g}{\partial y}$ leading to the system of equations

$$2x + 3y = \lambda$$
$$3x + 2y = \lambda$$
$$x + y = 1{,}000.$$

From the first two equations we obtain $2x + 3y = 3x + 2y$, which reduces to $y = x$. Combining this with the third equation, we get $x = y = 500$. Therefore, if there is local minimum or maximum, subject to the constraint, it must occur at $(500, 500)$.

In fact, $f(x, y)$ has a maximum at $(500, 500)$, subject to the constraint. To see that, we first note that $f(500, 500) = 5 \cdot 500^2$. Any other point on the line $x + y = 1{,}000$ can be written as $(500 - t, 500 + t)$ with $t \neq 0$, and $f(500 - t, 500 + t) = (500 - t)^2 + 3(500 - t)(500 + t) + (500 + t)^2 = 5 \cdot 500^2 - t^2$, which is less than $5 \cdot 500^2$.

5. Given $f(x, y) = (x - 1)^2 + (y - 2)^2 - 4$ and $g(x, y) = 3x + 5y - 47$, we define $F(x, y, \lambda) = f(x, y) - \lambda g(x, y)$ and set $\frac{\partial F}{\partial x} = 0$, $\frac{\partial F}{\partial y} = 0$, and $\frac{\partial F}{\partial \lambda} = 0$. These are equivalent to $\frac{\partial f}{\partial x} = \lambda \frac{\partial g}{\partial x}$, $\frac{\partial f}{\partial y} = \lambda \frac{\partial g}{\partial y}$, and $g(x) = 0$, and lead to the system of equations

$$2(x - 1) = 3\lambda$$
$$2(y - 2) = 5\lambda$$
$$3x + 5y = 47.$$

From the first two equations, we obtain $10(x-1) = 6(y-2)$, which reduces to $5x - 3y = -1$. Combining that with the third equation leads to the solution $x = 4$, $y = 7$.

We next apply the second derivative test. The partial derivatives are

$$\frac{\partial F}{\partial x} = 2(x-1) - 3\lambda, \quad \frac{\partial F}{\partial y} = 2(y-2) - 5\lambda, \quad \frac{\partial^2 F}{\partial x^2} = 2, \quad \frac{\partial^2 F}{\partial x \partial y} = 0, \quad \frac{\partial^2 F}{\partial y^2} = 2.$$

Since $D(x,y) = 2 \cdot 2 - 0^2 = 4 > 0$ and $\frac{\partial^2 F}{\partial x^2} = 2 > 0$, $f(x,y)$ has a local minimum at $(4,7)$, subject to constraint $g(x,y) = 0$.

We can also reach the same conclusion from geometric considerations similar to that of exercise 1 involving the level sets, which are circles centered at $(1,2)$. In fact, we reach the stronger conclusion that there is a global minimum at $(4,7)$.

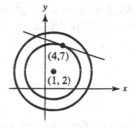

7. Given $f(x,y) = xy$ with the constraint $g(x,y) = x^2 + y^2 - 1 = 0$, we set $\frac{\partial f}{\partial x} = \lambda \frac{\partial g}{\partial x}$ and $\frac{\partial f}{\partial y} = \lambda \frac{\partial g}{\partial y}$, leading to the system of equations

$$y = 2x\lambda$$
$$x = 2y\lambda$$
$$x^2 + y^2 = 1.$$

By multiplying the first equation by y and the second by x, we obtain $y^2 = 2xy\lambda = x^2$. That reduces the third equation to $2x^2 = 1$, or $x = \pm\sqrt{2}/2$. Thus, there are four possible extreme points: $(\sqrt{2}/2, \sqrt{2}/2)$, $(\sqrt{2}/2, -\sqrt{2}/2)$, $(-\sqrt{2}/2, \sqrt{2}/2)$, and $(-\sqrt{2}/2, -\sqrt{2}/2)$.

The constraint curve is the circle of radius 1 about the origin, and the level curves of f are of the form $xy = c$. If $c > 0$, each curve has two branches, one in the first quadrant and the other in the third. Among those curves that intersect the constraint circle, the largest value of c is $1/2$, occuring at the two tangency points $(\sqrt{2}/2, \sqrt{2}/2)$ and $(-\sqrt{2}/2, -\sqrt{2}/2)$.

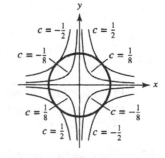

Thus, f has a global maximum, subject to the constraint, at each of those points. Similarly, f has a global minimum, subject to the constraint, at each of the points $(\sqrt{2}/2, -\sqrt{2}/2)$ and $(-\sqrt{2}/2, \sqrt{2}/2)$, where $c = -1/2$.

9. Given $f(x,y) = x^2 + y^2$ with constraint $g(x,y) = xy - 1 = 0$, we set $\frac{\partial f}{\partial x} = \lambda \frac{\partial g}{\partial x}$ and $\frac{\partial f}{\partial y} = \lambda \frac{\partial g}{\partial y}$, leading to the system of equations

$$2x = \lambda y$$
$$2y = \lambda x$$
$$xy = 1.$$

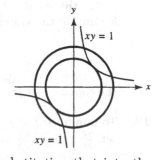

Multiplying the first equation by x and the second by y leads to $2x^2 = \lambda xy = 2y^2$, so that $y = \pm x$. However, $y \neq -x$, otherwise the third equation would become $-x^2 = 1$. Therefore, $y = x$, and by substituting that into the third equation, we obtain $x^2 = 1$, or $x = \pm 1$.

Thus, the possible extreme points are $(1,1)$ and $(-1,-1)$. By considering the relation of the level curves of f to the constraint curve as we have done in several previous exercises, we conclude that, subject to the constraint, f has a global minimum at each of the points $(1, 1)$ and $(-1, -1)$, with value $f(\pm 1, \pm 1) = 2$.

11. **(a)** The trail reaches its elevation at point C, where the height is 2100 feet.

　　(b) The trail curve meets the level curve tangentially.

　　(c) The function (x, y) has a local maximum under the constraint $g(x, y)$ at (x_0, y_0), which means that for some constant λ, $\frac{\partial f}{\partial x}(x_0, y_0) = \lambda \frac{\partial g}{\partial x}(x_0, y_0)$ and $\frac{\partial f}{\partial y}(x_0, y_0) = \lambda \frac{\partial g}{\partial y}(x_0, y_0)$.

13. With $f(x, y) = x^2 + y^2$ and $g(x, y) = 3x + y - 1$, we set $\frac{\partial f}{\partial x} = \lambda \frac{\partial g}{\partial x}$ and $\frac{\partial f}{\partial y} = \lambda \frac{\partial g}{\partial y}$, leading to the system of equations

$$2x = 3\lambda$$
$$2y = \lambda$$
$$3x + y = 1.$$

From the first two equations we get $y = \frac{1}{3}x$, and substituting that into the third equation gives $x = \frac{3}{10}$. Thus, the only extreme point is $\left(\frac{3}{10}, \frac{1}{10}\right)$.

　　An examination of the level curves, similar to that in exercise 1, shows that f has a global minimum, subject to the constraint $g(x, y) = 0$, at the point $\left(\frac{3}{10}, \frac{1}{10}\right)$, which is therefore the point on the line closest to the origin.

　　As an alternative, we can apply the second derivative test to $F(x, y, \lambda) = (x^2 + y^2) - \lambda(3x + y - 1)$ at the point $(x, y, \lambda) = \left(\frac{3}{10}, \frac{1}{10}, \frac{1}{5}\right)$.

15. With $f(x, y, z) = x^2 + y^2 + z^2$ and $g(x, y, z) = 2x - 2y + z - 9$, we set $\frac{\partial f}{\partial x} = \lambda \frac{\partial g}{\partial x}$, $\frac{\partial f}{\partial y} = \lambda \frac{\partial g}{\partial y}$, and $\frac{\partial f}{\partial z} = \lambda \frac{\partial g}{\partial z}$, leading to the system of equations

$$2x = 2\lambda$$
$$2y = -2\lambda$$
$$2z = \lambda$$
$$2x - 2y + z = 9.$$

From the first three equations we get $y = -x$ and $z = x/2$, and putting those into the last equation gives $x = 2$. Thus, the only possible extreme point is $(2, -2, 1)$. Therefore, the minimum value is $f(2, -2, 1) = 9$.

17. The question may be reformulated as follows: find the minimum value of $f(x, y, z) = x + y + z$ under the constraint $xyz = 8$. Let $g(x, y, z) = xyz - 8$ and set $\frac{\partial f}{\partial x} = \lambda \frac{\partial g}{\partial x}$, $\frac{\partial f}{\partial y} = \lambda \frac{\partial g}{\partial y}$, $\frac{\partial f}{\partial z} = \frac{\partial g}{\partial z}$, leading to the system of equations

$$1 = \lambda yz, \quad 1 = \lambda xz, \quad 1 = \lambda xy, \quad xyz = 8.$$

This system of equations has the unique solution $x = 2$, $y = 2$, $z = 2$, and $\lambda = 1/4$, Thus, under the constraint $xyz = 8$, $f(x, y, z)$ has a minimum at $(2, 2, 2)$ with the minimum value $f(2, 2, 2) = 6$.

19. To minimize $f(x, y, z) = x^2 + y^2 + z^2$ subject to the constraint $g(x, y, z) = x + y - 2z - 5 = 0$, we set $\frac{\partial f}{\partial x} = \lambda \frac{\partial g}{\partial x}$, $\frac{\partial f}{\partial y} = \lambda \frac{\partial g}{\partial y}$, and $\frac{\partial f}{\partial z} = \lambda \frac{\partial g}{\partial z}$, leading to the system of equations

$$2x = \lambda, \quad 2y = \lambda, \quad 2z = -2\lambda, \quad \text{and} \quad x + y - 2z = 5.$$

This system of equations has the solution $x = \frac{5}{6}$, $y = \frac{5}{6}$, $z = -\frac{5}{3}$, $\lambda = \frac{5}{3}$. Thus, $\left(\frac{5}{6}, \frac{5}{6}, -\frac{5}{3}\right)$ is the point on the plane closest to the origin.

21. Find the maximum and minimum of the function $f(x, y, z) = z$ for $z > 0$, subject to the constraints $x^2 + y^2 = 8$ and $x + y + z = 5$. Write $g(x, y, z) = x^2 + y^2 - 8$ and $h(x, y, z) = x + y + z - 5$, and set

$$\frac{\partial f}{\partial x} = \lambda \frac{\partial g}{\partial x} + \mu \frac{\partial h}{\partial x} \quad \text{or} \quad 0 = 2x\lambda + \mu$$

$$\frac{\partial f}{\partial y} = \lambda \frac{\partial g}{\partial y} + \mu \frac{\partial h}{\partial y} \quad \text{or} \quad 0 = 2y\lambda + \mu$$

$$\frac{\partial f}{\partial z} = \lambda \frac{\partial g}{\partial z} + \mu \frac{\partial h}{\partial z} \quad \text{or} \quad 1 = \mu$$

$$g(x, y, z) = 0 \quad \text{or} \quad 8 = x^2 + y^2$$

$$h(x, y, z) = 0 \quad \text{or} \quad 5 = x + y + z.$$

From the first two equations we obtain $x = y$, and combining that with the fourth equation, we get $x = y = \pm 2$. Combining this with the last equation, we obtain the points $(2, 2, 1)$ and $(-2, -2, 9)$ as the extreme points. The first has minimum height above the xy-plane, with $z = 1$, and the second has maximum height, with $z = 9$.

23. We need to maximize $P(x, y)$ subject to $Q(x, y) = x + y - 100 = 0$. Setting $\frac{\partial P}{\partial x} = \lambda \frac{\partial Q}{\partial x}$ and $\frac{\partial P}{\partial y} = \lambda \frac{\partial Q}{\partial y}$, we obtain the system of equations

$$10 - 0.2x = \lambda$$
$$20 - 0.2y = \lambda$$
$$x + y = 100,$$

whose unique solution is $x = 25$, $y = 75$, and $\lambda = 5$. Applying the second derivative test with $F(x, y, \lambda) = 10x + 20y - 0.1(x^2 + y^2) - \lambda(x + y - 100)$, we find that

$$\frac{\partial F}{\partial x} = 10 - 0.2x - \lambda, \quad \frac{\partial F}{\partial y} = 20 - 0.2y - \lambda, \quad \frac{\partial^2 F}{\partial x^2} = -0.2, \quad \frac{\partial^2 F}{\partial x \partial y} = 0, \quad \frac{\partial^2 F}{\partial y^2} = -0.2.$$

Then $D(x, y) = (-0.2)(-0.2) - 0^2 = 0.04 > 0$ and $\frac{\partial^2 F}{\partial x^2} = -0.2 < 0$, from which we conclude that $P(x, y)$ has a maximum at $(25, 75)$. When the company manufectures 25 units of X and 75 units of Y, it achieves its maximum profit $P(25, 75) = 1{,}125$.

25. We need to maximize $Q(x, y)$ under the constraint $70K + 65L = 4{,}000{,}000$. Letting $g(K, L) = 70K + 65L - 4{,}000{,}000$ and setting $\frac{\partial Q}{\partial x} = \lambda \frac{\partial g}{\partial x}$ and $\frac{\partial Q}{\partial y} = \lambda \frac{\partial g}{\partial y}$, we obtain the system of equations

$$14K^{-0.65} L^{0.65} = 70\lambda$$
$$26K^{0.35} L^{-0.35} = 65\lambda$$
$$70K + 65L = 4{,}000{,}000.$$

By the method of the last exercise, we obtain $K = 20{,}000$ and $L = 40{,}000$. Thus, the combination of 20,000 units of capital and 40,000 units of labor gives the maximum output. (As before, we refer to exercise 10 for the justification that we have indeed found the maximum.)

27. We want to maximize $V(x, y, z) = xyz$ (the volume of the package) under the constraint $W(x, y, z) = 2x + 2y + z - 108 = 0$ (which gives the maximum girth of 108). Setting $\frac{\partial V}{\partial x} = \lambda \frac{\partial W}{\partial x}$, $\frac{\partial V}{\partial y} = \lambda \frac{\partial W}{\partial y}$, and $\frac{\partial V}{\partial z} = \lambda \frac{\partial W}{\partial z}$, we obtain the system of equations

$$yz = 2\lambda$$
$$xz = 2\lambda$$
$$xy = \lambda$$
$$2x + 2y + z = 108.$$

From the first three equations we get $yz = xz = 2xy$, and since we can assume that none of the variables are zero (otherwise the volume would be zero), we obtain $y = x$ and $z = 2x$. Combining that with the fourth equation yields the solution $x = 18$, $y = 18$, $z = 36$, which are the dimensions (in inches) that maximize the volume of a rectangular package with the combined dimensions of 108 inches.

29. (a) To maximize $V(r, h) = \frac{1}{3}\pi r^2 h$ under the constraint $W(r, h) = r + h - 15 = 0$, we set $\frac{\partial V}{\partial r} = \lambda \frac{\partial W}{\partial r}$ and $\frac{\partial V}{\partial h} = \lambda \frac{\partial W}{\partial h}$, leading to the system of equations

$$\frac{2}{3}\pi r h = \lambda$$
$$\frac{1}{3}\pi r^2 = \lambda$$
$$r + h = 15.$$

From the first two equations we get $\frac{2}{3}\pi r h = \frac{1}{3}\pi r^2$, which reduces to $r = 2h$ (assuming, as we may, that $r \neq 0$). Substituting that into the third equation gives $3h = 15$. Thus, $h = 5$ feet and $r = 10$ feet will maximize the volume.

(b) The largest volume possible under the constraint is $V(10, 5) = 500\pi/3$ cubic feet.

(c) Suppose we change the radius from 10 to $10 + t$ for some t with $0 < t < 10$. Then, since $r + h = 15$, the height must change from 5 to $5 - t$, and the volume changes to

$$V(10 + t, 5 - t) = \frac{1}{3}\pi(10 + t)^2(5 - t) = \frac{500\pi}{3} - \frac{1}{3}\pi t^2(15 + t) < \frac{500\pi}{3}.$$

Therefore, the maximimum of $500\pi/3$ is attained at $t = 0$, or, in other words, at $(r, h) = (10, 5)$.

31. Assume that the rectangular tank is x cm long, y cm wide, and z cm high. We need to minimize the surface area $S(x, y, z) = xy + 2xz + 2yz$ under the constraint $T(x, y, z) = xyz - 32,000 = 0$ (the volume is 32,000). Setting $\frac{\partial S}{\partial x} = \lambda \frac{\partial T}{\partial x}$, $\frac{\partial S}{\partial y} = \lambda \frac{\partial T}{\partial y}$, and $\frac{\partial S}{\partial z} = \lambda \frac{\partial T}{\partial z}$, we obtain

$$y + 2z = yz\lambda$$
$$x + 2z = xz\lambda$$
$$2x + 2y = xy\lambda$$
$$xyz = 32,000.$$

From the first two equations we obtain $(y + 2z)/yz = \lambda = (x + 2z)/xz$. Clearing denominators and simplifying give $2xz^2 = 2yz^2$, which reduces to $x = y$ (if we assume, as we can, that $z \neq 0$). Next, from the second and third equations, with y replaced by x, we obtain $(x + 2z)/xz = \lambda = 4/x$. Clearing denominators and simplifying give $x = 2z$ (provided $x \neq 0$). Thus, $x = y = 2z$, and by combining that with the fourth equation we get $x = y = 40$ and $z = 20$, which are the dimensions (in centimeters) that minimize the surface area.

33. We need to minimize $f(x, y, z) = x^2 + y^2 + z^2$ under the constraints $x + 2y + z = 1$ and $x - y + z = 3$. Write $g(x, y, z) = x + 2y + z - 1$ and $h(x, y, z) = x - y + z - 3$, and set

$$\frac{\partial f}{\partial x} = \lambda \frac{\partial g}{\partial x} + \mu \frac{\partial h}{\partial x} \quad \text{or} \quad 2x = \lambda + \mu$$

$$\frac{\partial f}{\partial y} = \lambda \frac{\partial g}{\partial y} + \mu \frac{\partial h}{\partial y} \quad \text{or} \quad 2y = 2\lambda - \mu$$

$$\frac{\partial f}{\partial z} = \frac{\partial g}{\partial z} + \mu \frac{\partial h}{\partial z} \quad \text{or} \quad 2z = \lambda + \mu$$

$$g(x, y, z) = 0 \quad \text{or} \quad x + 2y + z = 1$$

$$h(x, y, z) = 0 \quad \text{or} \quad x - y + z = 3.$$

From the first and third equation we see that $x = z$. Substituting that into the last two equations leads to $2x + 2y = 1$ and $2x - y = 3$, whose solution is $x = \frac{7}{6}$ and $y = -\frac{2}{3}$. Thus, $(\frac{7}{6}, -\frac{2}{3}, \frac{7}{6})$ is the closest point to the origin on the line formed by $x + 2y + z = 1$ and $x - y + z = 3$.

35. Assume that the box is x units long, y units wide, and z units high. Let S_0 be the positive real number representing the fixed surface area. We need to maximize the volume $V(x, y, z) = xyz$ under the constraint $2xy + 2yz + 2xz = S_0$. Writing $S(x, y, z) = 2xy + 2yz + 2xz - S_0$, and setting $\frac{\partial V}{\partial x} = \lambda \frac{\partial S}{\partial x}$, $\frac{\partial V}{\partial y} = \lambda \frac{\partial S}{\partial y}$, and $\frac{\partial V}{\partial z} = \lambda \frac{\partial S}{\partial z}$, we obtain the system of equations

$$yz = \lambda(2y + 2z)$$

$$xz = \lambda(2x + 2z)$$

$$xy = \lambda(2x + 2y)$$

$$2xy + 2yz + 2xz = S_0.$$

Dividing the first equation by the second and simplifying the result give $x = y$. Dividing the second by the third and simplifying the result give $y = z$. Thus, $x = y = z = \sqrt{S_0/6}$. The cube has the maximum volume amount rectangular boxes with a fixed surface area.

CHAPTER 7: REVIEW EXERCISES

1. The distance from a point (x, y, z) to the yz-plane is $|x|$. The distances from the given points A, B, and C to the yz-plane are 3, 6, and 6, respectively, and A is the closest, and B lies on the xy-plane, since its third coordinate is zero.

3. The equation is $z - 3 = -5(x - 1) + \frac{2}{7}(y - 2)$ or $z = -5x + \frac{2}{7}y + \frac{52}{7}$.

5. (a) $f(x, y)$ is linear.
   (b) $f(x, y)$ is not linear, since $x^{1/2}$ is not linear.
   (c) $f(x, y)$ is not linear, since xy is not linear.

7. Assume that $f(x, y) = Ax + By + C$. Substituting the given numbers, we obtain the system of equations

$$f(1, 2) = \quad A + 2B + C = 12$$

$$f(1, 3) = \quad A + 3B + C = 14$$

$$f(4, 5) = 4A + 5B + C = 33.$$

It has the solution $A = 5$, $B = 2$, and $C = 3$. Therefore, $f(x, y) = 5x + 2y + 3$.

9. If $z = -9$, the level curve is given by the equation $x^2 + y^2 = 9$. It is the circle of radius 3 centered at the origin. If $z = 0$, the level "curve" is given by the equation $x^2 + y^2 = 0$. It is the single point $(0, 0)$. If $z = 4$ or $z = 9$, the level curve is empty, since $-x^2 - y^2 \leq 0$ for all (x, y).

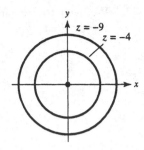

11. $\dfrac{\partial f}{\partial x} = 2xy - 2y^3 + 1, \qquad \dfrac{\partial f}{\partial y} = x^2 + 6y - 6xy^2 - 5$

13. $\dfrac{\partial g}{\partial x} = \dfrac{1}{x^2 + xy} \cdot \dfrac{\partial}{\partial x}(x^2 + xy) = \dfrac{2x + y}{x^2 + xy}$,

$\dfrac{\partial g}{\partial y} = \dfrac{1}{x^2 + xy} \cdot \dfrac{\partial}{\partial y}(x^2 + xy) = \dfrac{x}{x^2 + xy}$

15. $\dfrac{\partial f}{\partial x} = 2x \ln(1 + y^2), \qquad \dfrac{\partial f}{\partial y} = x^2 \dfrac{1}{1 + y^2} \cdot \dfrac{\partial}{\partial y}(1 + y^2) = \dfrac{2x^2 y}{1 + y^2}$

17. $\dfrac{\partial f}{\partial x} = e^{y+2} - \dfrac{2y}{x^3}$ and $\dfrac{\partial f}{\partial y} = xe^{y+2} + \dfrac{1}{x^2}$. Thus,

$$\frac{\partial f}{\partial x}(1, -2) = e^{-2+2} - \frac{2 \cdot (-2)}{1^3} = 5 \quad \text{and} \quad \frac{\partial f}{\partial y}(1, -2) = 1 \cdot e^{-2+2} + \frac{1}{1^2} = 2.$$

19. From the definition of the partial derivative,

$$\lim_{h \to 0} \frac{e^{x/(y+h)} - e^{x/y}}{h} = \frac{\partial}{\partial y}(e^{x/y}) = -\frac{x}{y^2} e^{x/y}.$$

21. First, $z_0 = f(1, 2) = 14$. Next, $\frac{\partial f}{\partial x} = 4x^3$, $\frac{\partial f}{\partial y} = 6y$, so that $\frac{\partial f}{\partial x}(1, 2) = 4$ and $\frac{\partial f}{\partial y}(1, 2) = 12$. Thus, the equation of the tangent plane is

$$z = z_0 + \frac{\partial f}{\partial x}(x_0, y_0)(x - x_0) + \frac{\partial f}{\partial y}(x_0, y_0)(y - y_0)$$

$$= 14 + 4(x - 1) + 12(y - 2) = 4x + 12y - 14.$$

23. Since, $\frac{\partial f}{\partial x} = y(-2x + 0.5)e^{-x^2+0.5x}$ and $\frac{\partial f}{\partial y} = e^{-x^2+0.5x}$, we have $\frac{\partial f}{\partial x}(0, 3) = 3 \cdot (0 + 0.5)e^0 = 1.5$ and $\frac{\partial f}{\partial y}(0, 3) = e^0 = 1$. In addition, $f(0, 3) = 3 \cdot e^0 = 3$. Thus, the linear approximation of f at $(0, 3)$ is

$$f(x, y) \approx f(0, 3) + \frac{\partial f}{\partial x}(0, 3)(x - 0) + \frac{\partial f}{\partial y}(0, 3)(y - 3) = 3 + 1.5x + 1 \cdot (y - 3) = 1.5x + y,$$

and $f(0.2, 3.1) \approx (1.5) \cdot (0.2) + 3.1 = 3.4$.

25. The equation of the tangent plane to the graph of f at $(10, 20)$ is

$$z = f(10, 20) + \frac{\partial f}{\partial x}(10, 20)(x - 10) + \frac{\partial f}{\partial y}(10, 20)(y - 20)$$

$$= 5 + 2(x - 10) + (-1)(y - 20) = 2x - y + 5.$$

The linear approximation of f at $(10, 20)$ is $f(x, y) \approx 2x - y + 5$. Thus, $f(9.9, 20.2) \approx 2 \cdot (9.9) - 20.2 + 5 = 4.6$.

27. First, $\frac{\partial f}{\partial x} = \frac{2x}{4+x^2+y^2}$ and $\frac{\partial f}{\partial y} = \frac{2y}{4+x^2+y^2}$. Then

$$\frac{\partial^2 f}{\partial x^2} = \frac{2(4 + x^2 + y^2) - (2x)(2x)}{(4 + x^2 + y^2)^2} = \frac{8 - 2x^2 + 2y^2}{(4 + x^2 + y^2)^2} \quad \text{and} \quad \frac{\partial^2 f}{\partial y^2} = \frac{8 - 2y^2 + 2x^2}{(4 + x^2 + y^2)^2}$$

$$\frac{\partial^2 f}{\partial y \partial x} = \frac{\partial f}{\partial y}\left(\frac{2x}{4 + x^2 + y^2}\right) = -\frac{2x}{(4 + x^2 + y^2)^2} \cdot \frac{\partial}{\partial y}(4 + x^2 + y^2) = -\frac{4xy}{(4 + x^2 + y^2)^2}$$

$$\frac{\partial^2 f}{\partial x \partial y} = \frac{\partial f}{\partial x}\left(\frac{2y}{4 + x^2 + y^2}\right) = -\frac{2y}{(4 + x^2 + y^2)^2} \cdot \frac{\partial}{\partial x}(4 + x^2 + y^2) = -\frac{4xy}{(4 + x^2 + y^2)^2}.$$

29. Setting $\frac{\partial f}{\partial x} = 6x - 4y - 20 = 0$, $\frac{\partial f}{\partial y} = 6y - 4x + 10 = 0$, we obtain the system of equations $3x - 2y = 10$ and $4x - 6y = 10$, which has the unique solution $x = 4$ and $y = 1$. Thus, $(4, 1)$ is the only critical point.

31. Set $\frac{\partial f}{\partial x} = 4x - 4x^3 = 4x(1 - x^2) = 0$, $\frac{\partial f}{\partial y} = -2y = 0$. This system of equations has three solutions: $(0, 0)$, $(-1, 0)$, and $(1, 0)$, which are the critical points. To use the second derivative test, we first compute

$$\frac{\partial^2 f}{\partial x^2} = 4 - 12x^2, \quad \frac{\partial^2 f}{\partial x \partial y} = 0, \quad \frac{\partial^2 f}{\partial y^2} = -2, \quad \text{and} \quad D(x, y) = 24x^2 - 8.$$

At both $(1, 0)$ and $(-1, 0)$, we have $D = 16 > 0$ and $\frac{\partial^2 f}{\partial x^2} = -8 < 0$, there is a local maximum at each of those points. Since $D(0, 0) = -8 < 0$, there is a saddle point at $(0, 0)$.

33. Since $\frac{\partial f}{\partial x}(1, 2) = \frac{\partial f}{\partial y}(1, 2) = 0$, the point $(1, 2)$ is a critical point; and since $D(1, 2) = (-2)(-1) - 1^2 = 1 > 0$ and $\frac{\partial^2 f}{\partial x^2}(1, 2) = -2 < 0$. Therefore, there is a local maximum at $(1, 2)$.

35. Set $\frac{\partial f}{\partial x} = 3x^2 + 3y = 0$ and $\frac{\partial f}{\partial y} = 3x + 3y^2 = 0$. Then $y = -x^2$ and $x = -y^2$. Substituting the second into the first gives $y = -y^4$, or $y(y^3 + 1) = 0$, with solutions $y = 0$ and $y = -1$. Thus, the critical points are $(0, 0)$ and $(-1, -1)$. Next, we compute

$$\frac{\partial^2 f}{\partial x^2} = 6x, \quad \frac{\partial^2 f}{\partial x \partial y} = 3, \quad \frac{\partial^2 f}{\partial y^2} = 6y, \quad \text{and} \quad D(x, y) = 36xy - 9.$$

Since $D(-1, -1) = 27 > 0$ and $\frac{\partial^2 f}{\partial x^2}(-1, -1) = -6 < 0$, f has a local maximum at $(-1, -1)$. Since $D(0, 0) = -9 < 0$, $(0, 0)$ is a saddle point of f.

37. The profit function is

$$P(x, y) = 3x + 5y - C(x, y)$$
$$= 3x + 5y - (2x^2 - 2xy + y^2 - 3x + y + 7)$$
$$= -2x^2 + 2xy - y^2 + 6x + 4y - 7.$$

Set $\frac{\partial P}{\partial x} = -4x + 2y + 6 = 0$, $\frac{\partial P}{\partial y} = 2x - 2y + 4 = 0$. This system of equations has the solution $x = 5$ and $y = 7$. Since $\frac{\partial^2 P}{\partial x^2} = -4$, $\frac{\partial^2 P}{\partial x \partial y} = 2$, and $\frac{\partial^2 P}{\partial y^2} = -2$, we have $D(x, y) = (-4)(-2) - 2^2 = 4 > 0$ and $\frac{\partial^2 P}{\partial x^2} = -4 < 0$. Therefore, $P(x, y)$ has a local maximum at $(5, 7)$, and also a global one since $P(x, y)$ is a quadratic function. Thus, $P(x, y)$ achieves its maximum value of \$22,000 when $x = 5$ thousand units and $y = 7$ thousand units.

39. Given $n = 3$ and
x_i	-1	0	2
y_i	2	1	3
, we obtain

$$x_1 y_1 + x_2 y_2 + x_3 y_3 = (-1) \cdot 2 + 0 \cdot 1 + 2 \cdot 3 = 4, \quad x_1 + x_2 + x_3 = -1 + 0 + 2 = 1,$$

$$y_1 + y_2 + y_3 = 2 + 1 + 3 = 6, \quad x_1^2 + x_2^2 + x_3^2 = (-1)^2 + 0^2 + 2^2 = 5.$$

Then $a = \frac{3 \cdot 4 - 1 \cdot 6}{3 \cdot 5 - 1^2} = \frac{3}{7}$ and $b = \frac{6 - (3/7) \cdot 1}{3} = \frac{13}{7}$, and the least square line has the equation $y = \frac{3}{7} x + \frac{13}{7}$.

41. From the formula $E(a, b) = (ax_1 + b - y_1)^2 + \cdots + (ax_n + b - y_n)^2$, we get

$$E(a, b) = (b - 2)^2 + (a + b - 1)^2 + (2a + b - 1)^2 + (4a + b - 2)^2 + (5a + b - 5)^2 + (6a + b - 4)^2.$$

43. Let x be the price and y the number of gallons sold. From the given data, we obtain $x_1 y_1 + \cdots + x_4 y_4 = 40.178$, $x_1 + \cdots + x_4 = 6.8$, $y_1 + \cdots + y_4 = 25$, and $x_1^2 + \cdots + x_4^2 = 11.9272$. Then

$$a = \frac{4 \cdot (40.178) - (6.8) \cdot 25}{4 \cdot (11.9272) - (6.8)^2} \approx -6.32353 \quad \text{and} \quad b = \frac{25 + (6.32353) \cdot (6.8)}{4} \approx 17.$$

Hence, the least-squares line is (approximately) $y = -6.32353x + 17$. The predicted amount sold when the price is \$1.50 per gallon is $y(1.5) \approx 7.5$ thousand gallons.

45. With $f(x, y) = 4x^2 + y^2$ and $g(x, y) = xy - 2$, set $\frac{\partial f}{\partial x} = \lambda \frac{\partial g}{\partial x}$ and $\frac{\partial f}{\partial y} = \lambda \frac{\partial g}{\partial y}$, leading to the system of equations

$$8x = \lambda y$$
$$2y = \lambda x$$
$$xy = 2.$$

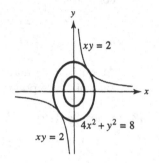

Multiplying the first equation by x and the second by y, we get $8x^2 = \lambda xy = 2y^2$. Therefore, $y = \pm 2x$, and by combining that with the third equation we find two solutions; $(1, 2)$ and $(-1, -2)$. By examining the level curves, we see that has a minimum value of 8 at each of those points.

47. Let $f(x, y, z) = x^2 + y^2 + z^2$ and $g(x, y, z) = 2x - 3y - z + 7$. Set $\frac{\partial f}{\partial x} = \lambda \frac{\partial g}{\partial x}$, or $2x = 2\lambda$, $\frac{\partial f}{\partial y} = \lambda \frac{\partial g}{\partial y}$, or $2y = -3\lambda$, $\frac{\partial f}{\partial z} = \lambda \frac{\partial g}{\partial z}$, or $2z = -\lambda$. This system of equations has the solution $x = -1$, $y = \frac{3}{2}$, $z = \frac{1}{2}$, and $\lambda = -1$. Thus, $(-1, \frac{3}{2}, \frac{1}{2})$ is the point on the plane $z = 2x - 3y + 7$ closest to the origin.

CHAPTER 7: PRACTICE EXAM

1. $\sqrt{(-1 - 1)^2 + (-2 - 0)^2 + (1 - 2)^2} = \sqrt{4 + 4 + 1} = 3$

3. The equation of this plane is $z - 4 = 1 \cdot (x - 2) + (-2)(y - 6)$, or $z = x - 2y + 14$.

5. $\frac{\partial f}{\partial x} = 3x^2 + 2xy + ye^{xy}$ and $\frac{\partial f}{\partial x}(2, 3) = 3 \cdot 2^2 + 2 \cdot 2 \cdot 3 + 3e^6 = 24 + 3e^6$

7. We have $\frac{\partial f}{\partial x} = 2x + 3y$ and $\frac{\partial f}{\partial y} = 1 + 3x$. Then $\frac{\partial f}{\partial x}(1, 2) = 8$ and $\frac{\partial f}{\partial y}(1, 2) = 4$. Also, $f(1, 2) = 1^2 + 2 + 3 \cdot 1 \cdot 2 = 9$. Thus near $(1,2)$ we have the linear approximation $f(x, y) \approx 9 + 8(x - 1) + 4(y - 2) = 8x + 4y - 7$.

9. The derivatives are

$$\frac{\partial f}{\partial x} = 3x^2 + 2y, \quad \frac{\partial f}{\partial y} = 2x - 12y, \quad \frac{\partial^2 f}{\partial x^2} = 6x, \quad \frac{\partial^2 f}{\partial y^2} = -12, \quad \frac{\partial^2 f}{\partial x \partial y} = 2.$$

Then $D(x, y) = 6x \cdot (-12) - 2^2 = -72x - 4$ and $D(0, 0) = -4 < 0$, and we conclude that $(0, 0)$ is a saddle point.

11. (a) In this case, $n = 3$ and

x_i	1	2	3
y_i	18	14	12

. Then

$$x_1 y_1 + x_2 y_2 + x_3 y_3 = 1 \cdot 18 + 2 \cdot 14 + 3 \cdot 12 = 82,$$

$$x_1 + x_2 + x_3 = 1 + 2 + 3 = 6, \quad y_1 + y_2 + y_3 = 18 + 14 + 12 = 44,$$

$$x_1^2 + x_2^2 + x_3^2 = 1 + 4 + 9 = 14.$$

Then $a = \frac{3 \cdot 82 - 6 \cdot 44}{3 \cdot 14 - 6^2} = -3$ and $b = \frac{44 - (-3) \cdot 6}{3} = \frac{62}{3}$. So the least squares line is $y = -3x + \frac{62}{3}$.
Another method is to minimize the squared-error function

$$E(a, b) = (a + b - 18)^2 + (2a + b - 14)^2 + (3a + b - 12)^2,$$

by solving the equations

$$\frac{\partial E}{\partial a} = 2(a + b - 18) + 4(2a + b - 14) + 6(3a + b - 12) = 0$$

$$\frac{\partial E}{\partial b} = 2(a + b - 18) + 2(2a + b - 14) + 2(3a + b - 12) = 0,$$

which reduce to the pair of equations

$$7a + 3b = 41 \quad \text{and} \quad 6a + 3b = 44.$$

Solving gives $a = -3$ and $b = \frac{62}{3}$.

(b) $y(4) = -3 \cdot 4 + \frac{62}{3} = \frac{26}{3} \approx 8.67$

13. We need to maximize $P(K, L) = 100 K^{0.36} L^{0.64}$ subject to the constraint $g(K, L) = 72K + 32L - 360{,}000 = 0$. Since $\frac{\partial P}{\partial K} = 36 K^{-0.64} L^{0.64}$, $\frac{\partial g}{\partial K} = 72$, $\frac{\partial P}{\partial L} = 64 K^{0.36} L^{-0.36}$, and $\frac{\partial g}{\partial L} = 32$, we must solve the system

$$36 K^{-0.64} L^{0.64} = \lambda \cdot 72, \quad 64 K^{0.36} L^{-0.36} = \lambda \cdot 32, \quad 72K + 32L = 360{,}000.$$

The first two equations give

$$\frac{36 K^{-0.64} L^{0.64}}{64 K^{0.36} L^{-0.36}} = \frac{72\lambda}{32\lambda} \quad \text{or} \quad L = 4K.$$

Then the constraint becomes $72K + 128K = 360{,}000$ or $200K = 360{,}000$, or $K = 1{,}800$. This gives $L = 7{,}200$. Thus $(K, L) = (1{,}800, 7{,}200)$ gives the maximum output.

CHAPTER 8
Trigonometric Functions

EXERCISES 8.1

1. $360° = 360 \cdot \dfrac{\pi}{180} = 2\pi$ radians, $\quad 450° = 450 \cdot \dfrac{\pi}{180} = \dfrac{5}{2}\pi$ radians,

$540° = 540 \cdot \dfrac{\pi}{180} = 3\pi$ radians

3. $135° = 135 \cdot \dfrac{\pi}{180} = \dfrac{3}{4}\pi$ radians, $\quad 225° = 225 \cdot \dfrac{\pi}{180} = \dfrac{5}{4}\pi$ radians

5. $\dfrac{5}{4}\pi = \dfrac{5}{4}\pi \cdot \dfrac{180}{\pi} = 225°, \quad \dfrac{7}{4}\pi = \dfrac{7}{4}\pi \cdot \dfrac{180}{\pi} = 315°$

7. $\dfrac{5}{6}\pi = \dfrac{5}{6}\pi \cdot \dfrac{180}{\pi} = 150°, \quad \dfrac{7}{6}\pi = \dfrac{7}{6}\pi \cdot \dfrac{180}{\pi} = 210°,$

$\dfrac{11}{6}\pi = \dfrac{11}{6}\pi \cdot \dfrac{180}{\pi} = 330°$

9. Each angle of an equilateral triangle is $60° = 60 \cdot \frac{\pi}{180} = \frac{\pi}{3}$ radians.

11.

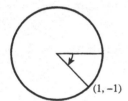

(1, −1)

13.

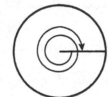

15.

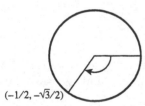

(−1/2, −√3/2)

17.

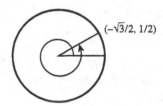

(−√3/2, 1/2)

19. $-\pi$

21. $\dfrac{3}{2}\pi$

23. -2π

25. $\dfrac{5}{4}\pi$

27. $\dfrac{13}{4}\pi$

29. $\dfrac{4}{3}\pi \cdot 5 = \dfrac{20}{3}\pi$ units

31.

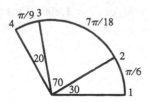

EXERCISES 8.2

1. The length of the hypotenuse is $\sqrt{2^2 + 1^2} = \sqrt{5}$. Therefore, $\sin\theta = \frac{1}{\sqrt{5}} = \frac{\sqrt{5}}{5}$, $\cos\theta = \frac{2}{\sqrt{5}} = \frac{2\sqrt{5}}{5}$, $\tan\theta = \frac{1}{2}$.

3. The length of the adjacent side is $\sqrt{3^2 - 1^2} = \sqrt{8}$. Therefore, $\sin\theta = \frac{1}{3}$, $\cos\theta = \frac{\sqrt{8}}{3} = \frac{2\sqrt{2}}{3}$, $\tan\theta = \frac{1}{\sqrt{8}} = \frac{\sqrt{2}}{4}$.

5. First, we convert $67°$ to $\frac{67\pi}{180}$ radians. If h is the height of the building, then $\tan(\frac{67\pi}{180}) = \frac{h}{20}$. Therefore, $h = 20\tan(\frac{67\pi}{180}) \approx 47.12$ feet.

7. (a) You multiply by one. The angle does not change if the sizes of the legs change.
   (b) Both of them equal $\cos\frac{\pi}{3} = \frac{1}{2}$. The cosine does not change if the lengths of both legs are multiplied by the same factor.

9. (a) If h is the length of the stick and l is the length of its shadow, then $\tan\theta = l/h$, or $h = l \cdot \tan\theta$.
   (b) Let s be the distance between Alexandria and Syene. Then r, s, θ (in radians) satisfy $s = r\theta$. Thus, the radius of the earth satisfies $r = s/\theta$.

11. $\sin\left(\frac{5\pi}{4}\right) = \sin\left(\pi + \frac{\pi}{4}\right) = \sin\pi\cos\left(\frac{\pi}{4}\right) + \cos\pi\sin\left(\frac{\pi}{4}\right) = -\frac{\sqrt{2}}{2}$

$\cos\left(\frac{5\pi}{4}\right) = \cos\left(\pi + \frac{\pi}{4}\right) = \cos\pi\cos\left(\frac{\pi}{4}\right) - \sin\pi\sin\left(\frac{\pi}{4}\right) = -\frac{\sqrt{2}}{2}$

$\tan\left(\frac{5\pi}{4}\right) = \frac{\sin\left(\frac{5\pi}{4}\right)}{\cos\left(\frac{5\pi}{4}\right)} = 1$

13. $\sin\left(-\frac{\pi}{3}\right) = -\sin\left(\frac{\pi}{3}\right) = -\frac{\sqrt{3}}{2}$, $\quad \cos\left(-\frac{\pi}{3}\right) = \cos\left(\frac{\pi}{3}\right) = \frac{1}{2}$, $\quad \tan\left(-\frac{\pi}{3}\right) = -\sqrt{3}$

15. $\tan\left(\frac{3\pi}{4}\right) = \tan\left(\frac{\pi}{2} + \frac{\pi}{4}\right) = \frac{\sin\left(\frac{\pi}{2} + \frac{\pi}{4}\right)}{\cos\left(\frac{\pi}{2} + \frac{\pi}{4}\right)} = \frac{\cos\left(\frac{\pi}{4}\right)}{-\sin\left(\frac{\pi}{4}\right)} = -1$

17. $\sin\left(-\frac{\pi}{6}\right) = -\sin\left(\frac{\pi}{6}\right) = -\frac{1}{2}$

19. $\cos\left(-\frac{2\pi}{3}\right) = \cos\left(\frac{2\pi}{3}\right) = -\frac{1}{2}$ (see exercise 12).

21. $\cos(3\pi) = \cos(2\pi + \pi) = \cos(\pi) = -1$.

23. If $0 \le \theta \le \frac{\pi}{2}$ and $\tan\theta = \frac{2}{3}$, then θ is the base angle of a right triangle with adjacent side of length 3 and opposite side of length 2. Therefore, the hypotenuse has length $\sqrt{3^2 + 2^2} = \sqrt{13}$, and $\sin\theta = \frac{2}{\sqrt{13}}$.

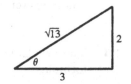

25. If $\frac{3\pi}{2} \le \theta \le 2\pi$, let $\mu = 2\pi - \theta$. Then $0 \le \mu \le \frac{\pi}{2}$, and $\sin\mu = -\sin\theta$, $\cos\mu = \cos\theta$, and $\tan\mu = -\tan\theta = \frac{4}{3}$. It follows that μ is the base angle of a right triangle with opposite side of length 4, adjacent side of length 3, and hypotenuse of length $\sqrt{3^2 + 4^2} = 5$.

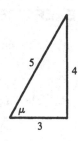

Therefore, $\sin\mu = \frac{4}{5}$ and $\cos\mu = \frac{3}{5}$, and we conclude that $\sin\theta = -\frac{4}{5}$ and $\cos\theta = \frac{3}{5}$.

27. $\sin(u - v) = \sin(u + (-v)) = \sin u \cdot \cos(-v) + \cos u \cdot \sin(-v) = \sin u \cos v - \cos u \sin v$
$\cos(u - v) = \cos(u + (-v)) = \cos u \cdot \cos(-v) - \sin u \cdot \sin(-v) = \cos u \cdot \cos v + \sin u \cdot \sin v$

29. We want $T(t) = a + b\sin(kt + d)$ to be cyclic over a 24-hour period; that is, $T(t + 24) = T(t)$. Therefore, we choose k so that $k \cdot 24 = 2\pi$, which gives $k = \pi/12$ and $T(t) = a + b\sin(\frac{\pi t}{12} + d)$.

At midnight, $t = 0$ and the temperature is 65 degrees. At noon, $t = 12$ and the temperature is 75 degrees. Thus,

$$T(0) = a + b\sin(d) = 65 \quad \text{and} \quad T(12) = a + b\sin(\pi + d) = a - b\sin(d) = 75.$$

Adding these equations gives $2a = 65 + 75 = 140$, or $a = 70$. Subtracting them gives $2b\sin(d) = -10$, or $b\sin(d) = -5$. The simplest choice of b and d satisfying this equation is $b = 5$ and $d = -\pi/2$. Thus, we obtain $T(t) = 70 + 5\sin(\frac{\pi t}{12} - \frac{\pi}{2})$.

There are other choices of b and d (though not of a) satisfying the given conditions, such as $b = -5$, $d = \pi/2$ (which gives the same function) and $b = -10$, $d = \pi/6$ (which gives a different function satisfying the same conditions).

31. If there are 70 heartbeats per minute, then the duration of one heartbeat cycle is 6/7 of a second. To make the function oscillate with that period, we take $c \cdot \frac{6}{7} = 2\pi$, or $c = \frac{7\pi}{3}$. Thus, $p(t) = a + b\sin(\frac{7\pi t}{3} + d)$.

Setting the maximum at $t = 0$ and the minimum at $t = \frac{3}{7}$ (halfway through the cycle), we get

$$p(0) = a + b\sin(d) = 120 \quad \text{and} \quad p\left(\frac{3}{7}\right) = a + b\sin(\pi + d) = a - b\sin(d) = 80.$$

Adding the equations gives $2a = 200$, or $a = 100$, and subtracting them gives $2b\sin(d) = 40$, or $b\sin(d) = 20$. We can take $b = 20$ and $d = \frac{\pi}{2}$. Thus,

$$p(t) = 100 + 20\sin\left(\frac{7\pi t}{3} + \frac{\pi}{2}\right).$$

(As in exercise 29, there is more than one possible choice.)

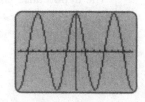

33. The graphs show are identical. We can verify that by the addition formula:

$$\sin\left(\frac{\pi}{2} + x\right) = \sin\left(\frac{\pi}{2}\right)\cos x + \cos\left(\frac{\pi}{2}\right)\sin x$$

$$= 1 \cdot \cos x + 0 \cdot \sin x = \cos x.$$

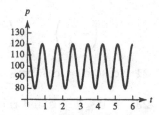

35. The graphs show that $\sin(\pi + x) = -\sin x$. We can verify that by the addition formula:

$$\sin(\pi + x) = \sin(\pi)\cos x + \cos(\pi)\sin x$$

$$= 0 \cdot \cos x - 1 \cdot \sin x = -\sin x.$$

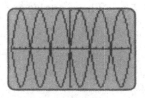

37. $\sin(2x) = \sin(x + x) = \sin x \cos x + \cos x \sin x = 2\sin x \cos x$

39. Using the addition formulas and the results of the previous two exercises:

$$\sin(3x) = \sin(2x + x) = \sin(2x)\cos x + \cos(2x)\sin x$$

$$= 2\sin x \cos x \cos x + (\cos^2 x - \sin^2 x)\sin x$$

$$= 3\sin x \cos^2 x - \sin^3 x$$

$$\cos(3x) = \cos(2x + x) = \cos(2x)\cos x - \sin(2x)\sin x$$

$$= (\cos^2 x - \sin^2 x)\cos x - 2\sin x \cos x \sin x$$

$$= \cos^3 x - 3\sin^2 x \cos x.$$

41. The graphs are identical—that is, $\tan(\pi + x) = \tan x$. That follows from the addition formula for the tangent (exercise 40), since

$$\tan(\pi + x) = \frac{\tan \pi + \tan x}{1 + \tan \pi \tan x} \frac{0 + \tan x}{1 + 0 \cdot \tan x}$$

$$= \tan x.$$

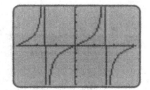

43. $\sin\left(-\dfrac{\pi}{12}\right) = \sin\left(\dfrac{\pi}{4} - \dfrac{\pi}{3}\right) = \sin\dfrac{\pi}{4}\cos\dfrac{\pi}{3} - \cos\dfrac{\pi}{4}\sin\dfrac{\pi}{3}$

$$= \frac{\sqrt{2}}{2} \cdot \frac{1}{2} - \frac{\sqrt{2}}{2} \cdot \frac{\sqrt{3}}{2} = \frac{1}{4}\left(\sqrt{2} - \sqrt{6}\right)$$

45. $\tan\left(\dfrac{7\pi}{12}\right) = \tan\left(\dfrac{\pi}{4} + \dfrac{\pi}{3}\right) = \dfrac{\tan\frac{\pi}{4} + \tan\frac{\pi}{3}}{1 - \tan\frac{\pi}{4}\tan\frac{\pi}{3}} = \dfrac{1 + \sqrt{3}}{1 - \sqrt{3}}$

47. $\cos(\frac{\pi}{4}) = \cos(\frac{\pi}{8} + \frac{\pi}{8}) = 1 - 2\sin^2(\frac{\pi}{8})$ (see exercise 38). Thus, $\sin^2(\frac{\pi}{8}) = \frac{1 - \cos(\pi/4)}{2} = \frac{2 - \sqrt{2}}{4}$. Since $\sin(\frac{\pi}{8}) > 0$, we obtain $\sin(\frac{\pi}{8}) = \frac{1}{2}\sqrt{2 - \sqrt{2}}$.

49. $\cos\left(\dfrac{\pi}{12}\right) = \cos\left(\dfrac{\pi}{24} + \dfrac{\pi}{24}\right) = 2\cos^2\left(\dfrac{\pi}{24}\right) - 1$ and $\dfrac{\pi}{12} = \dfrac{\pi}{3} - \dfrac{\pi}{4}$. Thus,

$$2\cos^2\left(\frac{\pi}{24}\right) - 1 = \cos\left(\frac{\pi}{3} - \frac{\pi}{4}\right) = \cos\frac{\pi}{3}\cos\frac{\pi}{4} + \sin\frac{\pi}{3}\sin\frac{\pi}{4} = \frac{\sqrt{6} + \sqrt{2}}{4}.$$

Therefore,

$$\cos\left(\frac{\pi}{24}\right) = \sqrt{\frac{1}{2} + \frac{\sqrt{6} + \sqrt{2}}{8}} = \sqrt{\frac{4 + \sqrt{2} + \sqrt{6}}{8}}.$$

51. $\tan^{-1} 3 \approx 1.2490$

53. These angles are $\frac{\pi}{2}$, $\sin^{-1}\frac{3}{5} \approx 0.6435$, and $\sin^{-1}\frac{4}{5} \approx 0.9273$.

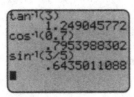

55. (a) We need to choose θ between 0 and $\pi/2$ so that $\tan\theta = \frac{1/8}{1/2} = 1/4$. A calculcator gives $\theta = \tan^{-1}(\frac{1}{4}) \approx 0.245$ radians.

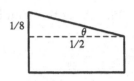

(b) We need to choose θ between 0 and $\pi/2$ so that $\cos\theta = 3/4$. A calculcator gives $\theta = \cos^{-1}(\frac{3}{4}) \approx 0.723$ radians.

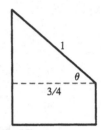

57.

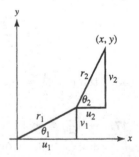

Guess: $\lim\limits_{t\to 0}\dfrac{\cos t - 1}{t} = 0$

59. We can think of each part of the arm as the hypotenuse of a right triangle, as shown. Then

$$\cos\theta_1 = \frac{u_1}{r_1} \quad \text{and} \quad \sin\theta_1 = \frac{v_1}{r_1},$$

so that $u_1 = r_1\cos\theta_1$ and $v_1 = r_1\sin\theta_1$. Similarly, $u_2 = r_2\cos\theta_2$ and $v_2 = r_2\sin\theta_1$. Moreover, $x = u_1 + u_2$ and $y = v_1 + v_2$. Thus, $x = r_1\cos\theta_1 + r_2\cos\theta_2$, $y = r_1\sin\theta_1 + r_2\sin\theta_2$.

61.
$$\begin{aligned}
x^2 + y^2 &= (r_1\cos\theta_1 + r_2\cos\theta_2)^2 + (r_1\sin\theta_1 + r_2\sin\theta_2)^2 \\
&= r_1^2\cos^2\theta_1 + 2r_1r_2\cos\theta_1\cos\theta_2 + r_2^2\cos^2\theta_2 \\
&\quad + r_1^2\sin^2\theta_1 + 2r_1r_2\sin\theta_1\sin\theta_2 + r_2^2\sin^2\theta_2 \\
&= r_1^2(\cos^2\theta_1 + \sin^2\theta_1) + r_2^2(\cos^2\theta_2 + \sin^2\theta_2) \\
&\quad + 2r_1r_2(\cos\theta_1\cos\theta_2 + \sin\theta_1\sin\theta_2) \\
&= r_1^2 + r_2^2 + 2r_1r_2\cos(\theta_2 - \theta_1)
\end{aligned}$$

EXERCISES 8.3

1. $\dfrac{d}{dx}(2\sin x - 3\cos x) = 2\dfrac{d}{dx}(\sin x) - 3\dfrac{d}{dx}(\cos x) = 2\cos x + 3\sin x$

3. $\dfrac{d}{dx}\left(\dfrac{1}{1+\cos x}\right) = -\dfrac{1}{(1+\cos x)^2}\cdot\dfrac{d}{dx}(1+\cos x) = \dfrac{\sin x}{(1+\cos x)^2}$

5. $\dfrac{d}{dx}(\sin^3 x - \cos^3 x) = 3\sin^2 x\cdot\dfrac{d}{dx}\sin x - 3\cos^2 x\cdot\dfrac{d}{dx}\cos x$

$$= 3\sin^2 x\cos x + 3\sin x\cos^2 x = 3\sin x\cos x(\sin x + \cos x)$$

7. $\dfrac{d}{dx}\tan(1+x^4) = \sec^2(1+x^4)\cdot\dfrac{d}{dx}(1+x^4) = 4x^3\sec^2(1+x^4)$

9. $\dfrac{d}{dt}[\cos^2(3t) - 2\sin^2(3t)] = 2\cos(3t)\dfrac{d}{dt}[\cos(3t)] - 4\sin(3t)\dfrac{d}{dt}[\sin(3t)]$

$$= 2\cos(3t)\cdot[-3\sin(3t)] - 4\sin(3t)\cdot[3\cos(3t)] = -18\sin(3t)\cos(3t)$$

11. $\dfrac{d}{dx}\left[\dfrac{1}{\tan(2x+1)}\right] = -\dfrac{\frac{d}{dx}\tan(2x+1)}{\tan^2(2x+1)}$

$$= -\dfrac{\sec^2(2x+1)}{\tan^2(2x+1)}\cdot\dfrac{d}{dx}(2x+1) = -\dfrac{2}{\sin^2(2x+1)}$$

13. $\dfrac{d}{dt}\{e^{2t}[\cos(3t) + \sin(3t)]\} = [\cos(3t) + \sin(3t)]\dfrac{d}{dt}e^{2t} + e^{2t}\dfrac{d}{dt}[\cos(3t) + \sin(3t)]$

$$= 2e^{2t}[\cos(3t) + \sin(3t)] + e^{2t}[-3\sin(3t) + 3\cos(3t)]$$

$$= e^{2t}[5\cos(3t) - \sin(3t)]$$

15. $\dfrac{d}{dx}(x\csc x) = \csc x + x\dfrac{d}{dx}\csc x = \csc x - x\csc x\cot x = \csc x(1 - x\cot x)$

17. $\dfrac{d}{dx}e^{\tan x} = e^{\tan x}\dfrac{d}{dx}\tan x = e^{\tan x}\sec^2 x$

19. (a) The vertical velocity of the sail boat is

$$D'(t) = 5[\sin^2 t\,\dfrac{d}{dt}\cos t + \cos t\,\dfrac{d}{dt}\sin^2 t]$$

$$= 5[-\sin^3 t + \cos t\cdot 2\sin t\,\dfrac{d}{dt}\sin t]$$

$$= 5[-\sin^3 t + 2\sin t\cos^2 t].$$

(b) $D'(k\pi) = 5[-\sin^3(k\pi) + 2\sin(k\pi)\cos^2(k\pi)] = 0$

21. By the chain rule

$$f'(x) = -2\cos\left(x + \frac{\pi}{4}\right)\sin\left(x + \frac{\pi}{4}\right).$$

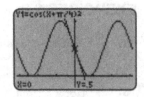

Thus, $f'(0) = -2\cos\left(\frac{\pi}{4}\right)\sin\left(\frac{\pi}{4}\right) = -1$. Also, $f(0) = \cos^2\left(\frac{\pi}{4}\right) = \frac{1}{2}$.
Therefore, the equation of the tangent line is $y = -x + \frac{1}{2}$.

23. $\dfrac{d}{dx}(\cot x) = \dfrac{d}{dx}\left(\dfrac{\cos x}{\sin x}\right) = \dfrac{(\sin x)\frac{d}{dx}\cos x - \cos x \frac{d}{dx}\sin x}{\sin^2 x}$

$$= \frac{-\sin^2 x - \cos^2 x}{\sin^2 x} = -\frac{\sin^2 x + \cos^2 x}{\sin^2 x} = -\frac{1}{\sin^2 x} = -\csc^2 x$$

25. Set $f'(x) = \frac{d}{dx}\sin^2 x = 2\sin x \frac{d}{dx}\sin x = 2\sin x \cos x = \sin(2x) = 0$. In the interval $(-\pi, \pi)$, this equation has the solutions $x = -\frac{\pi}{2}$, $x = 0$, and $x = \frac{\pi}{2}$, which are the critical points of $f(x)$. Checking the sign of $f'(x)$:

	$(-\pi, -\pi/2)$	$(-\pi/2, 0)$	$(0, \pi/2)$	$(\pi/2, \pi)$
$f'(x)$	$+$	$-$	$+$	$-$
$f(x)$	↗	↘	↗	↘

Thus, $f(x)$ has a local maximum at both $x = -\frac{\pi}{2}$ and $x = \frac{\pi}{2}$, and $f(x)$ has a local minimum at $x = 0$.

27. Set $f'(x) = \cos x + \sin x = 0$. This equation has the solutions $x = 3\pi/4$ and $7\pi/4$ in $(0, 2\pi)$. Checking the sign of $f'(x)$, we get:

	$(0,\ 3\pi/4)$	$(3\pi/4,\ 7\pi/4)$	$(7\pi/4,\ 2\pi)$
$f'(x)$	$+$	$-$	$+$

Thus, $f(x)$ is increasing in $(0, \frac{3\pi}{4})$ and $(\frac{7\pi}{4}, 2\pi)$, and decreasing on $(\frac{3\pi}{4}, \frac{7\pi}{4})$.

29. Set $f'(x) = 1 - 2\cos x = 0$. This equation has the solutions $x = \pi/3$ and $5\pi/3$ in $(0, 2\pi)$. Checking the sign of $f'(x)$, we get:

	$(0,\ \pi/3)$	$(\pi/3,\ 5\pi/3)$	$(5\pi/3,\ 2\pi)$
$f'(x)$	$-$	$+$	$-$

Thus, $f(x)$ is increasing in $(\frac{\pi}{3}, \frac{5\pi}{3})$, and $f(x)$ is decreasing in $(0, \frac{\pi}{3})$ and $(\frac{5\pi}{3}, 2\pi)$.

31. $y' = \dfrac{d}{dx}(x + \cos x) = 1 - \sin x$ and $y'' = \dfrac{d}{dx}(1 - \sin x) = -\cos x$

The solutions of $y'' = 0$ in $(0, 2\pi)$ are $x = \pi/2$ and $x = 3\pi/2$. The sign is:

	$(0, \pi/2)$	$(\pi/2, 3\pi/2)$	$(3\pi/2, 2\pi)$
y''	$-$	$+$	$-$

Thus, the graph has inflection points at $x = \pi/2$ and $x = 3\pi/2$.

33. **(a)** By the chain rule,

$$M'(t) = 2.4\sin\left(\frac{\pi t}{14}\right)\cos\left(\frac{\pi t}{14}\right) \cdot \frac{\pi}{14} = \frac{6\pi}{35}\sin\left(\frac{\pi t}{14}\right)\cos\left(\frac{\pi t}{14}\right).$$

Thus, $M'(t) = 0$ when $\frac{\pi t}{14} = \frac{k\pi}{2}$ for $k = 0, 1, 2, 3, \ldots$, or, in other words, when $t = 0, 7, 14, 21, 28, \ldots$. The sign of $M'(t)$ is:

	(0,7)	(7,14)	(14,21)	(21,28)	(28,35)	$\cdots$
$M'(t)$	+	−	+	−	+	$\cdots$

Mark is most happy when $t = 7, 21, 35, \ldots$, and he is least happy when $t = 0, 14, 28, \ldots$.

(b) $M(7) = M(21) = M(35) = \cdots = 0.95$
$M(0) = M(14) = M(28) = \cdots = -0.25$

(c) The frequency has increased but the range has decreased. That is, Mark's moods swing more often and he is not quite as happy at his high points, but he is a lot less sad at his low points.

35. (a) We first graph the function in a $[0, \frac{2\pi}{7}] \times [70, 130]$ window, use the TRACE to get as close as possible to the maximum (or minimum) point, then ZOOM in twice. We get a maximum value of 121.6333 and a minimum of 78.366704.

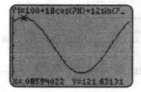 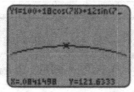 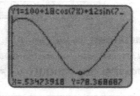 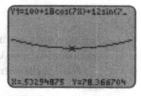

(b) Using the equation solver to solve $p'(t) = -126\sin(7t) + 84\cos(7t) = 0$, we get $t \approx 0.53279932244$, with a minimum value of $p(0.53279932244) \approx 78.36669235$, and $t \approx 0.08400037193$, with a maximum value of $p(0.08400037193) \approx 121.6333077$.

37. $f'(x) = 2\cos(2x) - \cos x$. When $f'(x)$ is negative (positive), $f(x)$ is decreasing (increasing). When $f'(x)$ is increasing (decreasing), the graph of $f(x)$ is concave up (down).

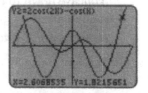

39. The relation between $f'(x)$ and $f(x)$ is as stated in exercise 37.

$$f'(x) = 1 - 2\sin x$$

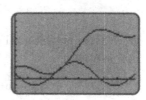

41. Let $x = x(t)$ be the position of the point of light on the shoreline at the end of t minutes. Then $x = 300\tan\theta$ and $x'(t) = 300\sec^2\theta\,\frac{d\theta}{dt}$. We are told that at the moment in question, when the hypotenuse of the right triangle is 500, $x'(t) = 2,500$. At that same moment $x = \sqrt{500^2 - 300^2} = 400$, and $\sec\theta = 5/3$.

We are also told that the light makes k rotations per hour, which means that it covers $\frac{2\pi k}{60} = \frac{k\pi}{30}$ radians per minute. Therefore $\frac{d\theta}{dt} = \frac{k\pi}{30}$. Substituting these values into the formula for $x'(t)$, we get $2,500 = 300 \cdot (\frac{5}{3})^2 \cdot \frac{k\pi}{30}$, and solving for k gives $k = \frac{2,500\cdot 30}{300\pi} \cdot \frac{9}{25} = \frac{90}{\pi}$.

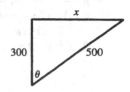

43. Let x be the distance between the deer and the base of the tree (see sketch). Then $x = 20 \tan \theta$ and $\frac{dx}{dt} = 20 \sec^2 \theta \frac{d\theta}{dt}$. When $x = 15$, the hypotenuse of the right triangle is $\sqrt{15^2 + 20^2} = 25$, so that $\sec \theta = \frac{25}{20} = \frac{5}{4}$. Thus,

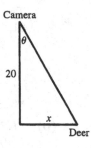

Camera

20

x

Deer

$$\frac{d\theta}{dt} = \frac{1}{20 \sec^2 \theta} \cdot \frac{dx}{dt} = \frac{1}{20 \left(\frac{5}{4}\right)^2} \cdot \frac{dx}{dt} = \frac{4}{125} \frac{dx}{dt}.$$

Since $\frac{dx}{dt} = \frac{22}{3}$ (as given in the problem), we get $\frac{d\theta}{dt} = \frac{4}{125} \cdot \frac{22}{3} = \frac{88}{375}$ radians per second.

45. With $y = \cos^2 x$, we have $y' = -2 \sin x \cos x = -\sin(2x)$, which is zero at $x = \frac{\pi}{2}, \pi, \frac{3\pi}{2}, 2\pi$. The sign of y' is:

x	$(0, \pi/2)$	$(\pi/2, \pi)$	$(\pi, 3\pi/2)$	$(3\pi/2, 2\pi)$
y'	$-$	$+$	$-$	$+$
y	↘	↗	↘	↗

Thus, $\cos^2 x$ is increasing on $(\frac{\pi}{2}, \pi)$ and $(\frac{3\pi}{2}, 2\pi)$ and decreasing on $(0, \frac{\pi}{2})$ and $(\pi, \frac{3\pi}{2})$, with local maxima at $x = 0, \pi, 2\pi$ and local minima at $x = \frac{\pi}{2}, \frac{3\pi}{2}$.

To check concavity, we compute $y'' = -2 \cos(2x)$, which is zero at $x = \frac{\pi}{4}, \frac{3\pi}{4}, \frac{5\pi}{4}, \frac{7\pi}{4}$. Its sign is:

x	$(0, \pi/4)$	$(\pi/4, 3\pi/4)$	$(3\pi/4, 5\pi/4)$	$(5\pi/4, 7\pi/4)$	$(7\pi/4, 2\pi)$
y''	$-$	$+$	$-$	$+$	$-$

Thus, the graph is concave up in $(\frac{\pi}{4}, \frac{3\pi}{4})$ and $(\frac{5\pi}{4}, \frac{7\pi}{4})$ and concave down in $(0, \frac{\pi}{4})$, $(\frac{3\pi}{4}, \frac{5\pi}{4})$, and $(\frac{7\pi}{4}, 2\pi)$, with inflection points at $x = \frac{\pi}{4}, \frac{3\pi}{4}, \frac{5\pi}{4}, \frac{7\pi}{4}$. There are no asymptotes. Also, note that $y \geq 0$ for all x, and its range is $[0, 1]$.

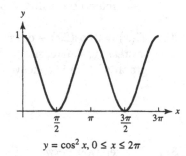

$y = \cos^2 x, 0 \leq x \leq 2\pi$

47. Over the interval $(-\pi, \pi)$, $\cot x$ is undefined at $x = 0$, and it has vertical asymptotes at $x = 0, \pm\pi$. In fact,

$$\lim_{x \to 0^+} \cot x = \lim_{x \to -\pi^+} \cot x = \infty \quad \text{and} \quad \lim_{x \to \pi^-} \cot x = \lim_{x \to 0^-} \cot x = -\infty.$$

$y' = -\csc^2 x$, which is positive throughout its domain but undefined at $x = 0$ (also at $x = \pm\pi, \pm2\pi$, etc.). Thus, $\cot x$ is decreasing in $(-\pi, 0)$ and $(0, \pi)$.

To check concavity, we compute $y'' = -2 \csc x \cdot (-\csc x \cdot \cot x) = 2 \csc^2 x \cot x$. Its sign is:

x	$(-\pi, -\pi/2)$	$(-\pi/2, 0)$	$(0, \pi/2)$	$(\pi/2, \pi)$
y''	$+$	$-$	$+$	$-$

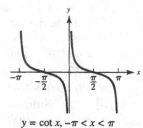

$y = \cot x, -\pi < x < \pi$

Thus, the graph is concave up in $(-\pi, -\frac{\pi}{2})$ and $(0, \frac{\pi}{2})$ and concave down in $(-\frac{\pi}{2}, 0)$ and $(\frac{\pi}{2}, \pi)$, with inflection points at $x = -\frac{\pi}{2}$ and $x = \frac{\pi}{2}$. Also, $y \geq 0$ for x in $(-\pi, \frac{\pi}{2}]$ and $(0, \frac{\pi}{2}]$, and $y \leq 0$ for x in $[-\frac{\pi}{2}, 0)$ and $[\frac{\pi}{2}, \pi)$.

49. With $y = x + \sin x$, $[0, 2\pi]$, we have $y' = 1 + \cos x$, which is zero at $x = \pi$. Its sign is:

x	$(0, \pi)$	$(\pi, 2\pi)$
y'	$+$	$+$
y	↗	↗

Thus, $x + \sin x$ is increasing throughout $(0, 2\pi)$, with a horizontal tangent (but no local extremum) at $x = \pi$. Over the interval $[0, 2\pi]$, it has a minimum of zero at $x = 0$ and a maximum of 2π at $x = 2\pi$.

To check concavity, we compute $y'' = -\sin x$, which is zero at $x = \pi$. Its sign is:

x	$(0, \pi)$	$(\pi, 2\pi)$
y''	$-$	$+$

Thus, the graph is concave up in $(\pi, 2\pi)$ and concave down in $(0, \pi)$, with an inflection point at $x = \pi$. There are no asymptotes.

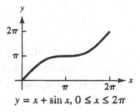

$y = x + \sin x, \ 0 \le x \le 2\pi$

51. (a) The height of the graph of $y = x \sin x$ is always less than or equal to that of $y = x$. That is because $\sin x \le 1$ for all x.

(b) The graphs interect at $x = 0$ and then at $x = \frac{\pi}{2}$ and $\frac{5\pi}{2}$, where $\sin x = 1$. The pattern continues at $\frac{9\pi}{2}, \frac{13\pi}{2}, \ldots$. Since $y' = \sin x + x \cos x$, we see that the slope of the graph equals 1 (the slope of the line) at the intersection points.

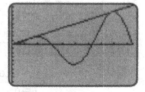

(c) Using the equation solver to solve $y' = \sin x + x \cos x = 0$, we find a local maximum at $x \approx 2.028758$ with (approximate) value 1.8197 and a second local maximum at $x \approx 7.9786657$ with (approximate) value 7.916727. In both cases, the maximum point is below the line $y = x$ and to the right of the point where the graph and the line intersect.

(d) The graph of $y = x \sin x$ lies above the line $y = -x$ except at $x = \frac{3\pi}{2}$, where they intersect and the graph has slope -1. They also intersect at $x = \frac{7\pi}{2}, \frac{11\pi}{2}, \ldots$, where $\sin x = -1$ and the slope equals -1.

Using the equation solver as before, we find that $x \sin x$ has minimum values of approximately -4.81447 at $x \approx 4.91318$ and approximately -11.04071 at $x \approx 11.08554$. Both are above the line and to the right of its intersection with the graph.

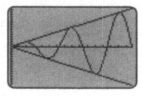

53. (a) Since $-1 \le \sin x \le 1$ and $f(x) > 0$, we have $-f(x) \le f(x) \sin x \le f(x)$.

(b) We want to solve $f(x) \sin x = f(x)$ for x. Since $f(x) \ne 0$, the equation reduces to $\sin x = 1$, whose solutions are $x = \frac{\pi}{2} + 2k\pi$ for $k = 0, 1, 2, 3, \ldots$. Since $y' = f'(x) \sin x + f(x) \cos x$, the slopes of $y = f(x) \sin x$ and $y = f(x)$ are both equal to $f'(2k\pi + \frac{\pi}{2})$ at $x = 2k\pi + \frac{\pi}{2}$, $k = 0, 1, 2, \ldots$.

(c) We want to solve $f(x) \sin x = -f(x)$ for x. Since $f(x) \ne 0$, the solutions are $x = 2k\pi + \frac{3\pi}{2}$, $k = 0, 1, 2, \ldots$. At those points, the slopes of $y = f(x) \sin x$ and $y = -f(x)$ are both equal to $-f'(2k\pi + \frac{3\pi}{2})$.

55. The system of equation $\frac{du}{dt} = 4v$ and $\frac{dv}{dt} = -u$ is a special case of (28), with $a = 2$ and $b = 1$. Thus, the general solution is

$$u = 2[M\sin(2t) + N\cos(2t)] \quad \text{and} \quad v = M\cos(2t) - N\sin(2t).$$

From the initial conditions $u(0) = -2$ and $v(0) = 1$, we get

$$-2 = 2[M\sin 0 + N\cos 0] = 2N \quad \text{and} \quad 1 = M\cos 0 - N\sin 0 = M.$$

Therefore, $M = 1$ and $N = -1$, and the solution is $u = 2\sin(2t) - 2\cos(2t)$ and $v = \sin(2t) + \cos(2t)$.

57. From equations (12) and (13) of section 8.2, we have

$$r_1\cos\theta_1 + r_2\cos\theta_2 = x \quad \text{and} \quad r_1\sin\theta_1 + r_2\sin\theta_2 = y,$$

where r_1 and r_2 are fixed and x, y, θ_1, and θ_2 are functions of t.

Since the arm moves along a vertical wall, x is constant, and, therefore, $\frac{dx}{dt} = 0$. And since the vertical speed is kept constant at 1 unit per second, $\frac{dy}{dt} = 1$. Therefore, by taking derivatives with respect to t is the equations above (and using the chain rule), we obtain $r_1\sin\theta_1\frac{d\theta_1}{dt} + r_2\sin\theta_2\frac{d\theta_2}{dt} = 0$ and $r_1\cos\theta_1\frac{d\theta_1}{dt} + r_2\cos\theta_2\frac{d\theta_2}{dt} = 1$.

59. $\displaystyle\frac{d}{dx}\cos x = \lim_{h\to 0}\frac{\cos(x+h) - \cos x}{h}$

$$= \lim_{h\to 0}\frac{\cos h\cos x - \sin h\sin x - \cos x}{h} \quad \text{(addition formula)}$$

$$= \lim_{h\to 0}\frac{\cos x(\cos h - 1) - \sin h\sin x}{h}$$

$$= \lim_{h\to 0}\left[\frac{\cos h - 1}{h}\cos x - \frac{\sin h}{h}\sin x\right]$$

$$= 0\cdot\cos x - 1\cdot\sin x = -\sin x$$

EXERCISES 8.4

1. Let $u = 3x$. Then $du = 3\,dx$ or $dx = \frac{1}{3}\,du$. Therefore,

$$\int\sin(3x)\,dx = \int\frac{1}{3}\sin u\,du = -\frac{1}{3}\cos u + c = -\frac{1}{3}\cos(3x) + c.$$

3. Let $u = x + 1$. Then $du = dx$, and

$$\int\sec^2(x+1)\,dx = \int\sec^2 u\,du = \tan u + c = \tan(x+1) + c.$$

5. Let $u = t$ and $dv = \sin t\,dt$. Then $du = dt$ and $v = -\cos t$. Therefore,

$$\int t\sin t\,dt = -t\cos t + \int\cos t\,dt = -t\cos t + \sin t + c.$$

7. Let $u = t^2$ and $dv = \cos t\, dt$. Then $du = 2t\, dt$ and $v = \sin t$. Therefore,

$$\int t^2 \cos t\, dt = t^2 \sin t - 2 \int t \sin t\, dt$$

$$= t^2 \sin t - 2(-t \cos t + \sin t) + c \quad \text{(by exercise 5)}$$

$$= t^2 \sin t + 2t \cos t - 2 \sin t + c.$$

9. Using the identity $\sin(2x) = 2 \sin x \cos x$, we obtain

$$\int \sin(2x) \cos x\, dx = 2 \int \sin x \cos^2 x\, dx$$

$$= -2 \int u^2\, du = -\frac{2}{3} u^3 + c = -\frac{2}{3} \cos^3 x + c,$$

where we have used the substitution $u = \cos x$ and $du = -\sin x\, dx$.

11. Let $u = \cot x$. Then $du = -\csc^2 x\, dx = -\frac{1}{\sin^2 x}\, dx$, so that

$$\int \frac{\cot^2 x}{\sin^2 x}\, dx = -\int u^2\, du = -\frac{1}{3} u^3 + c = -\frac{1}{3} \cot^3 x + c.$$

13. The area equals $\int_0^\pi \sin x\, dx = -\cos x \big|_0^\pi = -\cos \pi + \cos 0 = 2$.

15. The area equals. $\int_0^\pi x \sin x\, dx$. Let $u = x$ and $dv = \sin x\, dx$. Then $du = dx$ and $v = -\cos x$, and

$$\int_0^\pi x \sin x\, dx = -x \cos x \Big|_0^\pi + \int_0^\pi \cos x\, dx = \pi + \sin x \Big|_0^\pi = \pi.$$

17. Let $u = 3x$. Then $du = 3\, dx$ or $dx = \frac{1}{3}\, du$. If $x = -\frac{\pi}{2}$, then $u = -\frac{3\pi}{2}$; and if $x = \frac{\pi}{2}$, then $u = \frac{3\pi}{2}$. Therefore,

$$\int_{-\pi/2}^{\pi/2} \sin(3x)\, dx = \int_{-3\pi/2}^{3\pi/2} \sin u \cdot \frac{1}{3}\, du = -\frac{1}{3} \cos u \Big|_{-3\pi/2}^{3\pi/2} = \frac{2}{3} \cos \left(\frac{3\pi}{2} \right) = 0.$$

19. Let $u = \pi t^2$. Then $du = 2\pi\, dt$ or $dt = \frac{1}{2\pi} du$. When $t = 0$, $u = 0$; and when $t = 1$, $u = \pi$. Therefore,

$$\int_0^1 t \cos(\pi t^2)\, dt = \int_0^\pi \frac{1}{2\pi} \cos u\, du = \frac{1}{2\pi} \sin u \Big|_0^\pi = \frac{1}{2\pi}(\sin \pi - \sin 0) = 0.$$

21. $\displaystyle\int_{-\pi/4}^{\pi/4} \frac{1}{\cos^2 x}\, dx = \int_{-\pi/4}^{\pi/4} \sec^2 x\, dx = \tan x \Big|_{-\pi/4}^{\pi/4} = \tan \frac{\pi}{4} - \tan \left(-\frac{\pi}{4} \right) = 2$

23. $\displaystyle\int_0^{\pi/4} \tan x\, dx = \ln |\sec x| \, \Big|_0^{\pi/4} = \ln \left| \sec \frac{\pi}{4} \right| - \ln |\sec 0| = \ln \sqrt{2} - \ln 1 = \frac{\ln 2}{2}$

25. Let $u = \sec x$. Then $du = \sec x \tan x\, dx$. When $x = 0$, $u = \sec 0 = 1$; and when $x = \frac{\pi}{3}$, $u = \sec \frac{\pi}{3} = 2$. Then

$$\int_0^{\pi/3} \sec^3 x \tan x\, dx = \int_0^{\pi/3} \sec^2 x \cdot \sec x \cdot \tan x\, dx = \int_1^2 u^2\, du = \frac{1}{3} u^3 \Big|_1^2 = \frac{7}{3}.$$

27. From example 8.3.8, $P = 2 + 0.5[1.25\sin(0.02t) - 0.6\cos(0.02t)]$. Then the average over the time interval $[0, 10]$ is

$$\frac{1}{10}\int_0^{10}[2 + 0.625\sin(0.02t) - 0.3\cos(0.02t)]\,dt$$

$$= \frac{1}{10}\left[\int_0^{10}2\,dt + 0.625\int_0^{10}\sin(0.02t)\,dt - 0.3\int_0^{10}\cos(0.02t)dt\right]$$

$$= \frac{1}{10}\left[20 - 31.25\cos(0.02t)\Big|_0^{10} - 15\sin(0.02t)\Big|_0^{10}\right]$$

$$= \frac{1}{10}[20 - 31.25\cos 0.2 + 31.25 - 15\sin 0.2]$$

$$= 5.125 - 3.125\cos 0.2 - 1.5\sin 0.2 \approx 1.764 \text{ thousand prey.}$$

29. (a) Since $R(y) = \sec(1 - y)$, $0 \le y \le 2$, has its minimum value at $y = 1$, the viral infection occurs at the end of the first year. (Recall that $\cos u$ takes its maximum over $(-\pi, \pi)$, and therefore also over $(-1, 1)$ at $u = 0$. Thus $\sec u$ takes its minimum over the same interval at $u = 0$.)

(b) The total number of birds hatched in this two-year period is

$$\int_0^2 \sec(1 - y)dy = \int_0^2 \sec(y - 1)dy = \ln|\tan(y - 1) + \sec(y - 1)|\Big|_0^2$$

$$= \ln|\tan 1 + \sec 1| - \ln|\tan(-1) + \sec(-1)|$$

$$= \ln\left|\frac{\sec 1 + \tan 1}{\sec 1 - \tan 1}\right| \approx 2.45.$$

31. Using the substitution $u = 1 + \sin t$, we obtain

$$\int_0^{\pi/2}\frac{\cos t}{1 + \sin t}\,dt = \int_1^2\frac{du}{u}$$

$$= \ln u\Big|_1^2 = \ln 2 \approx 0.6931471806.$$

33. Using exercise 5, we obtain

$$\int_0^\pi x\sin x\,dx = (-x\cos x + \sin x)\Big|_0^\pi$$

$$= \pi \approx 3.141592654.$$

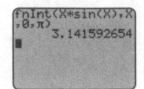

35. (a) Let $u = e^x$. Then $du = e^x\,dx$. The level of the chemical in the subject's blood is given by

$$y = \int e^x\cos(e^x)\,dx = \int\cos u\,du = \sin u + c = \sin(e^x) + c.$$

(b) Since $y(0) = 4$, we have $4 = \sin(e^0) + c = \sin 1 + c$, or $c = 4 - \sin 1$. Thus, $y = \sin(e^x) + 4 - \sin 1$.

(c) The level oscillates with greater and greater frequency and the time between peaks and lows approaching zero.

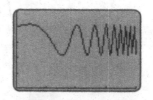

37. We use the equation solver (with an initial guess between 0 and $\pi/2$, such as $\pi/4$) to solve $x - \cos x = 0$. The result, as shown, is $x \approx 0.739$.

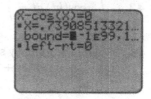

To the left of the intersection point, $\cos x > x$, and to the right $x > \cos x$. That follows, for instance, from the observation that $\cos 0 = 1$ and $\cos(\frac{\pi}{2}) = 0$. Therefore, the area between the graphs is (approximately)

$$\int_0^{0.739} (\cos x - x)\, dx + \int_{0.739}^{\pi/2} (x - \cos x)\, dx$$

$$= \left(\sin x - \frac{x^2}{2}\right)\Big|_0^{0.739} + \left(\frac{x^2}{2} - \sin x\right)\Big|_{0.739}^{\pi/2}$$

$$= \sin(0.739) - \frac{(0.739)^2}{2} + \left(\frac{\pi^2}{8} - 1\right) - \left[\frac{(0.739)^2}{2} - \sin(0.739)\right]$$

$$= 2\sin(0.739) - (0.739)^2 + \frac{\pi^2}{8} - 1 \approx 1.035.$$

Note: This integral may also be estimated by using the numerical integrator on your calculator, applied to the absolute value of $x - \cos x$, as shown.

39. The equations $\sec x = 3$ and $\cos x = \frac{1}{3}$ have the same solutions, so we use the equation solver, as shown, to to find the positive solution. We use symmetry to find the negative solution. Thus, the area is approximately

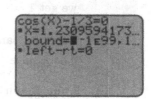

$$\int_{-1.231}^{1.231} (3 - \sec x)\, dx = 6 \cdot (1.231) - \ln|\sec x + \tan x|\Big|_{-1.231}^{1.231}$$

$$= 7.386 - \ln\left|\frac{\sec(1.231) + \tan(1.231)}{\sec(1.231) - \tan(1.231)}\right| \approx 3.860.$$

41. We first write

$$\int \csc x\, dx = \int \csc x \cdot \frac{\csc x + \cot x}{\csc x + \cot x}\, dx = \int \frac{\csc^2 x + \cot x \csc x}{\csc x + \cot x}\, dx.$$

Next, let $u = \csc x + \cot x$. Then $du = (-\csc^2 x - \cot x \csc x)\, dx$, and

$$\int \csc x \, dx = \int \csc x \cdot \frac{\csc x + \cot x}{\csc x + \cot x} \, dx = \int \frac{\csc^2 x + \cot x \csc x}{\csc x + \cot x} \, dx$$

$$= -\int \frac{1}{u} \, du = -\ln|u| = -\ln|\csc x + \cot x| + c.$$

43. $\int_0^\pi \cos(kt) \, dt = \frac{1}{k} \sin(kt)\Big|_0^\pi = \frac{1}{k} \sin(k\pi) - \frac{1}{k}\sin 0 = 0$

45. Using the identity $\sin^2(kt) = \frac{1}{2} - \frac{1}{2}\cos(2kt)$ and the substitution $x = kt$ we obtain that $\int \sin^2 kt \, dt = \frac{1}{2}t - \frac{1}{4k}\sin(2kt) + c.$ Then

$$\int_0^\pi \sin^2(kt)\,dt = \left[\frac{1}{2}t - \frac{1}{4k}\sin(2kt)\right]\Big|_0^\pi = \left[\frac{\pi}{2} - \frac{1}{4k}\sin(2k\pi)\right] - \left[0 - \frac{1}{4k}\sin 0\right] = \frac{\pi}{2}.$$

47. $c_1 = \frac{2}{\pi}\int_0^\pi f(t)\sin(2t)\,dt = \frac{2}{\pi}.\ c_2 = \frac{2}{\pi}\int_0^\pi f(t)\sin(4t)\,dt = -\frac{2}{\pi}.\ c_3 = \frac{2}{\pi}\int_0^\pi f(t)\sin(6t)\,dt = \frac{4}{\pi}$

49. $y' = \frac{\cos t}{y}$. Separating variables gives $y\,dy = \cos t\,dt$, or $\int y\,dy = \int \cos t\,dt$, or $\frac{1}{2}y^2 = \sin t + c$, or $y^2 = 2\sin t + 2c$. The initial condition $y(0) = 2$ gives $2^2 = 2\sin 0 + 2c$, or $2c = 4$. So $y^2 = 2\sin t + 4$.

51. $\frac{dy}{dt} = \frac{\cos t}{\cos y}$. Separating variables gives $\cos y\,dy = \cos t\,dt$, or $\int \cos y\,dy = \int \cos t\,dt$, or $\sin y = \sin t + c$. The initial condition $y(0) = 2\pi$ gives $\sin 2\pi = \sin 0 + c$, or $c = 0$. So $\sin y = \sin t$ or $y = t + 2k\pi$, $k = 0, \pm 1, \pm 2$. Again, the initial condition gives $y = t + 2\pi$.

53. If $y = M\sin(at) + N\cos(at)$, then $y' = Ma\cos(at) - Na\sin(at)$ and

$$y'' = -Ma^2\sin(at) - Na^2\cos(at) = -a^2[M\sin(at) + N\cos(at)].$$

So, $y'' + a^2 y = -a^2[M\sin(at) + N\cos(at)] + a^2[M\sin(at) + N\cos(at)] = 0.$

55. **(a)** Taking the derivative with respect to x on both sides of $x = \tan y$ and using the chain rule, we get $1 = \sec^2 y \frac{dy}{dx}$. Thus $\frac{dy}{dx} = \frac{1}{\sec^2 y}$.

(b) First, $1 + \tan^2 y = 1 + \frac{\sin^2 y}{\cos^2 y} = \frac{\sin^2 y + \cos^2 y}{\cos^2 y} = \frac{1}{\cos^2 y} = \sec^2 y.$
Then, since $x = \tan y$, we get $\frac{dy}{dx} = \frac{1}{\sec^2 y} = \frac{1}{1 + \tan^2 y} = \frac{1}{1 + x^2}.$

(c) From $\frac{d}{dx}\tan^{-1} x = \frac{1}{1+x^2}$, we get $\int \frac{1}{1+x^2}\,dx = \tan^{-1} x + c.$

(d) $\int_0^1 \frac{1}{1+x^2}\,dx = \tan^{-1} x\Big|_0^1 = \tan^{-1}(1) - \tan^{-1}(0) = \frac{\pi}{4} - 0 = \frac{\pi}{4}.$

CHAPTER 8: REVIEW EXERCISES

1. **(a)** $6° = 6 \cdot \frac{\pi}{180} = \frac{\pi}{30}$ radians
  (b) $150° = 150 \cdot \frac{\pi}{180} = \frac{5\pi}{6}$ radians

  (c) $345° = 345 \cdot \frac{\pi}{180} = \frac{23\pi}{12}$ radians

3. (a) 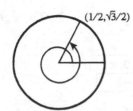 (1/2,√3/2)   (b)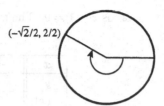
(−√2/2, 2/2)

5. The hypotenuse equals $\sqrt{3^2 + 4^2} = 5$. Therefore,

$$\sin\theta = \frac{3}{5}, \quad \cos\theta = \frac{4}{5}, \quad \tan\theta = \frac{3}{4}.$$

7. (a) $\sec\left(\dfrac{\pi}{6}\right) = \dfrac{1}{\cos(\pi/6)} = \dfrac{1}{\sqrt{3}/2} = \dfrac{2}{\sqrt{3}} = \dfrac{2\sqrt{3}}{3}$ (b) $\csc\left(\dfrac{5\pi}{4}\right) = \dfrac{1}{\sin(5\pi/4)} = -\dfrac{1}{\sqrt{2}/2} = -\sqrt{2}$

(c) $\cot\left(\dfrac{3\pi}{2}\right) = \dfrac{\cos(3\pi/2)}{\sin(3\pi/2)} = 0$ (d) $\sec\left(\dfrac{2\pi}{3}\right) = \dfrac{1}{\cos(2\pi/3)} = -\dfrac{1}{1/2} = -2$

9. $1 + \cot^2 t = 1 + \dfrac{\cos^2 t}{\sin^2 t} = \dfrac{\sin^2 t + \cos^2 t}{\sin^2 t} = \dfrac{1}{\sin^2 t} = \csc^2 t$

11. $\cot(u+v) = \dfrac{\cos(u+v)}{\sin(u+v)} = \dfrac{\cos u\cos v - \sin u\sin v}{\cos v\sin u + \sin v\cos u}$, and dividing the numerator and denominator by $\sin u\sin v$ gives

$$\frac{\dfrac{\cos u\cos v}{\sin u\sin v} - 1}{\dfrac{\cos v}{\sin v} + \dfrac{\cos u}{\sin u}} = \frac{\cot u\cot v - 1}{\cot u + \cot v}.$$

13. $\dfrac{dy}{dx} = \tan x\,\dfrac{d}{dx}x + x\,\dfrac{d}{dx}\tan x = \tan x + x\sec^2 x$

15. $\dfrac{dy}{dx} = -\dfrac{\frac{d}{dx}(1 + \cos^2 x)}{(1 + \cos^2 x)^2} = -\dfrac{2\cos x\,\frac{d}{dx}\cos x}{(1 + \cos^2 x)^2} = \dfrac{2\sin x\cos x}{(1 + \cos^2 x)^2}$

17. First, we compute

$$f'(x) = (\sin x + \cos x)\,\frac{d}{dx}e^x + e^x\,\frac{d}{dx}(\sin x + \cos x)$$

$$= e^x(\sin x + \cos x) + e^x(\cos x - \sin x) = 2e^x\cos x.$$

Then

$$f''(x) = 2\left(\cos x\,\frac{d}{dx}e^x + e^x\,\frac{d}{dx}\cos x\right) = 2(e^x\cos x - e^x\sin x).$$

19. $f'(x) = 2\cos(2x) - 4\cos x$, and

$$f''(x) = -4\sin(2x) + 4\sin x = -8\sin x\cos x + 4\sin x = -8\sin x\left(\cos x - \frac{1}{2}\right).$$

The only solution in $\left(0, \frac{\pi}{2}\right)$ is where $\cos x = \frac{1}{2}$, that is, at $x = \pi/3$. The sign of $f''(x)$ is:

x	$(0, \pi/3)$	$(\pi/3, \pi/2)$
$f''(x)$	$-$	$+$

We conclude that $f(x)$ has an inflection point at $x = \frac{\pi}{3}$.

21. First, set $y' = 2\cos x - 1 = 0$. The solutions in $(-\pi, \pi)$ are $x = -\frac{\pi}{3}$ and $x = \frac{\pi}{3}$. The sign of y' is as follows:

x	$(-\pi, -\pi/3)$	$(-\pi/3, \pi/3)$	$(\pi/3, \pi)$
y'	$-$	$+$	$-$
$f(x)$	$\searrow$	$\nearrow$	$\searrow$

Thus, $2\sin x - x$ is increasing in $(-\frac{\pi}{3}, \frac{\pi}{3})$ and decreasing in $(-\pi, -\frac{\pi}{3})$, $(\frac{\pi}{3}, \pi)$, with a local maximum at $x = \frac{\pi}{3}$ and local minimum at $x = -\frac{\pi}{3}$ To check concavity, we compute $y'' = -2\sin x$, which is zero at $x = 0$. Then

x	$(-\pi, 0)$	$(0, \pi)$
y''	$+$	$-$
y	$\cup$	$\cap$

Thus, the graph is concave up in $(-\pi, 0)$ and concave down in $(0, \pi)$, with an inflection point at $x = 0$. There are no asymptotes. Also $y(0) = 0$.

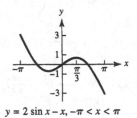

$y = 2\sin x - x,\ -\pi < x < \pi$

23. From $\tan\theta = \frac{x}{20}$, $-\frac{\pi}{2} < \theta < \frac{\pi}{2}$, we obtain $x = 20\tan\theta$ and, therefore, $\frac{dx}{dt} = 20\sec^2\theta\frac{d\theta}{dt}$. Since the spotlight is rotating at a constant rate, $\frac{d\theta}{dt}$ is constant, which means that $\frac{dx}{dt}$ achieves a minimum when $\sec^2\theta$ does. That occurs at $\theta = 0$, since $\sec 0 = 1$ and $\sec^2\theta > 1$ for all other θ. But $\theta = 0$ means that $x = 0$, so that the velocity of the point of light on the wall is at a minimum when it passes through point A.

25. First method: using the identity $\sin(2t) = 2\sin t\cos t$, we have

$$\int \sin(3x)\cos(3x)\,dx = \int \frac{1}{2}\sin(6x)\,dx = -\frac{1}{12}\cos(6x) + c.$$

Second method: using the substitution $u = \sin(3x)$, with $du = 3\cos(3x)\,dx$,

$$\int \sin(3x)\cos(3x)\,dx = \frac{1}{3}\int u\,du = \frac{u^2}{6} + c = \frac{\sin^2(3x)}{6} + c.$$

The answers are seen to be equivalent using the identity $\sin^2\theta = \frac{1}{2}[1 + \cos(2\theta)]$.

27. Let $u = x$ and $dv = \sec^2 x\,dx$. Then $du = dx$ and $v = \tan x$. Therefore,

$$\int x\sec^2 x\,dx = x\tan x - \int \tan x\,dx = x\tan x - \ln|\sec x| + c.$$

29. Let $u = 1 + \sin t$. Then $du = \cos t\,dt$. If $t = 0$, then $u = 1 + \sin 0 = 1$; and if $t = \frac{\pi}{2}$, then $u = 1 + \sin\frac{\pi}{2} = 2$. Thus,

$$\int_0^{\pi/2} \frac{\cos t}{1 + \sin t}\,dt = \int_1^2 \frac{1}{u}\,du = \ln|u|\Big|_1^2 = \ln 2 - \ln 1 = \ln 2.$$

31. The area under the graph of $y = \cos^3 x$ from $-\frac{\pi}{2}$ to $\frac{\pi}{2}$ is $\int_{-\pi/2}^{\pi/2} \cos^3 x \, dx$. Let $u = \sin x$. Then $du = \cos x \, dx$. If $x = -\frac{\pi}{2}$, then $u = \sin(-\frac{\pi}{2}) = -1$; and if $x = \frac{\pi}{2}$, then $u = \sin(\frac{\pi}{2}) = 1$. Therefore,

$$\int_{-\pi/2}^{\pi/2} \cos^3 x \, dx = \int_{-\pi/2}^{\pi/2} \cos^2 x \cos x \, dx = \int_{-\pi/2}^{\pi/2} (1 - \sin^2 x) \cos x \, dx$$

$$= \int_{-1}^{1} (1 - u^2) \, du = \left(u - \frac{1}{3} u^3 \right) \Big|_{-1}^{1}$$

$$= \left(1 - \frac{1}{3} \right) - \left(-1 + \frac{1}{3} \right) = \frac{4}{3}.$$

33. Note that $\sec x < 2$ on the interval $(-\frac{\pi}{3}, \frac{\pi}{3})$, with $\sec(\pm\frac{\pi}{3}) = 2$. Therefore, the area between the graph and the line is given by

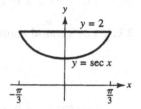

$$\int_{-\pi/3}^{\pi/3} (2 - \sec x) \, dx = 2x - \ln|\sec x + \tan x| \Big|_{-\pi/3}^{\pi/3}$$

$$= \frac{4\pi}{3} - \ln\left|\sec\frac{\pi}{3} + \tan\frac{\pi}{3}\right| + \ln\left|\sec\left(-\frac{\pi}{3}\right) + \tan\left(-\frac{\pi}{3}\right)\right|$$

$$= \frac{4\pi}{3} - \ln|2 + \sqrt{3}| - \ln|2 - \sqrt{3}| = \frac{4\pi}{3} - \ln\left|\frac{2 + \sqrt{3}}{2 - \sqrt{3}}\right| \approx 1.555.$$

35. $\int_0^\pi \cos^2 t \, dt = \int_0^\pi \frac{\cos 2t + 1}{2} \, dt = \frac{1}{4} \sin(2t) + \frac{t}{2} \Big|_0^\pi = \frac{1}{4} \sin(2\pi) + \frac{1}{2}\pi = \frac{1}{2}\pi$

37. $\lim\limits_{t \to 0} \frac{\sin(2t)}{t} = \lim\limits_{t \to 0} \frac{2 \sin t \cos t}{t} = 2 \lim\limits_{t \to 0} \frac{\sin t}{t} \lim\limits_{t \to 0} \cos t = 2$

39. $\lim\limits_{h \to 0} \frac{\cos h - 1}{\sin h} = \lim\limits_{h \to 0} \frac{\cos h - 1}{h} \frac{h}{\sin h} = \lim\limits_{h \to 0} \frac{\cos h - 1}{h} \cdot \lim\limits_{h \to 0} \frac{h}{\sin h} = 0 \cdot 1 = 0$

CHAPTER 8: PRACTICE EXAM

1. We apply two facts: the sum of the angles of a triangle equals π radians (180 degrees), and a right angle has $\pi/2$ radians (90 degrees). Thus, if θ is one of the acute angles of a right triangle, with the other being $2\pi/5$ radians, then $\theta + \frac{2\pi}{5} + \frac{\pi}{2} = \pi$. Therefore, $\theta = \frac{\pi}{2} - \frac{2\pi}{5} = \frac{\pi}{10}$ radians.

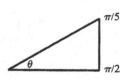

3. If $\frac{3\pi}{2} < \theta < 2\pi$, then $\sin \theta < 0$. Therefore,

$$\sin \theta = -\sqrt{1 - \cos^2 \theta} = -\sqrt{1 - (2/3)^2} = \frac{-\sqrt{5}}{3},$$

$\tan \theta = \frac{\sin \theta}{\cos \theta} = \frac{-\sqrt{5}/3}{2/3} = \frac{-\sqrt{5}}{2}$, and $\sec \theta = \frac{1}{\cos \theta} = \frac{3}{2}$.

5. Since $\cos(\frac{5\pi}{6}) = -\frac{\sqrt{3}}{2}$ and $0 \le \frac{5\pi}{6} \le \pi$, we have $\cos^{-1}(-\frac{\sqrt{3}}{2}) = \frac{5\pi}{6}$.

7. $\dfrac{d}{dt}\tan(\cos t) = \sec^2(\cos t)\dfrac{d}{dt}(\cos t) = -\sec^2(\cos t)\sin t$

9. Set $f'(x) = 1 - 2\cos(2x) = 0$. Then $\cos(2x) = 1/2$, so that $x = \frac{\pi}{6}, \frac{5\pi}{6}, \frac{7\pi}{6}$, or $\frac{11\pi}{6}$ (under the restriction $0 \le x \le 2\pi$). Comparing the values of $f(x)$ at these critical points and the endpoints, we get the following table (with the precise values in the second row and one-place decimal approximations in the second):

x	0	$\pi/6$	$5\pi/6$	$7\pi/6$	$11\pi/6$	2π
$f(x)$	0	$\frac{\pi}{6} - \frac{\sqrt{3}}{2}$	$\frac{5\pi}{6} + \frac{\sqrt{3}}{2}$	$\frac{7\pi}{6} - \frac{\sqrt{3}}{2}$	$\frac{11\pi}{6} + \frac{\sqrt{3}}{2}$	2π
$\approx$	0	-0.3	3.5	2.8	6.6	6.3

Thus, we see that the maximum occurs at $x = 11\pi/6$ and the minimum at $x = \pi/6$.

11. The general solution to this pair of differential equations is

$$u = M\sin\left(\frac{t}{2}\right) + N\cos\left(\frac{t}{2}\right) \quad \text{and} \quad v = \frac{1}{2}\left[M\cos\left(\frac{t}{2}\right) - N\sin\left(\frac{t}{2}\right)\right]$$

(see example 8.3.7). From the initial conditions $u(0) = 1$ and $v(0) = -2$, we get $N = 1$ and $\frac{1}{2}M = -2$, or $M = -4$. Thus, $u = -4\sin(\frac{t}{2}) + \cos(\frac{t}{2})$ and $v = -2\cos(\frac{t}{2}) - \frac{1}{2}\sin(\frac{t}{2})$.

13. The average fish population over the next 5 years is

$$\frac{1}{5}\int_0^5 \left[200 + 30\sin\left(\frac{\pi t}{10}\right) - 12\cos\left(\frac{\pi t}{10}\right)\right]dt$$

$$= \frac{1}{5}\left[200t - \frac{300}{\pi}\cos\left(\frac{\pi t}{10}\right) - \frac{120}{\pi}\sin\left(\frac{\pi t}{10}\right)\right]\Big|_0^5$$

$$= \frac{1}{5}\left[1,000 - \frac{300}{\pi}\cos\left(\frac{\pi}{2}\right) - \frac{120}{\pi}\sin\left(\frac{\pi}{2}\right)\right] - \frac{1}{5}\left[0 - \frac{300}{\pi}\cos 0 - \frac{120}{\pi}\sin 0\right]$$

$$= 200 + \frac{36}{\pi} \approx 211.46.$$

15. $\displaystyle\lim_{h\to 0}\frac{\sin^2 h}{h} = \lim_{h\to 0}\frac{\sin h}{h}\cdot\sin h = \lim_{h\to 0}\frac{\sin h}{h}\cdot\lim_{h\to 0}\sin h = 1\cdot 0 = 0$

CHAPTER 9
Differential Equations

EXERCISES 9.1

1. (a) 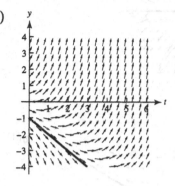   (b)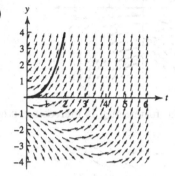

(c) 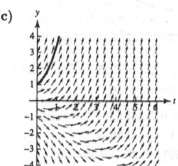   (d)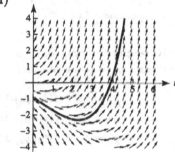

Solution (a) is a linear function.

3. The matching direction field is (d). Reasons: (i) the slope is positive for $x < 0$, negative for $x > 0$, and zero for $x = 0$; (ii) all points with the same x coordinate have the same slope.

5. The matching direction field is (a). Reasons: (i) the slope is positive for $y < 1$, negative for $y > 1$, and zero for $y = 1$; (ii) all points with the same y coordinate have the same slope.

7.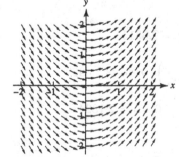

9.

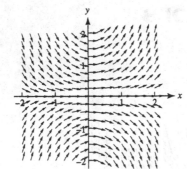 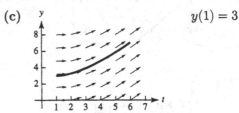

11. (a) The function y is increasing, since $y' > 0$ for $t > 0$.

(b)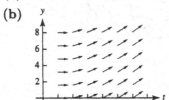

(c)

$y(1) = 3$

13. (a) To divide the interval $[0, 1]$ into $n = 4$ segments, we use $\Delta t = 0.25$. With $f(t, y) = 1 + y$, $t_0 = 0$, and $y_0 = 0$, we obtain

$$y_1 = y_0 + f(t_0, y_0)\Delta t = 0 + (1 + 0) \cdot 0.25 = 0.25$$
$$y_2 = y_1 + f(t_1, y_1)\Delta t = 0.25 + (1 + 0.25) \cdot 0.25 = 0.5625$$
$$y_3 = y_2 + f(t_2, y_2)\Delta t = 0.5625 + (1 + 0.5625) \cdot 0.25 = 0.953125$$
$$y_4 = y_3 + f(t_3, y_3)\Delta t = 0.953125 + (1 + 0.953125) \cdot 0.25 \approx 1.44141.$$

Thus, $y(1) \approx 1.44141$.

(b) Separating variables, we get $dy/(1 + y) = dt$, and integrating both sides gives $\ln |1 + y| = t + c$, or $y = Ce^t - 1$, where $C = \pm e^c$. From the initial condition $y(0) = 0$, we obtain $0 = C - 1$, or $C = 1$. Therefore, $y = e^t - 1$ and $y(1) = e - 1 \approx 1.71828$.

15. (a) Given $f(x, y) = x/y$, $x_0 = 0$, $y_0 = 2$, and $\Delta x = 0.25$, we have $x_0 = 0$, $x_1 = 0.25$, $x_2 = 0.5$, $x_3 = 0.75$, $x_4 = 1$. Then

$$y_1 = y_0 + f(x_0, y_0)\Delta x = 2 + \left(\frac{0}{2}\right)(0.25) = 2$$

$$y_2 = y_1 + f(x_1, y_1)\Delta x = 2 + \left(\frac{0.25}{2}\right)(0.25) = 2.03125$$

$$y_3 = y_2 + f(x_2, y_2)\Delta x = 2.03125 + \left(\frac{0.5}{2.03125}\right)(0.25) \approx 2.0928$$

$$y_4 = y_3 + f(x_3, y_3)\Delta x \approx 2.0928 + \left(\frac{0.75}{2.0928}\right)(0.25) \approx 2.1824.$$

Therefore, $y(1) \approx 2.1824$.

(b) Separating variables, we get $y\,dy = x\,dx$, and integrating both sides gives $\frac{y^2}{2} = \frac{x^2}{2} + c$, or $y^2 = x^2 + 2c$. Since $y(0) > 0$, we choose the positive square root and obtain $y = \sqrt{x^2 + 2c}$. Next, setting $x = 0$ and $y(0) = 2$, we find that $2 = \sqrt{2c}$, which gives $c = 2$. Therefore, $y = \sqrt{x^2 + 4}$ and $y(1) = \sqrt{5} \approx 2.236$.

17. We are given $f(t, M) = 0.07M - 14{,}000$, $t_0 = 0$, $M_0 = 120{,}000$, and $\Delta t = 1$. Then $t_0 = 0$, $t_1 = 1$, $t_2 = 2, \ldots, t_5 = 5$, and we obtain

$$M_1 = M_0 + f(t_0, M_0)\Delta t = 120{,}000 + 0.07 \cdot (120{,}000) - 14{,}000 = 114{,}400$$
$$M_2 = M_1 + f(t_1, M_1)\Delta t = 114{,}400 + 0.07 \cdot (114{,}400) - 14{,}000 = 108{,}408$$
$$M_3 = M_2 + f(t_2, M_2)\Delta t = 108{,}408 + 0.07 \cdot (108{,}408) - 14{,}000 = 101{,}996.56$$
$$M_4 = M_3 + f(t_3, M_3)\Delta t = 101{,}996.56 + 0.07 \cdot (101{,}996.56) - 14{,}000 \approx 95{,}136.32$$
$$M_5 = M_4 + f(t_4, M_4)\Delta t \approx 95{,}136.32 + 0.07 \cdot (95{,}136.32) - 14{,}000 \approx 87{,}795.86.$$

Therefore, $M(5) \approx 87{,}795.86$ dollars.

19. We are given $f(t, p) = 0.02p(1 - p)$, $t_0 = 0$, $p_0 = 0.28$, and $\Delta t = 4$. Then $t_0 = 0$, $t_1 = 4$, $t_2 = 8$, $t_3 = 12$, $t_4 = 16$, $t_5 = 20$, and we obtain

$$p_1 = p_0 + 0.02p_0(1 - p_0)\Delta t = 0.28 + (0.08)(0.28)(0.72) = 0.296128$$
$$p_2 = p_1 + 0.02p_1(1 - p_1)\Delta t = 0.296128 + 0.08(0.296128)(0.703872) \approx 0.312803$$
$$p_3 = p_2 + 0.02p_2(1 - p_2)\Delta t \approx 0.312803 + 0.08(0.312803)(0.687197) \approx 0.33000$$
$$p_4 = p_3 + 0.02p_3(1 - p_3)\Delta t \approx 0.33000 + 0.08(0.33000)(0.67000) = 0.347688$$
$$p_5 = p_4 + 0.02p_4(1 - p_4)\Delta t \approx 0.347688 + 0.08(0.347688)(0.652312) \approx 0.365832.$$

According to this model, the U.S. population in the year 2020 will be approximately 0.366 billion.

21. (a) We are given $f(t, r) = 3t^{0.2} - 2tr$ and $n = 8$. Dividing the interval $[0, 2]$ into 8 equal segments means taking $\Delta t = 0.25$. Using Euler's method, we compute

$$r_k = r_{k-1} + f(t_{k-1}, r_{k-1}) \cdot \Delta t = r_{k-1} + (3t_{k-1}^{0.2} - 2t_{k-1}r_{k-1}) \cdot (0.25)$$

for $k = 1, 2, \ldots, 8$, starting with $t_0 = 0$ and $r_0 = 0.25$. We obtain the following:

k	1	2	3	4	5	6	7	8
t_k	0.25	0.5	0.75	1	1.25	1.5	1.75	2
r_k	0.25	0.787	1.243	1.485	1.493	1.344	1.149	0.982

(b) The total sales are approximately $0.25 \cdot (r_1 + r_2 + \cdots + r_8) \approx 2.18325$ million dollars (using the Riemann sum to approximate $\int_0^2 r(t)\, dt$).

23. We are given $f(t, y) = 0.2y(e^{1-(y/5)} - 1)$, $t_0 = 0$, and $y_0 = 1.5$. The interval is $[0, 20]$ and $n = 20$, so that $\Delta t = 1$. Using Euler's method, we compute

$$y_k = y_{k-1} + f(t_{k-1}, y_{k-1}) \cdot 1 = y_{k-1} + 0.2y_{k-1}(e^{1-(y_{k-1}/5)} - 1)$$

for $k = 1, 2, 3, \ldots, 20$. The result is:

k	1	2	3	4	5	6	7
y_k	1.80413	2.12703	2.45732	2.78309	3.09367	3.38084	3.63942

8	9	10	11	12	13	14
3.86706	4.06375	4.23112	4.37178	4.48884	4.58548	4.66475

15	16	17	18	19	20
4.72945	4.78204	4.82465	4.85909	4.88687	4.90924

The deer population 20 years later is approximately equal to 491.

EXERCISES 9.2

1. Let $g(y) = -0.1y$. Solving $g(y) = 0$ gives $y = 0$ as the only equilibrium solution. Since $g'(y) = -0.1 < 0$, the equilibrium solution $y = 0$ is asymptotically stable.

   To find the general solution, we separate variables to obtain $dy/y = -0.1\,dt$. Integrating gives $\ln|y| = -0.1t + c$, and therefore $y(t) = Ce^{-0.1t}$, where $C = \pm e^c$. Four solutions are graphed with initial conditions $y(0) = \pm 0.25$ and $y(0) = \pm 0.5$.

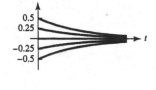

3. Let $g(y) = 1 - y$. Solving $g(y) = 0$ gives $y = 1$ as the only equilibrium solution. Since $g'(y) = -1 < 0$, the equilibrium solution $y = 1$ is asymptotically stable.

   To find the general solution, we separate variables to obtain $\frac{dy}{1-y} = dt$. Integrating gives $-\ln|1 - y| = t + c$, and therefore $y = Ce^{-t} + 1$, where $C = \pm e^{-c}$. Four solutions are graphed with initial conditions $y(0) = 0.5$, $y(0) = 0.75$, $y(0) = 1.25$, and $y(0) = 1.5$.

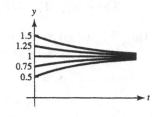

5. Let $g(y) = y - 3$. Solving $g(y) = 0$ gives $y = 3$ as the only equilibrium solution. Since $g'(y) = 1 > 0$, the equilibrium solution $y = 3$ is unstable.

   To find the general solution, we separate variables to obtain $\frac{dy}{y-3} = dt$. Integrating gives $\ln|y - 3| = t + c$, and therefore $y = Ce^t + 3$, where $C = \pm e^c$. Four solutions are graphed with initial conditions $y(0) = 2$, $y(0) = 2.5$, $y(0) = 3.5$, and $y(0) = 4$.

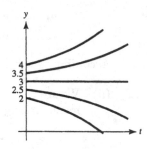

7. Let $g(p) = 0.1p(1 - \frac{p}{10})$. Solving $g(p) = 0$ gives $p = 0$ and $p = 10$ as the equilibrium solutions. Next, we compute $g'(p) = 0.1 - 0.02p$. Since $g'(0) = 0.1 > 0$, the equilibrium solution $p = 0$ is unstable; and since $g'(10) = -0.1 < 0$, the equilibrium solution is $p = 10$ is asymptotically stable.

   We separate variables to obtain $\frac{dp}{p(p-10)} = -0.01\,dt$. Integrating by the method of partial fractions, we get $\frac{1}{10}\ln\left|\frac{p-10}{p}\right| = -0.01t + c$, and solving for p gives, $p = \frac{10}{1 - Ce^{-0.1t}}$, where $C = \pm e^{10c}$. Four solutions are graphed with initial conditions $y(0) = -2$, $y(0) = 2$, $y(0) = 8$, and $y(0) = 12$.

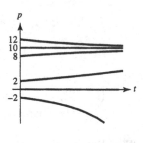

9. Let $g(y) = y^3$. Solving $g(y) = 0$ gives $y = 0$ as the only equilibrium solution. Since $g'(y) = 3y^2$ and $g'(0) = 0$, we cannot apply Theorem 9.2.1. To find the general solution, we separate variables to get $\frac{dy}{y^3} = dt$. Integrating gives $-\frac{1}{2y^2} = t + c$, and therefore $y^2 = \frac{1}{C - 2t}$, where $C = -2c$. Then

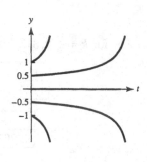

$$y = \begin{cases} \dfrac{1}{\sqrt{C - 2t}} & \text{if } y(0) > 0 \\[2ex] -\dfrac{1}{\sqrt{C - 2t}} & \text{if } y(0) < 0. \end{cases}$$

In both cases, $C = \frac{1}{y_0^2} > 0$, and $\lim_{t \to C} y = \infty$. Therefore, the equilibrium solution is unstable. Four solutions are graphed.

11. Let $g(y) = 1 - e^{2-y}$. Solving $g(y) = 0$, we find that the only equilibrium solution is $y = 2$. Since $g'(y) = e^{2-y}$ and $g'(2) = 1 > 0$, the equilibrium solution $y = 2$ is unstable.

13. Let $g(y) = 4y - y^2 = y(4 - y)$. Solving $g(y) = 0$, we find that the equilibrium solutions are $y = 0$ and $y = 4$. Next, we compute $g'(y) = 4 - 2y$. Since $g'(0) = 4 > 0$, the equilibrium solution $y = 0$ is unstable. Since $g'(4) = -4 < 0$, the equilibrium solution $y = 4$ is asymptotically stable.

15. Let $g(y) = -y^2 + y + 2 = -(y - 2)(y + 1)$. Solving $g(y) = 0$, we find that the equilibrium solutions are $y = 2$ and $y = -1$. Next, we compute $g'(y) = -2y + 1$. Since $g'(2) = -3 < 0$, the equilibrium solution $y = 2$ is asymptotically stable. Since $g'(-1) = 3 > 0$, the equilibrium solution $y = -1$ is unstable.

17. Solving $g(y) = -y^3 + 6y^2 - 8y = -y(y - 2)(y - 4) = 0$, we find that the equilibrium solutions are $y = 0$, $y = 2$, and $y = 4$. Next, we compute $g'(y) = -3y^2 + 12y - 8$. Since $g'(0) = -8 < 0$, the equilibrium solution $y = 0$ is asymptotically stable. Since $g'(2) = 4 > 0$, the equilibrium solution $y = 2$ is unstable. Since $g'(4) = -8 < 0$, the equilibrium solution $y = 4$ is asymptotically stable.

19. Setting $g(x) = 0.4\sqrt{k} - 0.1k = 0.1\sqrt{k}(4 - \sqrt{k}) = 0$, leads to either $\sqrt{k} = 0$ or $\sqrt{k} = 4$. Therefore, $k = 0$ and $k = 16$ are the equilibrium solutions. Next, $g'(k) = 0.2k^{-1/2} - 0.1$. Since $g'(16) = -0.05 < 0$, the equilibrium solution $k = 16$ is asymptotically stable. However, $g'(0)$ is undefined, and we cannot apply Theorem 9.2.1 to the equilibrium solution $k = 0$.

To determine whether $k = 0$ is asymptotically stable, we plot a direction field for the differential equation, shown here for $0 \leq t \leq 4$ and $0 \leq k \leq 1$. It strongly suggests that the solutions move away from the line $k = 0$ and, therefore the equilibrium solution is unstable.

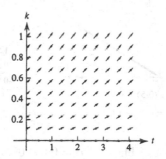

21. Setting $g(p) = -rp(1 - \frac{p}{T})(1 - \frac{p}{K}) = 0$, we obtain the equilibrium solutions $p = 0$, $p = T$, and $p = K$. By repeated applications of the product rule, we get $g'(p) = -r + 2r(\frac{1}{T} + \frac{1}{K})p - \frac{3rp^2}{TK}$.

Since $g'(0) = -r < 0$, the equilibrium solution $p = 0$ is asymptotically stable. Next, recalling that $0 < T < K$, we get $g'(T) = r(K - T)/K > 0$, and we conclude that the equilibrium solution $p = T$ is unstable. Also, $g'(K) = r(T - K)/T < 0$, and we conclude that the equilibrium solution $p = K$ is asymptotically stable.

23. We are given $g(y) = 1 - \frac{2}{3}y$.

(a) Setting $1 - \frac{2}{3}y = 0$, we see that the only equilibrium solution is $y = \frac{3}{2}$.

(b) Since $g'(y) = -\frac{2}{3} < 0$, the equilibrium solution $y = \frac{3}{2}$ is asymptotically stable.

(c) We first observe that $g(y)$ is a decreasing function with no local maximum or minimum. Therefore, any solution $y(t)$ of the differential equation has no inflection point and keeps the same concavity for all $t > 0$.

If $y(0) = -1$, then $y'(0) = g(-1) > 0$, which means that $y(t)$ is increasing for $t > 0$. Also, $y''(0) = g'(-1) \cdot y'(0) = -\frac{2}{3}y'(0) < 0$, which means that the graph is concave down.

If $y(0) = 3$, then $y'(0) = g(3) < 0$, which means that $y(t)$ is decreasing for $t > 0$. Also, $y''(0) = g'(3) \cdot y'(0) = -\frac{2}{3}y'(0) > 0$, which means that the graph is concave up.

In both cases $\lim_{t \to \infty} y(t) = \frac{3}{2}$.

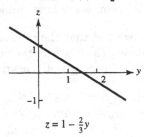

$z = 1 - \frac{2}{3}y$

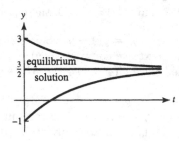

25. We are given $g(y) = 0.2y - 0.1$.

    (a) Setting $0.2y - 0.1 = 0$, we see that the only equilibrium solution is $y = 0.5$.

    (b) Since $g'(y) = 0.2 > 0$, the equilibrium solution $y = 0.5$ is unstable.

    (c) We first observe that $g(y)$ is an increasing function with no local maximum or minimum. Therefore, any solution $y(t)$ of the differential equation has no inflection point and keeps the same concavity for all $t > 0$.

        If $y(0) = -0.5$, then $y'(0) = g(-0.5) < 0$, which means that $y(t)$ is decreasing for $t > 0$, and $y(t) \to -\infty$ as $t \to \infty$. Also, $y''(0) = g'(-0.5)y'(0) = 0.2y'(0) < 0$, which means that the graph is concave down.

        If $y(0) = 1.5$, then $y'(0) = g(1.5) > 0$, which means that $y(t)$ is increasing for $t > 0$ and $y(t) \to \infty$ as $t \to \infty$. Also, $y''(0) = g'(1.5) \cdot y'(0) = 0.2y'(0) > 0$, which means that the graph is concave up.

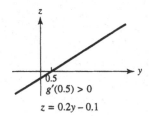

$z = 0.2y - 0.1$

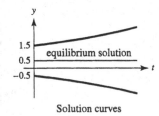

Solution curves

27. We are given $g(y) = 0.1y(6 - y)$. Then $g'(y) = 0.6 - 0.2y$.

    (a) Setting $0.1y(6 - y) = 0$, we see that the equilibrium solutions are $y = 0$ and $y = 6$.

    (b) Since $g'(0) = 0.6 > 0$, the equilibrium solution $y = 0$ is unstable. Since $g'(6) = -0.6 < 0$, the equilibrium solution $y = 6$ is asymptotically stable.

    (c) We first observe that $g(y)$ has a local maximum at $y = 3$. Therefore, any solution $y(t)$ is concave up for $y < 3$, concave down for $y > 3$ and has an inflection point where it crosses the line $y = 3$.

        If $y(0) = -0.5$, then $y'(0) = g(-0.5) < 0$, which means that $y(t)$ is decreasing for $t > 0$, and $y(t) \to -\infty$. Also, $y''(0) = g'(-0.5) \cdot y'(0) = 0.7y'(0) < 0$, which means that the graph is concave down for all t.

        If $y(0) = 2$, then $y'(0) = g(2) > 0$, which means that $y(t)$ is increasing for $t > 0$ and $y(t) \to 6$. Also, $y''(0) = g'(2) \cdot y'(0) = 0.2 \cdot y'(0) > 0$, which means that the graph is concave up for t near zero and changes to concave down after crossing the line $y = 3$.

        If $y(0) = 4$, then $y'(0) = g(4) > 0$, which means that $y(t)$ is increasing for $t > 0$ and $y(t) \to 6$. Also, $y''(0) = g'(4) \cdot y'(0) = -0.2y'(0) < 0$, which means that the graph is concave down for all t.

        If $y(0) = 8$, then $y'(0) = g(8) < 0$, which means that $y(t)$ is decreasing for $t > 0$ and $y(t) \to 6$. Also, $y''(0) = g'(8) \cdot y'(0) = -y'(0) > 0$, which means that the graph is concave up for all t.

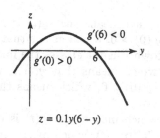

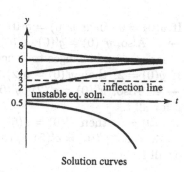

Solution curves

29. We are given $g(y) = 0.2y(y - 5)$. Then $g'(y) = 0.4y - 1$.

(a) Setting $0.2y(y - 5) = 0$, we see that the equilibrium solutions are $y = 0$ and $y = 5$.

(b) Since $g'(0) = -1 < 0$, the equilibrium solution $y = 0$ is asymptotically stable. Since $g'(5) = 1 > 0$, the equilibrium solution $y = 5$ is unstable.

(c) We first observe that $g(y)$ has a local minimum at $y = 2.5$. Therefore, any solution $y(t)$ is concave down for $y < 2.5$, concave up for $y > 2.5$ and has an inflection point where it crosses the line $y = 2.5$.

　　If $y(0) = -0.5$, then $y'(0) = g(-0.5) > 0$, which means that $y(t)$ is increasing for $t > 0$, and $y(t) \to 0$. Also, $y''(0) = g'(-0.5) \cdot y'(0) = -1.2y'(0) < 0$, which means that the graph is concave down for all t.

　　If $y(0) = 1$, then $y'(0) = g(1) < 0$, which means that $y(t)$ is decreasing for $t > 0$ and $y(t) \to 0$. Also, $y''(0) = g'(1) \cdot y'(0) = -0.6 \cdot y'(0) > 0$, which means that the graph is concave up for all t.

　　If $y(0) = 3$, then $y'(0) = g(3) < 0$, which means that $y(t)$ is decreasing for $t > 0$ and $y(t) \to 0$. Also, $y''(0) = g'(3) \cdot y'(0) = 0.2y'(0) < 0$, which means that the graph is concave down for t near zero and changes to concave up when it crosses the line $y = 2.5$.

　　If $y(0) = 7$, then $y'(0) = g(7) > 0$, which means that $y(t)$ is increasing for $t > 0$ and $y(t) \to \infty$. Also, $y''(0) = g'(7) \cdot y'(0) = 1.8y'(0) > 0$, which means that the graph is concave up for all t.

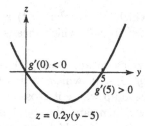

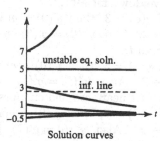

Solution curves

31. We are given $g(y) = 4 + 3y - y^2 = -(y - 4)(y + 1)$. Then $g'(y) = 3 - 2y$.

(a) Setting $4 + 3y - y^2 = 0$, we see that the equilibrium solutions are $y = 4$ and $y = -1$.

(b) Since $g'(4) = -5 < 0$, the equilibrium solution $y = 4$ is asymptotically stable. Since $g'(-1) = 5 > 0$, the equilibrium solution $y = -1$ is unstable.

(c) We first observe that $g(y)$ has a local maximum at $y = 3/2$. Therefore, any solution $y(t)$ is concave up for $y < 3/2$, concave down for $y > 3/2$ and has an inflection point where it crosses the line $y = 3/2$.

　　If $y(0) = -2$, then $y'(0) = g(-2) < 0$, which means that $y(t)$ is decreasing for $t > 0$ and $y(t) \to -\infty$. Also, $y''(0) = g'(-2) \cdot y'(0) = 7 \cdot y'(0) < 0$, which means that the graph is concave down for all t.

If $y(0) = 0$, then $y'(0) = g(0) > 0$, which means that $y(t)$ is increasing for $t > 0$, and $y(t) \to 4$. Also, $y''(0) = g'(0) \cdot y'(0) = 3y'(0) > 0$, which means that the graph is concave up for t near zero and changes to concave down when it crosses the line $y = 3/2$.

If $y(0) = 2$, then $y'(0) = g(2) > 0$, which means that $y(t)$ is increasing for $t > 0$ and $y(t) \to 4$. Also, $y''(0) = g'(2) \cdot y'(0) = -y'(0) < 0$, which means that the graph is concave down for all t.

If $y(0) = 5$, then $y'(0) = g(5) < 0$, which means that $y(t)$ is decreasing for $t > 0$ and $y(t) \to 4$. Also, $y''(0) = g'(5) \cdot y'(0) = -7y'(0) > 0$, which means that the graph is concave up for all t.

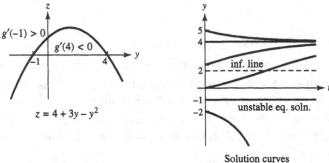

33. We are given $g(y) = y^2 - 8y + 12 = (y - 6)(y - 2)$. Then $g'(y) = 2y - 8$.

(a) Setting $y^2 - 8y + 12 = 0$, we see that the equilibrium solutions are $y = 2$ and $y = 6$.

(b) Since $g'(2) = -4 < 0$, the equilibrium solution $y = 2$ is asymptotically stable. Since $g'(6) = 4 > 0$, the equilibrium solution $y = 6$ is unstable.

(c) We first observe that $g(y)$ has a local minimum at $y = 4$. Therefore, any solution $y(t)$ is concave down for $y < 4$, concave up for $y > 4$ and has an inflection point where it crosses the line $y = 4$.

If $y(0) = 1/2$, then $y'(0) = g(1/2) > 0$, which means that $y(t)$ is increasing for $t > 0$ and $y(t) \to 2$. Also, $y''(0) = g'(1/2) \cdot y'(0) = -7 \cdot y'(0) < 0$, which means that the graph is concave down for all t.

If $y(0) = 3$, then $y'(0) = g(3) < 0$, which means that $y(t)$ is decreasing for $t > 0$, and $y(t) \to 2$. Also, $y''(0) = g'(3) \cdot y'(0) = -2y'(0) > 0$, which means that the graph is concave up for all t.

If $y(0) = 5$, then $y'(0) = g(5) < 0$, which means that $y(t)$ is decreasing for $t > 0$ and $y(t) \to 2$. Also, $y''(0) = g'(5) \cdot y'(0) = 2y'(0) < 0$, which means that the graph is concave down for t near zero and changes to concave up where it crosses the line $y = 4$.

If $y(0) = 7$, then $y'(0) = g(7) > 0$, which means that $y(t)$ is increasing for $t > 0$ and $y(t) \to \infty$. Also, $y''(0) = g'(7) \cdot y'(0) = 6y'(0) > 0$, which means that the graph is concave up for all t.

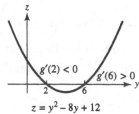

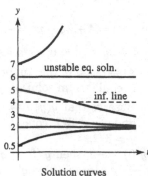

Solution curves

35. We are given $g(y) = -0.2y(1 - 2y)(2 - y) = -0.4y^3 + y^2 - 0.4y$. Then $g'(y) = -1.2y^2 + 2y - 0.4$.

(a) Setting $-0.2y(1 - 2y)(2 - y) = 0$, we see that the equilibrium solutions are $y = 0$, $y = 0.5$ and $y = 2$.

(b) Since $g'(0) = -0.4 < 0$, the equilibrium solution $y = 0$ is asymptotically stable. Since $g'(0.5) = 0.3 > 0$, the equilibrium solution $y = 0.5$ is unstable. And, since $g'(2) = -1.2 < 0$, the equilibrium solution $y = 2$ is asymptotically stable.

(c) The solutions of $g'(y) = 0$ are $\frac{5}{6} - \frac{1}{6}\sqrt{13} \approx 0.2324$ and $\frac{5}{6} + \frac{1}{6}\sqrt{13} \approx 1.43426$. The second derivative test shows that $g(y)$ has a local minimum at the first of these and a local maximum at the second. Therefore, any solution $y(t)$ is concave down for $y < \frac{5}{6} - \frac{1}{6}\sqrt{13}$ or $y > \frac{5}{6} + \frac{1}{6}\sqrt{13}$ and concave up for $\frac{5}{6} - \frac{1}{6}\sqrt{13} < y < \frac{5}{6} + \frac{1}{6}\sqrt{13}$. It has an inflection point where it crosses either of the lines $y = \frac{5}{6} - \frac{1}{6}\sqrt{13}$ or $y = \frac{5}{6} + \frac{1}{6}\sqrt{13}$.

If $y(0) = 0.2$, then $y'(0) = g(0.2) < 0$, which means that $y(t)$ is decreasing for $t > 0$ and $y(t) \to 0$. Also, $y''(0) = g'(0.2) \cdot y'(0) = -0.048 \cdot y'(0) > 0$, which means that the graph is concave up for all t.

If $y(0) = 0.75$, then $y'(0) = g(0.75) > 0$, which means that $y(t)$ is increasing for $t > 0$, and $y(t) \to 2$. Also, $y''(0) = g'(0.75) \cdot y'(0) = 0.425y'(0) > 0$, which means that the graph is concave up for t near zero and changes to concave down when it crosses the line $y = \frac{5}{6} + \frac{1}{6}\sqrt{13}$.

If $y(0) = 1.5$, then $y'(0) = g(1.5) > 0$, which means that $y(t)$ is increasing for $t > 0$ and $y(t) \to 2$. Also, $y''(0) = g'(1.5) \cdot y'(0) = -0.1y'(0) < 0$, which means that the graph is concave down for all t.

If $y(0) = 3$, then $y'(0) = g(3) < 0$, which means that $y(t)$ is decreasing for $t > 0$ and $y(t) \to 2$. Also, $y''(0) = g'(3) \cdot y'(0) = -5.2y'(0) > 0$, which means that the graph is concave up for all t.

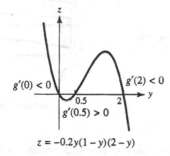

$z = -0.2y(1-y)(2-y)$

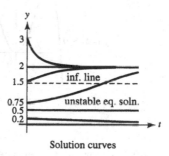

Solution curves

37. We are given $g(y) = y^2(16 - y^2) = 16y^2 - y^4$. Then $g'(y) = 32y - 4y^3$.

(a) Setting $y^2(16 - y^2) = 0$, we see that the equilibrium solutions are $y = 0$, $y = 4$, and $y = -4$.

(b) Since $g'(4) = -128 < 0$, the equilibrium solution $y = 4$ is asymptotically stable. Since $g'(-4) = 128 > 0$, the equilibrium solution $y = -4$ is unstable. However, we cannot apply Theorem 9.2.1 to $y = 0$ because $g'(0) = 0$ By plotting a direction field (as shown), we conclude that the equilibrium solution $y = 0$ is unstable.

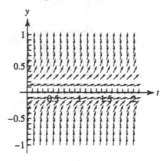

(c) The solutions of $g'(y) = 0$ are $y = 0$ and $y = \pm 2\sqrt{2}$. Checking the sign of $g''(y)$ shows that $g(y)$ has a local minimum at $y = 0$ and local maxima at $x = \pm 2\sqrt{2}$. Therefore, any solution $y(t)$ is concave down for $-2\sqrt{2} < y < 0$ or $y > 2\sqrt{2}$ and concave up for $y < -2\sqrt{2}$ and $0 < y < 2\sqrt{2}$. It has an inflection point where it crosses either of the lines or $y = \pm 2\sqrt{2}$. (Since $y = 0$ is an equilibrium solution, no other solution crosses it.)

If $y(0) = -5$, then $y'(0) = g(-5) < 0$, which means that $y(t)$ is decreasing for $t > 0$ and $y(t) \to -\infty$. Also, $y''(0) = g'(-5) \cdot y'(0) = 340 \cdot y'(0) < 0$, which means that the graph is concave down for all t.

If $y(0) = -3$, then $y'(0) = g(-3) > 0$, which means that $y(t)$ is increasing for $t > 0$, and $y(t) \to 0$. Also, $y''(0) = g'(-3) \cdot y'(0) = 12y'(0) > 0$, which means that the graph is concave up for t near zero and changes to concave down when it crosses the line $y = -2\sqrt{2}$.

If $y(0) = 1$, then $y'(0) = g(1) > 0$, which means that $y(t)$ is increasing for $t > 0$ and $y(t) \to 4$. Also, $y''(0) = g'(1) \cdot y'(0) = 28y'(0) > 0$, which means that the graph is concave up for t near zero and changes to concave down when it crosses the line $y = 2\sqrt{2}$.

If $y(0) = 3$, then $y'(0) = g(3) > 0$, which means that $y(t)$ is increasing for $t > 0$ and $y(t) \to 4$. Also, $y''(0) = g'(3) \cdot y'(0) = -12y'(0) < 0$, which means that the graph is concave down for all t.

If $y(0) = 5$, then $y'(0) = g(5) < 0$, which means that $y(t)$ is decreasing for $t > 0$ and $y(t) \to 4$. Also, $y''(0) = g'(5) \cdot y'(0) = -340y'(0) > 0$, which means that the graph is concave up for all t.

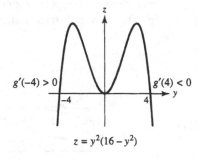

$$z = y^2(16 - y^2)$$

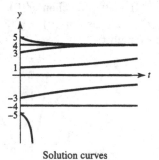

Solution curves

39. Since $g(y) \geq 1 > 0$ for all y over the domain shown, the differential equation $y' = g(y)$ has no equilibrium solution, and any solution $y(t)$ is an increasing function with $\lim_{t \to \infty} y(t) = \infty$. Thus, the solution with initial condition $y(0) = -1.5$ must cross the t-axis.

Since $y' = g(y) > 0$ and $y'' = g'(y) \cdot y' = g'(y) \cdot g(y)$, we see from the graph of $z = g(y)$ that $y'' < 0$ (and the solution curve is concave down) if $y < -1$ or $1 < y < 3$ and $y'' > 0$ (and the solution curve is concave up) if $-1 < y < 1$ or $y > 3$. The solution curve $y(t)$ has inflection points where it crosses the lines $y = -1$, $y = 1$, or $y = 3$.

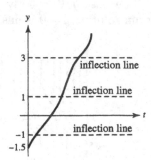

41. (a) The variables involved are the time t, measured in suitable units (such as years), and the density $y(t)$, measured in number of plants per unit area.

(b) There are three equilibrium solutions $y = 0$, $y = m$, and $y = M$, where m is the minimum density required for cross-pollination, and M is the optimum density for the plants.

(c) $\frac{dy}{dt} = -ry(1 - \frac{y}{m})(1 - \frac{y}{M})$, where r is the intrinsic growth rate.

(d) The threshold is m and carrying capacity is M. Write $g(y) = -ry(1 - \frac{y}{m})(1 - \frac{y}{M})$. We can check that $g'(0) < 0$ and $g'(M) < 0$, which means that both equilibrium solutions $y = M$ and $y = 0$ are asymptotically stable; and $y'(m) > 0$, which means that $y = m$ is unstable. We can also check that $g(y)$ has a local minimum in the interval $(0, m)$ and a local maximum in (m, M). Thus there are two inflection lines, as shown.

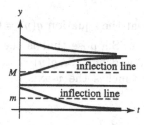

43. We are given $g(y) = 0.4y^{1/3}(15.3y - 36.9 - y^2) = -0.4y^{1/3}(y - 3)(y - 12.3)$. Expanding, we get $g(y) = 6.12y^{4/3} - 14.76y^{1/3} - 0.4y^{7/3}$, so that $g'(y) = 8.16y^{1/3} - 4.92y^{-2/3} - \frac{14}{15}y^{4/3}$. Solving $g(y) = 0$ gives the equilibrium solutions $y = 0$, $y = 3$, and $y = 12.3$.

Since $g'(3) > 0$, the equilibrium solution $y = 3$ is unstable; and since $g'(12.3) < 0$, the equilibrium solution $y = 12.3$ is asymptotically stable. Thus, if the initial condition satisfies $3 < y(0) < 12.3$, the solution $y(t)$ will increase toward 12.3 as $t \to \infty$; and if $y(0) > 12.3$, the solution will decrease toward 12.3. If $0 < y(0) < 3$, the solution will decrease toward zero as $t \to \infty$.

(a) Since y is in tens of birds, we conclude that the largest sustainable population is 123 birds.

(b) We must have $y > 3$, which means the number of birds must be greater than 30. In other words, the ranger must release at least 31 birds.

45. (a) Setting $g(p) = 0.01p(p - 1)(1 - \frac{p}{20}) = -0.0005(p^3 - 21p^2 + 20p) = 0$, we find that the equilibrium solutions are $p = 0$, $p = 1$, and $p = 20$. Next, we compute $g'(p) = -0.0005(3p^2 - 42p + 20)$. Then $g'(0) = -0.01$, $g'(1) = 0.0095$, and $g'(20) = -0.19$. Thus the equilibrium solutions $p = 0$ and $p = 20$ are asymptotically stable, and $p = 1$ is unstable. It follows that if $p(t)$ is the solution with initial value $p(0) = 8$, then it is an increasing function, and $p(t) \to 20$ as $t \to \infty$.

To check the concavity of $p(t)$, we look for the relative extrema of $g(p)$. Solving $g'(p) = 0$, we obtain $p = \frac{1}{6}(42 \pm \sqrt{42^2 - 240}) \approx 0.4936, 13.5064$.

The first is a local minimum and the second is a local maximum. We conclude that the graph of $p(t)$ (with $p(0) = 8$) is concave up until it crosses the line $p = \frac{1}{6}(42 + \sqrt{42^2 - 240})$, where it changes to concave down. That is inflection point is the value where the rate of growth (i.e., the slope) of $p(t)$ is maximum.

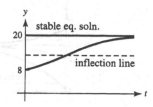

(b) The graph approaches the equilibrium solution $p = 20$, which means that $\lim_{t\to\infty} p(t) = 20$.

(c) The threshold level is $p = 1$ since any solution with initial value below that tends toward the equilibrium solution $p = 0$. The carrying capacity is 20, since any solution with initial value greater than one tends toward the line $p = 20$.

47. Let a be the size of the annual catch. Then the whale population is modeled by the differential equation

$$\frac{dy}{dt} = 0.02y \ln \frac{45}{y} - a.$$

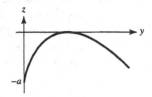

We want to determine a so that the equation $g(y) = 0.02y \ln \frac{45}{y} - a = 0$ has exactly one solution. Suppose that is the case. Then, observing that $g(0) = -a < 0$ and $\lim_{y\to\infty} g(y) = -\infty$, we conclude that $g(y)$ have a local maximum at the point where it touches the y-axis (as shown). We thus obtain the simultaneous equations

$$g(y) = 0.02y \ln \frac{45}{y} - a = 0 \quad \text{and} \quad g'(y) = 0.02 \ln \frac{45}{y} - 0.02 = 0.$$

Solving the second equation for y, we get $y = 45e^{-1}$. By substituting that into the first equation and solving for a, we obtain $a = 0.9e^{-1} \approx 0.331$.

Since y is in thousands of whales per year, we see that, if the model is correct, an annual catch of 331 whales will doom them to extinction.

EXERCISES 9.3

1. (a) Given $\alpha = 0.3$, $s = 0.22$, $\delta = 0.1$, the corresponding Solow differential equation is $\frac{dk}{dt} = 0.22k^{0.3} - 0.1k$.

(b) $k_s = \left(\frac{s}{\delta}\right)^{1/1-\alpha} = \left(\frac{0.22}{0.1}\right)^{1/1-0.3} = (2.2)^{1/0.7} \approx 3.0844$

(c) $k_i = \alpha^{1/1-\alpha} k_s = (0.3)^{1/0.7} \cdot (2.2)^{1/0.7} = (0.66)^{1/0.7} \approx 0.5523$

(d) Letting $g(k) = 0.22k^{0.3} - 0.1k$, we have $g'(k) = 0.066k^{-0.7} - 0.1$. Since $g'((2.2)^{1/0.7}) = -0.07 < 0$, the equilibrium solution $k = (2.2)^{1/0.7}$ is asymptotically stable.

(e) If $k_0 = 0.2$, then $k(t)$ is increasing and $\lim_{t\to\infty} k(t) = (2.2)^{1/0.7}$. The graph is concave up for $0.2 < k < (0.66)^{1/0.7}$ and concave down for $(0.66)^{1/0.7} < k < (2.2)^{1/0.7}$ with inflection value $k = (0.66)^{1/0.7}$.

If $k_0 = 2.2$, then $k(t)$ is increasing and $\lim_{t\to\infty} y(t) = (2.2)^{1/0.7}$. The graph is concave down for all t.

If $k_0 = 4.2$, then $k(t)$ is decreasing and $\lim_{t\to\infty} y(t) = (2.2)^{1/0.7}$. The graph is concave up for all t.

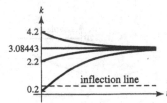

(f)   Using Mathematica with commands

```
In[13] := f1[t_, k_] = (2.2 + (k^(0.7 - 2.2) Exp[-0.07 t]) ^ (10/7)

In[30] := Plot[Evaluate[Table[f1[t, k], {k, 0.5, 6, 0.5}]], {t, 0, 25},
            PlotRange → {0, 6.5}, AxesLabel → {"t", "k"}, AspectRatio → .75]
```

we obtain

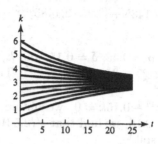

3. (a)   Given $\alpha = 0.3$, $s = 0.34$, $\delta = 0.1$, the corresponding Solow differential equation is $\frac{dk}{dt} = 0.34k^{0.3} - 0.1k$.

(b) $k_s = \left(\frac{s}{\delta}\right)^{1/1-\alpha} = \left(\frac{0.34}{0.1}\right)^{1/1-0.3} = (3.4)^{1/0.7} \approx 5.7445$

(c) $k_i = \alpha^{1/1-\alpha} k_s = (0.3)^{1/0.7} \cdot (3.4)^{1/0.7} = (1.02)^{1/0.7} \approx 1.0287$

(d)   Letting $g(k) = 0.34k^{0.3} - 0.1k$, we have $g'(k) = 0.102k^{-0.7} - 0.1$. Since $g'((3.4)^{1/0.7}) = -0.07 < 0$, the equilibrium solution $k = (3.4)^{1/0.7}$ is asymptotically stable.

(e)   If $k_0 = 0.5$, then $k(t)$ is increasing and $\lim_{t \to \infty} k(t) = (3.4)^{1/0.7}$. The graph is concave up for $0.5 < k < (1.02)^{1/0.7}$ and concave down for $(1.02)^{1/0.7} < y < (3.4)^{1/0.7}$ with inflection value $k = (1.02)^{1/0.7} \approx 1.0287$.

　　　If $k_0 = 3$, then $k(t)$ is increasing and $\lim_{t \to \infty} k(t) = (3.4)^{1/0.7}$. The graph is concave down for all t.

　　　If $k_0 = 8$, then $k(t)$ is decreasing and $\lim_{t \to \infty} y(t) = (3.4)^{1/0.7}$. The graph is concave up for all t.

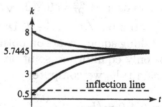

(f)   Using Mathematica with commands similar to those in the previous exercise, we obtain

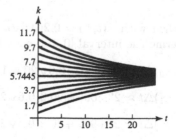

5. (a) We are given $f(t, k) = 0.25k^{0.45} - 0.15k$ with $k_0 = 2$ and $\Delta t = 1$. To estimate $k(5)$ we consider the interval $[0, 5]$, so that $n = 5$. Then Euler's method gives

$$k_1 = 2 + [0.25 \cdot 2^{0.45} - 0.15(2)] \cdot 1 \approx 2.0415$$
$$k_2 \approx 2.0415 + [0.25 \cdot 2.0415^{0.45} - 0.15(2.0415)] \cdot 1 \approx 2.0800$$
$$k_3 \approx 2.0800 + [0.25 \cdot 2.0800^{0.45} - 0.15(2.0800)] \cdot 1 \approx 2.1156$$
$$k_4 \approx 2.1156 + [0.25 \cdot 2.1156^{0.45} - 0.15(2.1156)] \cdot 1 \approx 2.1485$$
$$k_5 \approx 2.1485 + [0.25 \cdot 2.1485^{0.45} - 0.15(2.1485)] \cdot 1 \approx 2.1789.$$

Thus, $k(5) \approx 2.1789$.

(b) In this case, $s = 0.25$, $\alpha = 0.45$, $\delta = 0.15$, and $k_0 = 2$. Then, from formula (25) the explicit solution is $k = [\frac{5}{3} + (2^{0.55} - \frac{5}{3})e^{-0.0825t}]^{1/0.55}$. Then $k(5) = [\frac{5}{3} + (2^{0.55} - \frac{5}{3})e^{-0.4125}]^{1/0.55} \approx 2.1733$.

(c) Setting $g(k) = 0.25k^{0.45} - 0.15k = 0$, we obtain $k = (\frac{5}{3})^{1/0.55} \approx 2.5314$. Next, computing $g'(k) = 0.1125k^{-0.55} - 0.15$, we find that $g'(2.5314) \approx -0.08 < 0$. Therefore, the equilibrium solution $k = 2.5314$ is stable.

7. Given $\frac{dk}{dt} = 0.3\sqrt{k} - 0.1k$, $k(0) = 25$, let $y = \sqrt{k}$. Then

$$\frac{dy}{dt} = \frac{1}{2\sqrt{k}} \cdot \frac{dk}{dt} = \frac{1}{2\sqrt{k}}(0.3\sqrt{k} - 0.1k)$$

$$= 0.15 - 0.05\sqrt{k} = 0.15 - 0.05y = -0.05(y - 3).$$

Separating variables, we get $\frac{dy}{y-3} = -0.05\,dt$, and integrating gives

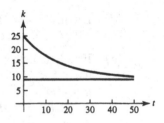

$$\ln|y - 3| = -0.05t + c \quad \text{or} \quad y = 3 + Ce^{-0.05t}$$

where $C = \pm e^c$. If $k(0) = 25$, then $y(0) = \sqrt{k(0)} = 5$, which gives $5 = 3 + C$, or $C = 2$. Thus, $y = 3 + 2e^{-0.05t}$, and, since $k = y^2$, we have the solution: $k = (3 + 2e^{-0.05t})^2$. Note that $\lim_{t\to\infty} k(t) = 9$.

9. (a) The Solow differential equation is $\frac{dk}{dt} = 0.2(0.6\sqrt{k} + 1)^2 - 0.1k$.

(b) We need to solve $g(k) = 0.2(0.6\sqrt{k} + 1)^2 - 0.1k = 0$. We rewrite this equation in the form $(0.6\sqrt{k} + 1)^2 = 0.5k$. Taking square roots on both sides and rearranging terms we get $(\sqrt{0.5} - 0.6)\sqrt{k} = 1$. Thus, the only equilibrium solution is $k = 1/(\sqrt{0.5} - 0.6)^2 \approx 87.17$.

   Next, using the chain rule, we compute

$$g'(k) = 0.4(0.6\sqrt{k} + 1) \cdot \frac{0.3}{\sqrt{k}} - 0.1 = 0.12\left(0.6 + \frac{1}{\sqrt{k}}\right) - 0.1.$$

Then $g'(\frac{1}{(\sqrt{0.5}-0.6)^2}) = 0.12\sqrt{0.5} - 0.1 \approx -0.015$, which shows that the equilibrium solution is stable.

(c) We use Euler's method with $f(t, k) = 0.2(0.6\sqrt{k} + 1)^2 - 0.1k$, $t_0 = 0$, $k_0 = 20$, and $n = 3$. Since we are considering the interval $[0, 15]$, we take $\Delta t = \frac{15-0}{3} = 5$. Then

$$k_1 = k_0 + f(t_0, k_0)\Delta t = 20 + [0.2(0.6\sqrt{20} + 1)^2 - 2] \cdot 5 \approx 23.5666$$

$$k_2 = k_1 + f(t_1, k_1)\Delta t \approx 23.5666 + [0.2(0.6\sqrt{23.5666} + 1)^2 - 2.35666] \cdot 5 \approx 27.0927$$

$$k_3 = k_2 + f(t_2, k_2)\Delta t \approx 27.0927 + [0.2(0.6\sqrt{27.0927} + 1)^2 - 2.70927] \cdot 5 \approx 30.5458.$$

Thus, $k(15) \approx 30.5458$.

(d) The following Mathematica commands apply Euler's method to the given initial value problem, with $n = 200$ and $\Delta t = 1$.

```
In[82] := eumeth[f_, {x0_, y0_}, h_, n_] :=
            (emstep[{x_, y_}] := {x + h, y + h * f[x, y]} // N;
            NestList[emstep, {x0, y0}, n])

In[91] := f[x_, y_] = 0.2 * (0.6 Sqrt[y] + 1)^2 - 0.1 y

Out[91]= 0.2 (1 + 0.6√y̅)² - 0.1 y

In[92] := k = eumeth[f, {0, 20}, 1, 200]
```

The resulting output consists of 200 pairs of numbers giving the approximations at each stage, starting with $(0, 20), (1, 20.7133), (2, 21.426), \ldots$ and ending with $(200, 83.1266)$ Thus, we obtain $k(200) \approx 83.1266$.

As an alternative, we can use the Mathematica's NDSolve command, which uses a much more accurate numerical method to approximate the solution of an initial value problem. The result is a function that interpolates all the approximate values over a given interval. The function can then be evaluated or graphed. In the present case, the commands are as follows:

```
In[5]:= z = NDSolve[{k'[t] = 0.2 * (0.6 Sqrt[k[t]] + 1)^2 - 0.1 k[t], k[0] == 20}, k, {t, 0, 500}]

Out[5]= {{k → InterpolatingFunction[{{0., 500.}}, <>]}}

In[8]:= k[t_] = k[t] /. First[z]

Out[8]= InterpolatingFunction[{{0., 500.}}, <>][t]

In[7]:= k[200]

Out[7]= 83.0512
```

The result, as shown, gives $k(200) \approx 83.0512$.

CHAPTER 9: REVIEW EXERCISES

1. The slope of the graph of $y(t)$ at $(3, 2)$ is equal to the value of $\frac{dy}{dt}$ at $t = 3$; that is, $y'(3) = 9 \cdot 2 - 5 \cdot 3 = 3$.

3. $y'(0) = e^{3-y(0)} - 1 = e^{3-2} - 1 = e - 1$

5. (a) $y'(0) = 4y(0) - y(0)^2 = 4 \cdot 1 - 1^2 = 3$ and $L(t) = y(0) + y'(0)(t - 0) = 1 + 3t$
   (b) $y(0.1) \approx L(0.1) = 1 + 3 \cdot 0.1 = 1.3$

7. (a) We are given $f(t, y) = 2y - 10t$, $t_0 = 0$, $y_0 = 3$. Our interval here is $[0, 1]$ and $n = 5$. Therefore, $\Delta t = \frac{1-0}{5} = 0.2$, and, by using the formula

$$y_k = y_{k-1} + f(t_{k-1}, y_{k-1}) \cdot \Delta t = y_{k-1} + (2y_{k-1} - 10t_{k-1}) \cdot (0.2),$$

for $k = 1, 2, 3, 4, 5$, we obtain the following:

k	1	2	3	4	5
t_k	0.2	0.4	0.6	0.8	1
y_k	4.2	5.48	6.872	8.4208	10.1891

(b) If $y(t) = \frac{1}{2}(5 + e^{2t} + 10t)$, then $y'(t) = \frac{1}{2}(2e^{2t} + 10) = e^{2t} + 5$ and

$$2y - 10t = 2 \cdot \frac{1}{2}(5 + e^{2t} + 10t) - 10t = 5 + e^{2t}.$$

Thus, we see that $y'(t) = 2y - 10t$.

We next check the initial condition: $y(0) = \frac{1}{2}(5 + e^0 + 0) = 3$. Thus, $y(t) = \frac{1}{2}(5 + e^{2t} + 10t)$ is the solution of the initial value problem.

(c) Comparing the actual value $y(t_k)$ to the approximation y_k at each step results in the following table, where the error equals $|y(t_k) - y_k|$.

k	1	2	3	4	5
y_k	4.2	5.48	6.872	8.4208	10.1891
$y(t_k)$	4.24591	5.61277	7.16006	8.97652	11.1945
error	0.04591	0.13277	0.28806	0.55572	1.0054

The largest error occurs at $t = 1$.

9. We are given $f(t, p) = 0.04p \ln(2/p)$, $t_0 = 0$, and $p_0 = 0.8$. Our interval here is $[0, 10]$ and $\Delta t = 2$. Therefore, $n = 5$, and Euler's method gives

$$p_1 = p_0 + f(t_0, p_0)\Delta t = 0.8 + 0.04 \cdot 0.8 \ln \frac{2}{0.8} \cdot 2 \approx 0.858643$$

$$p_2 = p_1 + f(t_1, p_1)\Delta t \approx 0.858643 + 0.04 \cdot 0.8 \ln \frac{2}{0.858643} \cdot 2 \approx 0.916725$$

$$p_3 = p_2 + f(t_2, p_2)\Delta t \approx 0.916728 + 0.04 \cdot 0.8 \ln \frac{2}{0.916725} \cdot 2 \approx 0.973935$$

$$p_4 = p_3 + f(t_3, p_3)\Delta t \approx 0.973935 + 0.04 \cdot 0.8 \ln \frac{2}{0.973935} \cdot 2 \approx 1.03$$

$$p_5 = p_4 + f(t_4, p_4)\Delta t \approx 1.03 + 0.04 \cdot 0.8 \ln \frac{2}{1.03} \cdot 2 \approx 1.08468.$$

Thus, the population of the city 10 years later will be approximately 1.08468 million.

11. Setting $g(y) = 0.1(10y - 16 - y^2) = -0.1(y - 8)(y - 2) = 0$, we find the equilibrium solutions $y = 2$ and $y = 8$. Next, we compute $g'(y) = 0.1(10 - 2y)$. Since $g'(2) = 0.6 > 0$, the equilibrium solution $y = 2$ is unstable. Since $g'(8) = -0.6 < 0$, the equilibrium solution $y = 8$ is asymptotically stable.

13. To find the equilibrium solutions, we solve $g(y) = 0.2ye^{1-(y/8)} - 0.3y = 0$. Rewriting the equation in the form $0.2y[e^{1-(y/8)} - \frac{3}{2}] = 0$, we see that either $y = 0$ or $e^{1-(y/8)} = \frac{3}{2}$. The second of these reduces to $y = 8(1 - \ln \frac{3}{2})$.

Next, using the product and chain rules, we compute

$$g'(y) = -\frac{0.2}{8}ye^{1-(y/8)} + 0.2e^{1-(y/8)} - 0.3 = (0.2 - 0.025y)e^{1-(y/8)} - 0.3.$$

Since $g'(0) = 0.2e - 0.3 > 0$, the equilibrium solution $y = 0$ is unstable. On the other hand, since $g'(8(1 - \ln \frac{3}{2})) = -0.3(1 - \ln \frac{3}{2}) < 0$, the equilibrium solution $y = 8(1 - \ln \frac{3}{2})$ is asymptotically stable.

15. If $y(t)$ is any solution of the differential equation, then

$$y'(t) = -1 - y(t)^2 \le -1 < 0.$$

It follows from the first derivative test that $y(t)$ is a decreasing function.

17. The only equilibrium solution of the differential equation is $y = 0$. If $y(t)$ is any solution, then $y'(t) = -y(t)^2$ for all t, which means that $y(t) \le y(0)$. In particular, if $y(0) = -1$, then $y'(t) = -y(t)^2 \le -1$ for all t. Thus, the graph is decreasing and, since $y(0)$ is below the only equilbrium solution, $\lim_{t \to \infty} y(t) = -\infty$.

To check concavity, we observe that $y''(t) = -2y(t)y'(t) = 2y(t)^3 \le -2$. Therefore, the graph is concave down. To get a further idea of the shape of the graph, we sketch a direction field of the differential equation and then superimpose the solution on it, as shown.

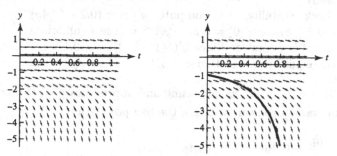

19. Setting $g(y) = 3 + y = 0$, we see that the only equilibrium solution is $y = -3$. Since $g'(y) = 1$, it is unstable. Also, $y'' = g'(y) \cdot y' = 3 + y$.

If $y(0) = -2$, then $y'(0) = g(-2) = 1 > 0$. Thus, $y(t)$ is increasing and $y(t) \to \infty$ as $t \to \infty$. Also, $y''(0) = 1$, and the graph is concave up for all t.

If $y(0) = -4$, then $y'(0) = g(-4) = -1 < 0$. Thus, $y(t)$ is decreasing and $y(t) \to -\infty$ as $t \to \infty$. Also, $y''(0) = 1$, and the graph is concave down for all t.

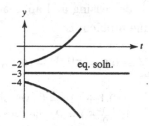

21. (a) Let $p(t)$ be the U.S. population in billions and t the time in years. The initial value problem is

$$\frac{dp}{dt} = 0.02p\left(1 - \frac{p}{0.5}\right) + 0.001, \quad p(0) = 0.28.$$

(b) Set $g(p) = 0.02p(1 - \frac{p}{0.5}) + 0.001 = -0.04p^2 + 0.02p + 0.001 = 0$. Solving this equation gives the equilibrium solutions $p = 0.25 - 0.05\sqrt{35}$ and $p = 0.25 + 0.05\sqrt{35}$. Under the restriction $p > 0$, the only equilibrium solution is $p = 0.25 + 0.05\sqrt{35} \approx 0.5458$. Since $g'(p) = -0.08p + 0.02$, we have $g'(0.25 + 0.05\sqrt{35}) = -0.004\sqrt{35} < 0$. Therefore, the equilibrium solution $p = 0.25 + 0.05\sqrt{35}$ is asymptotically stable.

(c) Since $\frac{dp}{dt} = -0.04(p - 0.25 + 0.05\sqrt{35})(p - 0.25 - 0.05\sqrt{35})$ > 0 for $0 < p < 0.25 + 0.05\sqrt{35}$, $p(t)$ is increasing, and $\lim_{t \to \infty} p(t) = 0.25 + 0.05\sqrt{35}$.

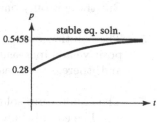

To check concavity, we observe that $g(p)$ has a local (and global) maximum at $p = 0.25$. Any solution $p(t)$ is concave up for $p < 0.25$ and concave down for $p > 0.25$. Therefore, the solution with initial value $p(0) = 0.28$ is concave down throughout.

23. **(a)** The initial value problem with y in hundreds of thousands, is

$$\frac{dy}{dt} = 0.2ye^{1-(y/5)} - 0.1y, \quad y(0) = 5.$$

Solving $g(y) = 0.2ye^{1-(y/5)} - 0.1y = 0.1y(2e^{1-(y/5)} - 1) = 0$, we find the equilibrium solutions are $y = 0$ and $y = 5(1 + \ln 2) \approx 8.4657$.

To check stability, we compute $g'(y) = (0.2 - 0.04y)e^{1-(y/5)} - 0.1$. Since $g'(0) = 0.2e - 0.1 > 0$, the equilibrium solution $y = 0$ is unstable. Since $g'(5(1 + \ln 2)) = -0.1\ln 2 - 0.1 < 0$, the equilibrium solution $y = 5(1 + \ln 2)$ is asymptotically stable.

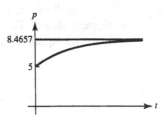

If $y(0) = 5$, then $y(t)$ is increasing and approaches $y = 5(1 + \ln 2)$ as $t \to \infty$.

(b) The initial value problem modeling the fish population is

$$\frac{dy}{dt} = 0.2ye^{1-(y/5)} - 0.4y, \quad y(0) = 5.$$

Solving $g(y) = 0.2ye^{1-(y/5)} - 0.4y = 0$, we obtain the equilibrium solutions $y = 0$ and $y = 5(1 - \ln 2) \approx 1.5343$.

To check stability, we compute $g'(y) = (0.2 - 0.04y)e^{1-(y/5)} - 0.4$. Since $g'(0) = 0.2e - 0.4 > 0$, the equilibrium solution $y = 0$ is unstable. Since $g'(5(1 - \ln 2)) = 0.4(\ln 2 - 1) < 0$, the equilibrium solution $y = 5(1 - \ln 2)$ is asymptotically stable.

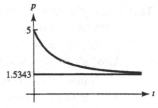

If $y(0) = 5$, then $y(t)$ is decreasing and approaches $5(1 - \ln 2)$ as $t \to \infty$.

(c) In this case, the initial value problem is

$$\frac{dy}{dt} = 0.2ye^{1-(y/5)} - (0.1 + k)y, \quad y(0) = 5.$$

Solving $g(y) = 0.2ye^{1-(y/5)} - (0.1 + k)y = 0$, we obtain the equilibrium solutions $y = 0$ and $y = 5(1 - \ln \frac{0.1+k}{0.2})$. If $5(1 - \ln \frac{0.1+k}{0.2}) = 0$, the two equilibrium solutions merge into one, and the fish will become extinct, no matter what the initial population is. That equation has the solution $k = 0.2e - 0.1 \approx 0.44$.

CHAPTER 9: EXAM

1. The derivative of the solution passing through $(1, 3)$ is positive, since the segment of the direction field through that point has positive slope.

3. There is no equilibrium solution with a value between 0 and 4. If there were, its graph would be a horizontal line, and the segments of the direction field along that line would all be horizontal. But there is only one horizontal segment shown in the direction field, through the point $(0, 2)$.

5. The solution curve through $(0, 2)$ would be superimposed on a field of short line segments with positive and increasing slope. Therefore, the solution curve itself has positive and increasing slope, and, therefore, it has no maximium. (See exercise 6.)

7. The equilibrium solutions are $y = 0$, $y = 2$, $y = 4$.

The equilibrium solution $y = 0$ is unstable, since each nonconstant solution in $0 < y < 2$ moves farther and farther away from the equilibrium solution $y = 0$ as t gets larger.

The equilibrium solution $y = 2$ is asymptotically stable, since each nonconstant solution in $0 < y < 2$ or $2 < y < 4$ moves closer and closer to the equilibrium solution $y = 2$ as t gets larger.

The equilibrium solution $y = 4$ is unstable, since each nonconstant solution in $2 < y < 4$ or $y > 4$ moves farther and farther away from the equilibrium solution $y = 4$ as t gets larger.

9. Following the direction field, we see that the solution corresponding to the initial condition $y(0) = 0.25$ goes from concave up to concave down. Similarly, the solution corresponding to the initial condition $y(0) = 3.75$ goes from concave down to concave up.

11. We are given $f(t, y) = 2y + t$, $t_0 = 0$, $y_0 = 1$, and $\Delta t = 0.1$. To cover the interval $[0, 0.2]$ with steps of size Δt, we need $n = 2$. Then Euler's method gives

$$y_1 = y_0 + f(t_0, y_0)\Delta t = 1 + (2 \cdot 1 + 0) \cdot 0.1 = 1.2$$

$$y_2 = y_1 + f(t_1, y_1)\Delta t = 1.2 + (2 \cdot 1.2 + 0.1) \cdot 0.1 = 1.45.$$

Thus, $y(0.2) \approx 1.45$.

13. (a) Let p be the population of the region in millions and t the time in years. The population is increasing according to a logistic model with $r = 0.02$ and $K = 8$, which means that the rate of increase is $0.02p(1 - \frac{p}{8})$. At the same time, the population is decreasing at a rate of $0.01p$ (one percent of the population) through emigration. Combining those, we get the differential equation

$$\frac{dp}{dt} = 0.02p\left(1 - \frac{p}{8}\right) - 0.01p = 0.01p - 0.0025p^2$$

with $p(0) = 6$ (the initial population) as the initial condition.

(b) Setting $g(p) = 0.01p - 0.0025p^2 = 0.0025p(4 - p) = 0$, we obtain the equilibrium solutions $p = 0$ and $p = 4$. Next, $g'(p) = 0.01 - 0.005p$. Then $g'(0) = 0.01 > 0$, which means that the equilibrium solution $p = 0$ is unstable; and $g'(4) = -0.01 < 0$, which means that the equilibrium solution $p = 4$ is asymptotically stable.

(c) Since $\frac{dp}{dt} < 0$ for $p > 4$, $p(t)$ is decreasing. Since $p = 4$ is an asymptotically stable equilibrium solution, $\lim_{t \to \infty} p(t) = 4$. To check concavity, we compute

$$\frac{d^2p}{dt^2} = g'(p)\frac{dp}{dt} = (0.01 - 0.005p)g(p) = 0.0000125p(p - 2)(p - 4).$$

Since $d^2p/dt^2 > 0$ for $p > 4$, $p(t)$ is concave down.

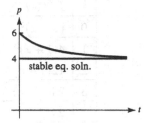

stable eq. soln.

CHAPTER 10
Higher-Order Approximations

EXERCISES 10.1

1. The coefficients are in the right-hand column of the following table:

k	$p^{(k)}(x)$	$p^{(k)}(0)$	$k!$	$p^{(k)}(0)/k!$
0	$1 + \frac{x}{2} + \frac{x^2}{3} + \frac{x^3}{4} + \frac{x^4}{5}$	1	1	1
1	$\frac{1}{2} + \frac{2}{3}x + \frac{3}{4}x^2 + \frac{4}{5}x^3$	$\frac{1}{2}$	1	$\frac{1}{2}$
2	$\frac{2}{3} + \frac{3}{2}x + \frac{12}{5}x^2$	$\frac{2}{3}$	2	$\frac{1}{3}$
3	$\frac{3}{2} + \frac{24}{5}x$	$\frac{3}{2}$	6	$\frac{1}{4}$
4	$\frac{24}{5}$	$\frac{24}{5}$	24	$\frac{1}{5}$

3. The coefficients are in the right-hand column of the following table:

k	$p^{(k)}(x)$	$p^{(k)}(0)$	$k!$	$p^{(k)}(0)/k!$
0	$1 - x + x^2 - x^3 + x^4 - x^5 + x^6$	1	1	1
1	$-1 + 2x - 3x^2 + 4x^3 - 5x^4 + 6x^5$	-1	1	-1
2	$2 - 6x + 12x^2 - 20x^3 + 30x^4$	2	2	1
3	$-6 + 24x - 60x^2 + 120x^3$	-6	6	-1
4	$24 - 120x + 360x^2$	24	24	1
5	$-120 + 720x$	-120	120	-1
6	720	720	720	1

5. (a) $n = 4$

k	$f^{(k)}(x)$	$f^{(k)}(0)$	$k!$	$f^{(k)}(0)/k!$
0	$(1+x)^4$	1	1	1
1	$4(1+x)^3$	4	1	4
2	$12(1+x)^2$	12	2	6
3	$24(1+x)$	24	6	4
4	24	24	24	1

$c_0 = 1$, $c_1 = 4$, $c_2 = 6$, $c_3 = 4$, $c_4 = 1$

(b) $n = 5$

k	$f^{(k)}(x)$	$f^{(k)}(0)$	$k!$	$f^{(k)}(0)/k!$
0	$(1+x)^5$	1	1	1
1	$5(1+x)^4$	5	1	5
2	$20(1+x)^3$	20	2	10
3	$60(1+x)^2$	60	6	10
4	$120(1+x)$	120	24	5
5	120	120	120	1

$c_0 = 1,\ c_1 = 5,\ c_2 = 10,\ c_3 = 10,\ c_4 = 5,\ c_5 = 1$

(c) $n = 6$

k	$f^{(k)}(x)$	$f^{(k)}(0)$	$k!$	$f^{(k)}(0)/k!$
0	$(1+x)^6$	1	1	1
1	$6(1+x)^5$	6	1	6
2	$30(1+x)^4$	30	2	15
3	$120(1+x)^3$	120	6	20
4	$360(1+x)^2$	360	24	15
5	$720(1+x)$	720	120	6
6	720	720	720	1

$c_0 = 1,\ c_1 - 6,\ c_2 - 15,\ c_3 = 20,\ c_4 = 15,\ c_5 = 6,\ c_6 = 1$

7. $(u+v)^n = u^n \left(1 + \dfrac{v}{u}\right)^n$

$$= u^n \left[c_0 + c_1\frac{v}{u} + c_2\left(\frac{v}{u}\right)^2 + \cdots + c_k\left(\frac{v}{u}\right)^k + \cdots + c_n\left(\frac{v}{u}\right)^n\right]$$

$$= c_0 u^n + c_1 u^{n-1}v + c_2 u^{n-2}v + \cdots + c_k u^{n-k}v^k + \cdots + c_n v^n,$$

where $c_k = \dfrac{n(n-1)(n-2)\cdots(n-k+1)}{k!}$.

9.

k	$f^{(k)}(x)$	$f^{(k)}(1)$	$k!$	$f^{(k)}(1)/k!$
0	$x^3 - 3x^2 + 2x + 1$	1	1	1
1	$3x^2 - 6x + 2$	-1	1	-1
2	$6x - 6$	0	2	0
3	6	6	6	1

Thus, $P_3(x) = 1 - (x-1) + (x-1)^3$. To check, we expand and collect terms:

$$P_3(x) = 1 - x + 1 + x^3 - 3x^2 + 3x - 1 = x^3 - 3x^2 + 2x + 1.$$

11.

k	$f^{(k)}(x)$	$f^{(k)}(-1)$	$k!$	$f^{(k)}(-1)/k!$
0	x^4	1	1	1
1	$4x^3$	-4	1	-4
2	$12x^2$	12	2	6
3	$24x$	-24	6	-4
4	24	24	24	1

$P_4(x) = 1 - 4(x+1) + 6(x+1)^2 - 4(x+1)^3 + (x+1)^4 = x^4$. To check:

$$P_4(x) = 1 - 4x - 4 + 6(x^2 + 2x + 1) - 4(x^3 + 3x^2 + 3x + 1)$$

$$+ (x^4 + 4x^3 + 6x^2 + 4x + 1)$$

$$= x^4 + (-4+4)x^3 + (6 - 12 + 6)x^2 + (-4 + 12 - 12 + 4)x$$

$$+ (1 - 4 + 6 - 4 + 1)$$

$$= x^4.$$

13. $f(x) = \ln x$, $a = 1$, $n = 6$

k	$f^{(k)}(x)$	$f^{(k)}(1)$	$k!$	$f^{(k)}(1)/k!$
0	$\ln x$	0	1	0
1	$\frac{1}{x}$	1	1	1
2	$-\frac{1}{x^2}$	-1	2	$-\frac{1}{2}$
3	$\frac{2}{x^3}$	2	6	$\frac{1}{3}$
4	$-\frac{6}{x^4}$	-6	24	$-\frac{1}{4}$
5	$\frac{24}{x^5}$	24	120	$\frac{1}{5}$
6	$-\frac{120}{x^6}$	-120	720	$-\frac{1}{6}$

$$P_6(x) = (x-1) - \frac{1}{2}(x-1)^2 + \frac{1}{3}(x-1)^3 - \frac{1}{4}(x-1)^4 + \frac{1}{5}(x-1)^5 - \frac{1}{6}(x-1)^6$$

15. $f(x) = e^{-x^2}$, $a = 0$, $n = 4$

k	$f^{(k)}(x)$	$f^{(k)}(0)$	$k!$	$f^{(k)}(0)/k!$
0	e^{-x^2}	1	1	1
1	$-2xe^{-x^2}$	0	1	0
2	$(4x^2 - 2)e^{-x^2}$	-2	2	-1
3	$(-8x^3 + 12x)e^{-x^2}$	0	6	0
4	$(16x^4 - 48x^2 + 12)e^{-x^2}$	12	24	$\frac{1}{2}$

$$P_4(x) = 1 - x^2 + \frac{1}{2}x^4$$

17. $f(x) = (1+x)^{1/3}$, $a = 0$, $n = 3$

k	$f^{(k)}(x)$	$f^{(k)}(0)$	$k!$	$f^{(k)}(0)/k!$
0	$(1+x)^{1/3}$	1	1	1
1	$\frac{1}{3}(1+x)^{-2/3}$	$\frac{1}{3}$	1	$\frac{1}{3}$
2	$-\frac{2}{9}(1+x)^{-5/3}$	$-\frac{2}{9}$	2	$-\frac{1}{9}$
3	$\frac{10}{27}(1+x)^{-8/3}$	$\frac{10}{27}$	6	$\frac{5}{81}$

$$P_3(x) = 1 + \frac{1}{3}x - \frac{1}{9}x^2 + \frac{5}{81}x^3$$

19. $f(x) = e^x + e^{-x}$, $a = 0$, $n = 6$

k	$f^{(k)}(x)$	$f^{(k)}(0)$	$k!$	$f^{(k)}(0)/k!$
0	$e^x + e^{-x}$	2	1	2
1	$e^x - e^{-x}$	0	1	0
2	$e^x + e^{-x}$	2	2	1
3	$e^x - e^{-x}$	0	6	0
4	$e^x + e^{-x}$	2	24	$\frac{1}{12}$
5	$e^x - e^{-x}$	0	120	0
6	$e^x + e^{-x}$	2	720	$\frac{1}{360}$

$$P_6(x) = 2 + x^2 + \frac{1}{12}x^4 + \frac{1}{360}x^6$$

21. $f(x) = \cos x$, $a = \pi$, $n = 6$

k	$f^{(k)}(x)$	$f^{(k)}(\pi)$	$k!$	$f^{(k)}(\pi)/k!$
0	$\cos x$	-1	1	-1
1	$-\sin x$	0	1	0
2	$-\cos x$	1	2	$\frac{1}{2}$
3	$\sin x$	0	6	0
4	$\cos x$	-1	24	$-\frac{1}{24}$
5	$-\sin x$	0	120	0
6	$-\cos x$	1	720	$\frac{1}{720}$

$$P_6(x) = -1 + 2(x-\pi)^2 - \frac{1}{24}(x-\pi)^4 + \frac{1}{720}(x-\pi)^6$$

23. $f(x) = e^{2x}$, $a = 0$

k	$f^{(k)}(x)$	$f^{(k)}(0)$	$k!$	$f^{(k)}(0)/k!$
0	e^{2x}	1	0!	1
1	$2e^{2x}$	2	1!	$\frac{2}{1!}$
2	$2^2 e^{2x}$	2^2	2!	$\frac{2^2}{2!}$
3	$2^3 e^{2x}$	2^3	3!	$\frac{2^3}{3!}$
4	$2^4 e^{2x}$	2^4	4!	$\frac{2^4}{4!}$
5	$2^5 e^{2x}$	2^5	5!	$\frac{2^5}{5!}$
6	$2^6 e^{2x}$	2^6	6!	$\frac{2^6}{6!}$
7	$2^7 e^{2x}$	2^7	7!	$\frac{2^7}{7!}$
8	$2^8 e^{2x}$	2^8	8!	$\frac{2^8}{8!}$

General pattern: $c_k = \dfrac{2^k}{k!}$.

25. $f(x) = \ln x$, $a = 1$

k	$f^{(k)}(x)$	$f^{(k)}(1)$	$k!$	$f^{(k)}(1)/k!$
0	$\ln x$	0	1	0
1	x^{-1}	1	1	1
2	$-x^{-2}$	-1	2!	$-\frac{1}{2}$
3	$2x^{-3}$	2	3!	$\frac{1}{3}$
4	$-3!x^{-4}$	$-3!$	4!	$-\frac{1}{4}$
5	$4!x^{-5}$	$4!$	5!	$\frac{1}{5}$
6	$-5!x^{-6}$	$-5!$	6!	$-\frac{1}{6}$
7	$6!x^{-7}$	$6!$	7!	$\frac{1}{7}$
8	$-7!x^{-8}$	$-7!$	8!	$-\frac{1}{8}$

General pattern:
$$c_k = \frac{(-1)^{k+1}}{k}, \quad k = 1, 2, 3, \ldots, \text{ and } c_0 = 0.$$

27. $f(x) = \dfrac{1}{(1+x)^2}$, $a = 0$

k	$f^{(k)}(x)$	$f^{(k)}(0)$	$k!$	$f^{(k)}(0)/k!$
0	$\frac{1}{(1+x)^2}$	1	1	1
1	$-\frac{2}{(1+x)^3}$	-2	1	-2
2	$\frac{3!}{(1+x)^4}$	$3!$	2!	3
3	$-\frac{4!}{(1+x)^5}$	$-4!$	3!	-4
4	$\frac{5!}{(1+x)^6}$	$5!$	4!	5
5	$-\frac{6!}{(1+x)^7}$	$-6!$	5!	-6
6	$\frac{7!}{(1+x)^8}$	$7!$	6!	7
7	$-\frac{8!}{(1+x)^9}$	$-8!$	7!	-8
8	$\frac{9!}{(1+x)^{10}}$	$9!$	8!	9

General pattern: $c_k = (-1)^k (k+1)$.

29. $f(x) = \sin x$, $a = 0$

k	$f^{(k)}(x)$	$f^{(k)}(0)$	$k!$	$f^{(k)}(0)/k!$
0	$\sin x$	0	0!	0
1	$\cos x$	1	1!	$\frac{1}{1!}$
2	$-\sin x$	0	2!	0
3	$-\cos x$	-1	3!	$-\frac{1}{3!}$
4	$\sin x$	0	4!	0
5	$\cos x$	1	5!	$\frac{1}{5!}$
6	$-\sin x$	0	6!	0
7	$-\cos x$	-1	7!	$-\frac{1}{7!}$
8	$\sin x$	0	8!	0

General pattern: $c_{2k} = 0$, $c_{2k+1} = \dfrac{(-1)^k}{(2k+1)!}$.

31. From example 10.1.3, we have $P_2(x) = 1 + x + \frac{1}{2}x^2$. Thus,

$$e^{-0.1} \approx P_2(-0.1) = 1 + (-0.1) + \frac{1}{2}(-0.1)^2 = 0.905.$$

The error satisfies $|R_2(x)| \leq \frac{M|x-0|^3}{3!}$, where M is the maximum value of $|f^{(3)}(x)| = e^x$ on the interval between x and 0. Since e^x is increasing, its maximum value is 1, achieved at $x = 0$. Therefore, $M = 1$, and

$$|R_2(-0.1)| \leq \frac{0.1^3}{3!} = \frac{0.001}{6} < 0.000167.$$

33. To find the Taylor polynomials of $f(x) = \sqrt{x}$ about 4, we first compute the following:

k	$f^{(k)}(x)$	$f^{(k)}(4)$	$k!$	$f^{(k)}(4)/k!$
0	$x^{1/2}$	2	1	2
1	$\frac{1}{2}x^{-1/2}$	$\frac{1}{4}$	1	$\frac{1}{4}$
2	$-\frac{1}{4}x^{-3/2}$	$-\frac{1}{32}$	2	$-\frac{1}{64}$
3	$\frac{3}{8}x^{-5/2}$	$\frac{3}{256}$	6	$\frac{1}{512}$
4	$-\frac{15}{16}x^{-7/2}$	$-\frac{15}{2,048}$	24	$-\frac{5}{16,384}$

Then $P_1(x) = 2 + \frac{1}{4}(x-4)$, $P_2(x) = 2 + \frac{1}{4}(x-4) - \frac{1}{64}(x-4)^2$, and $P_3(x) = 2 + \frac{1}{4}(x-4) - \frac{1}{64}(x-4)^2 + \frac{1}{512}(x-4)^3$.

For $n = 1$, $\sqrt{5} \approx P_1(5) = 2 + \frac{1}{4} = 2.25$, and $|R_1(5)| \leq \frac{M_1|5-4|^2}{2!}$, where M_1 is the maximum of $\frac{1}{4}x^{-3/2}$ on the interval $[4,5]$. Since it is decreasing, it achieves a maximum of $\frac{1}{32}$ at $x = 4$. Therefore, $|R_1(5)| \leq \frac{1/32}{2!} = \frac{1}{64}$.

For $n = 2$, $\sqrt{5} \approx P_2(5) = 2 + \frac{1}{4} - \frac{1}{64} = 2.234375$, and we have $|R_2(5)| \leq \frac{M_2|5-4|^3}{3!}$, where M_2 is the maximum of $\frac{3}{8}x^{-5/2}$ on the interval $[4,5]$. It achieves a maximum of $\frac{3}{256}$ at $x = 4$. Therefore, $|R_2(5)| \leq \frac{3/256}{3!} = \frac{1}{512}$.

For $n = 3$, $\sqrt{5} \approx P_3(5) = 2 + \frac{1}{4} - \frac{1}{64} + \frac{1}{512} = 2.236328125$, and we have $|R_3(5)| \leq \frac{M_3|5-4|^4}{4!}$, where M_3 is the maximum of $\frac{15}{16}x^{-7/2}$ on the interval $[4,5]$. It achieves a maximum of $\frac{15}{2,048}$ at $x = 4$. Therefore, $|R_3(5)| \leq \frac{1}{4!} \cdot \frac{15}{2,048} = \frac{5}{16,384}$.

35. **(a)** From the table in Exercise 34, we infer that $c_0 = 0$ and $c_k = \frac{(-1)^{k+1}}{k}$ for $k = 1, 2, \ldots$. In fact, if, as the table suggests, $f^{(k)}(x) = (-1)^{k-1}\frac{(k-1)!}{(1+x)^k}$ with $f^{(k)}(0) = (-1)^{k-1}(k-1)!$, then

$$f^{(k+1)}(x) = \frac{d}{dx}\left[(-1)^{k-1}\frac{(k-1)!}{(1+x)^k}\right] = (-1)^{k-1} \cdot (-k)\frac{(k-1)!}{(1+x)^{k+1}} = (-1)^k \frac{k!}{(1+x)^{k+1}},$$

and $f^{(k+1)}(0) = (-1)^k k!$. Therefore, $c_{k+1} = \frac{f^{(k+1)}(0)}{(k+1)!} = \frac{(-1)^k}{k+1}$, and the pattern continues. Thus, the nth-degree Taylor polynomial is

$$P_n(x) = x - \frac{1}{2}x^2 + \frac{1}{3}x^3 - \cdots + (-1)^{n+1}\frac{1}{n}x^n.$$

(b) In approximating $\ln(1.1) \approx 0.1 - \frac{1}{2}(0.1)^2 + \frac{1}{3}(0.1)^3 - \cdots + (-1)^{n+1}\frac{1}{n}(0.1)^n$, the error satisfies $|R_n(0.1)| \leq \frac{M_n|0.1-0|^{n+1}}{(n+1)!}$, where M_n is the maximum of $|f^{(n+1)}(x)| = \frac{n!}{(1+x)^{n+1}}$ on the interval $[0, 0.1]$. Since it is decreasing over that interval, $\frac{n!}{(1+x)^{n+1}}$ achieves a maximum of $n!$ at $x = 0$. Therefore, $M_n = n!$, and we want n so that $|R_n(0.1)| \leq \frac{n!(0.1)^{n+1}}{(n+1)!} = \frac{(0.1)^{n+1}}{n+1} \leq 10^{-6}$. Direct calculation gives the following table:

n	1	2	3	4	5
$\frac{(0.1)^{n+1}}{n+1}$	0.005	0.00033	0.000025	0.000002	0.000000167

Thus, $n \geq 5$.

37. The Taylor coefficients are given by the following table:

k	$f^{(k)}(x)$	$f^{(k)}(0)$	$k!$	$f^{(k)}(0)/k!$
0	$e^{x/2}$	1	1	1
1	$\frac{1}{2}e^{x/2}$	$\frac{1}{2}$	1	$\frac{1}{2}$
2	$\frac{1}{4}e^{x/2}$	$\frac{1}{4}$	2	$\frac{1}{8}$
3	$\frac{1}{8}e^{x/2}$	$\frac{1}{8}$	6	$\frac{1}{48}$

Thus, $P_3(x) = 1 + \frac{1}{2}x + \frac{1}{8}x^2 + \frac{1}{48}x^3$, and

$$\int_0^1 e^{x/2}\, dx \approx \int_0^1 \left(1 + \frac{1}{2}x + \frac{1}{8}x^2 + \frac{1}{48}x^3\right) dx = \left(x + \frac{1}{4}x^2 + \frac{1}{24}x^3 + \frac{1}{192}x^4\right)\Big|_0^1$$

$$= 1 + \frac{1}{4} + \frac{1}{24} + \frac{1}{192} = \frac{83}{64} \approx 1.296875.$$

The fundamental theorem of calculus gives

$$\int_0^1 e^{x/2}\, dx = 2e^{x/2}\Big|_0^1 = 2e^{1/2} - 2 \approx 1.29744.$$

39. By Exercise 30, $P_4(x) = x + x^2 + \frac{1}{2}x^3 + \frac{1}{6}x^4$. Therefore,

$$\int_0^1 xe^x\, dx \approx \int_0^1 \left(x + x^2 + \frac{1}{2}x^3 + \frac{1}{6}x^4\right) dx$$

$$= \frac{1}{2}x^2 + \frac{1}{3}x^3 + \frac{1}{8}x^4 + \frac{1}{30}x^5\Big|_0^1$$

$$= \frac{1}{2} + \frac{1}{3} + \frac{1}{8} + \frac{1}{30} = \frac{119}{120} \approx 0.991667.$$

Using integration by parts, we get

$$\int_0^1 xe^x\, dx = (xe^x - e^x)\Big|_0^1 = (e - e) - (-1) = 1.$$

41. The Taylor coefficients are given by the following table:

k	$f^{(k)}(x)$	$f^{(k)}(0)$	$k!$	$f^{(k)}(0)/k!$
0	$\frac{1}{1+x^2}$	1	1	1
1	$-\frac{2x}{(1+x^2)^2}$	0	1	0
2	$\frac{6x^2-2}{(1+x^2)^3}$	-2	2	-1

Therefore, $P_2(x) = 1 - x^2$, and

$$\int_0^{0.1} \frac{1}{1+x^2}\,dx \approx \int_0^{0.1} (1-x^2)\,dx = x - \frac{1}{3}x^3\Big|_0^{0.1} = 0.1 - \frac{1}{3}(0.1)^3 \approx 0.09967.$$

43. We want to estimate $\int_0^1 h(t)\,dt$, where $h(t) = \frac{40}{(t^2+5)^2}$. The Taylor coefficients are given by the following table:

k	$h^{(k)}(t)$	$h^{(k)}(0)$	$k!$	$h^{(k)}(0)/k!$
0	$\frac{40}{(t^2+5)^2}$	$\frac{8}{5}$	1	$\frac{8}{5}$
1	$-\frac{160t}{(t^2+5)^3}$	0	1	0
2	$\frac{800(t^2-1)}{(t^2+5)^4}$	$-\frac{32}{25}$	2	$-\frac{16}{25}$

Therefore, $P_2(t) = \frac{8}{5} - \frac{16}{25}t^2$, and

$$\int_0^1 h(t)\,dt \approx \int_0^1 \left(\frac{8}{5} - \frac{16}{25}t^2\right) dt = \left(\frac{8}{5}t - \frac{16}{75}t^3\right)\Big|_0^1 = \frac{8}{5} - \frac{16}{75} = \frac{104}{75} \approx 1.3867.$$

The area of the parcel is approximately 13,867 square feet.

45. If $y' - y + e^t - t$, then $y'' = y' + e^t - 1$ and $y''' - y'' + e^t$. Using the initial condition $y(0) = 0$, we get $y'(0) = y(0) + e^0 - 0 = 1$, $y''(0) = y'(0) + e^0 - 1 = 1$, and $y'''(0) = y''(0) + e^0 = 2$. Then we obtain the Taylor coefficients:

k	$y^{(k)}(0)$	$k!$	$y^{(k)}(0)/k!$
0	0	1	0
1	1	1	1
2	1	2	$\frac{1}{2}$
3	2	6	$\frac{1}{3}$

$P_3(t) = t + \frac{1}{2}t^2 + \frac{1}{3}t^3$, and $y(0.1) \approx P_3(0.1) = 0.1 + \frac{1}{2}(0.1)^2 + \frac{1}{3}(0.1)^3 \approx 0.1053.$

47. If $y' = t + y^2$, then $y'' = 1 + 2yy'$. Using the initial condition $y(0) = -1$, we get $y'(0) = y(0)^2 = 1$ and $y''(0) = 1 + 2y(0)y'(0) = -1$. Then we obtain:

k	$y^{(k)}(0)$	$k!$	$y^{(k)}(0)/k!$
0	-1	1	-1
1	1	1	1
2	-1	2	$-\frac{1}{2}$

$P_2(t) = -1 + t - \frac{1}{2}t^2$, and $y(0.1) \approx P_2(0.1) = -1 + 0.1 - \frac{1}{2}(0.1)^2 = -0.905.$

49. (a) If $B'(y) = 0.4(y + 0.2)^{-0.6}$, then $B''(y) = -0.24(y + 0.2)^{-1.6}$, $B'''(y) = 0.384(y + 0.2)^{-2.6}$, and $B^{(4)}(y) = -0.9984(y + 0.2)^{-3.6}$.

Using these derivatives and the initial condition $B(0) = 38$, we get the Taylor coefficients as follows:

k	$B^{(k)}(0)$	$k!$	$B^{(k)}(0)/k!$
0	38	1	38
1	$0.4(0.2)^{-0.6}$	1	$0.4(0.2)^{-0.6} \approx 1.05$
2	$-0.24(0.2)^{-1.6}$	2	$-0.12(0.2)^{-1.6} \approx -1.58$
3	$0.384(0.2)^{-2.6}$	6	$0.064(0.2)^{-2.6} \approx 4.20$
4	$-0.9984(0.2)^{-3.6}$	24	$-0.0416(0.2)^{-3.6} \approx 13.66$

Then $P_4(y) \approx 38 + 1.05y - 1.58y^2 + 4.20y^3 - 13.66y^4$.

(b)

y	0	0.1	0.2	0.3	0.4	0.5
$P_4(y) \approx$	38	38.09	38.16	38.18	38.09	37.80

(c) $B(y) = \int 0.4(y + 0.2)^{-0.6}\, dy = (y + 0.2)^{0.4} + c$. The initial condition $B(0) = 38$ gives $38 = 0.2^{0.4} + c$, or $c = 38 - 0.2^{0.4}$. So $B(y) = (y + 0.2)^{0.4} + 38 - 0.2^{0.4}$.

y	0	0.1	0.2	0.3	0.4	0.5
$B(y)$	38	38.09	38.17	38.23	38.29	38.34

The accuracy of the estimate decreases as y increases.

51. If $p' = 0.02p(1 - p)$, then, by the product rule,

$$p'' = 0.02[(1 - p)p' - pp'] = 0.02(1 - 2p)p'.$$

Taking t to be the number of years after 2010, we have the initial condition $p(0) = 0.31$. From it, we obtain

$$p'(0) = 0.02p(0)[1 - p(0)] = 0.02 \cdot 0.31 \cdot (1 - 0.31) = 0.004278$$

and $p''(0) = 0.02(1 - 2 \cdot 0.31) \cdot 0.004278 = 0.0000325128$. Then

$$P_2(t) = p(0) + p'(0)t + \frac{p''(0)}{2}t^2 = 0.31 + 0.004278t + 0.0000162564t^2,$$

and $p(1) \approx P_2(1) = 0.31 + 0.004278 + 0.0000162564 = 0.3142942564$. Thus, the population will be approximately 0.3143 billion in 2011.

53. The Taylor coefficients are given by the following table:

k	$f^{(k)}(x)$	$f^{(k)}(0)$	$k!$	$f^{(k)}(0)/k!$
0	$e^x - x$	1	1	1
1	$e^x - 1$	0	1	0
2	e^x	1	2	$1/2$
3	e^x	1	6	$1/6$

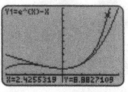

[-3, 3] × [-1, 10] window

55. The Taylor coefficients are given by the following table:

k	$f^{(k)}(x)$	$f^{(k)}(0)$	$k!$	$f^{(k)}(0)/k!$
0	xe^{-x}	0	1	0
1	$(1-x)e^{-x}$	1	1	1
2	$(x-2)e^{-x}$	-2	2	-1
3	$(3-x)e^{-x}$	3	6	$\frac{1}{2}$
4	$(x-4)e^{-x}$	-4	24	$-\frac{1}{6}$
5	$(5-x)e^{-x}$	5	120	$\frac{1}{24}$

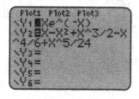 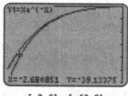 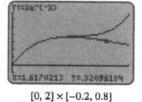

$[-3, 0] \times [-60, 0]$ $[0, 2] \times [-0.2, 0.8]$

57. The Taylor coefficients are given by the following table:

k	$f^{(k)}(x)$	$f^{(k)}(0)$	$k!$	$f^{(k)}(0)/k!$
0	$\frac{x-1}{x+1}$	-1	1	-1
1	$2(x+1)^{-2}$	2	1	2
2	$-4(x+1)^{-3}$	-4	2	-2
3	$12(x+1)^{-4}$	12	6	2

 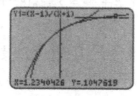

$[-0.75, 1] \times [-2, 1]$

59. (a) If $u = f''(t)$ and $dv = (t - x)\, dt$, then $du = f^{(3)}(t)\, dt$, and $v = \frac{1}{2}(t - x)^2$. Then applying integration by parts to (22) gives

$$R_1(x) = \int_a^x (x - t)f''(t)\, dt = -\int_a^x (t - x)f''(t)\, dt$$

$$= -\frac{1}{2}f''(t)(t - x)^2 \Big|_a^x + \int_a^x \frac{1}{2}f^{(3)}(t)(t - x)^2\, dx$$

$$= \frac{1}{2}f''(a)(x - a)^2 + \frac{1}{2}\int_a^x (x - t)^2 f^{(3)}(t)\, dt.$$

Therefore,

$$f(x) = f(a) + f'(a)(x - a) + R_1(x)$$

$$= f(a) + f'(a)(x - a) + \frac{1}{2}f''(a)(x - a)^2 + \frac{1}{2}\int_a^x (x - t)^2 f^{(3)}(t)\, dt$$

$$= f(a) + f'(a)(x - a) + \frac{1}{2}f''(a)(x - a)^2 + R_2(x),$$

where $R_2(x) = \frac{1}{2}\int_a^x (x - t)^2 f^{(3)}(t)\, dt$.

(b) For simplicity, we assume that $x > a$. Let M be the maximum value of $|f^{(3)}(t)|$ for $a \le t \le x$. Then

$$|R_2(x)| = \frac{1}{2}\left|\int_a^x (x - t)^2 f^{(3)}(t)\, dt\right| \le \frac{1}{2}\int_a^x |x - t|^2 |f^{(3)}(t)|\, dt$$

$$\le \frac{M}{2}\int_a^x (x - t)^2\, dt = \frac{M}{2}\frac{1}{3}(t - x)^3\Big|_a^x$$

$$= \frac{M(x - a)^3}{3!} = \frac{M|x - a|^3}{3!}.$$

EXERCISES 10.2

1. Since $|\frac{2}{3}| = \frac{2}{3} < 1$, $\sum_{k=0}^{\infty}(\frac{2}{3})^k$ coverges to $\frac{1}{1-2/3} = 3$.

3. $1 - \frac{1}{5} + (\frac{1}{5})^2 - (\frac{1}{5})^3 + (\frac{1}{5})^5 - \cdots = \sum_{k=0}^{\infty}(-\frac{1}{5})^k$. Since $|-\frac{1}{5}| = \frac{1}{5} < 1$, $\sum_{k=0}^{\infty}(-\frac{1}{5})^k$ converges to $\frac{1}{1-(-1/5)} = \frac{5}{6}$.

5. $1 + e^{-1} + e^{-2} + e^{-3} + \cdots = \sum_{k=0}^{\infty}(e^{-1})^k$. Since $|e^{-1}| = \frac{1}{e} < 1$, $\sum_{k=0}^{\infty}(e^{-1})^k$ coverges to $\frac{1}{1-e^{-1}} = \frac{e}{e-1}$.

7. $\frac{3}{2} - \frac{3}{4} + \frac{3}{8} - \frac{3}{16} + \cdots = \frac{3}{2}(1 - \frac{1}{2} + \frac{1}{4} - \frac{1}{8} + \cdots) = \frac{3}{2}\sum_{k=0}^{\infty}(-\frac{1}{2})^k$. Since $|-\frac{1}{2}| = \frac{1}{2} < 1$, $\frac{3}{2}\sum_{k=0}^{\infty}(-\frac{1}{2})^k$ converges to $\frac{3}{2} \cdot \frac{1}{1-(-1/2)} = 1$.

9. $0.\overline{21} = \frac{21}{100}\sum_{k=0}^{\infty}\left(\frac{1}{100}\right)^k = \frac{21}{100} \cdot \frac{1}{1 - 1/100} = \frac{21}{99} = \frac{7}{33}$

11. $2.\overline{13} = 2 + 0.\overline{13} = 2 + \frac{13}{100}\sum_{k=0}^{\infty}\left(\frac{1}{100}\right)^k$

$$= 2 + \frac{13}{100} \cdot \frac{1}{1 - 1/100} = 2 + \frac{13}{99} = \frac{211}{99}$$

13. If you invest A dollars at 5% annual interest, compounded monthly, the amount of money D at the end of k years satisfies

$$D = A\left(1 + \frac{0.05}{12}\right)^{12k} = A\left(\frac{241}{240}\right)^{12k}.$$

Equivalently, $A = D(\frac{240}{241})^{12k}$. Therefore, to have $1,000 at the end of k years, you must invest $1,000\left(\frac{240}{241}\right)^{12k}$ dollars now. Taking the sum for $k = 1, 2, \ldots$ gives the the total investment needed to have $1,000 at the end of every year:

$$\sum_{k=1}^{\infty} 1,000 \left(\frac{240}{241}\right)^{12k} = 1,000 \sum_{k=1}^{\infty} \left[\left(\frac{240}{241}\right)^{12}\right]^{k}$$

$$= 1,000 \cdot \frac{(240/241)^{12}}{1 - (240/241)^{12}}$$

$$\approx 19545.80 \text{ dollars.}$$

15. (a) The ball drops 27 feet, hits the driveway and bounces back up $(0.82) \cdot 27$ feet. It then falls back down $(0.82) \cdot 27$ feet, hits the driveway and bounces back up $(0.82)^2 \cdot 27$ feet. It then falls back down $(0.82)^2 \cdot 27$ and hits the driveway a third time. The total distance traveled is

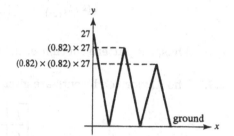

$$27 + 2(0.82) \cdot 27 + 2(0.82)^2 \cdot 27 \approx 107.59 \text{ feet.}$$

(b) After it hits the driveway for the kth time, the ball rebounds up to $(0.82)^k \cdot 27$ feet and then falls back down that same amount to hit the driveway for the $(k + 1)$st time. If it continues bouncing indefinitely, the total distance traveled (in feet) will be

$$27 + \sum_{k=1}^{\infty} 2(0.82)^k \cdot 27 = 27 + 2 \cdot (0.82) \cdot 27 \cdot \sum_{k=0}^{\infty} (0.82)^k$$

$$= 27 + (1.64) \cdot 27 \cdot \frac{1}{1 - 0.82} = 273.$$

17. (a) $10e^{-0.8} + 10e^{-2 \cdot (0.8)} + 10e^{-3 \cdot (0.8)} = 10(e^{-0.8} + e^{-1.6} + e^{-2.4})$

(b) $10e^{-0.8} + 10e^{-2 \cdot (0.8)} \cdots + 10e^{-n \cdot (0.8)} = 10 \sum_{k=1}^{n} (e^{-0.8})^k$

(c) $\lim_{n \to \infty} 10 \sum_{k=1}^{n} (e^{-0.8})^k = 10 \sum_{k=1}^{\infty} (e^{-0.8})^k = 10 \cdot \frac{e^{-0.8}}{1 - e^{-0.8}} \approx 8.16 \text{ mg}$

19. If the initial population of fish is A, then the population of the fish D after k years satisfies $D = Ae^{0.22k}$, or $A = De^{-0.22k}$. Therefore, in order to harvest 1,000 at the end of k six-month periods (that is, $k/2$ years), there must be $1,000e^{-0.11k}$ fish in the initial population. In order to harvest that amount at the end of *every* six-month period, we add those initial amounts for $k = 1, 2, 3, \ldots$, which yields

$$1,000 \sum_{k=1}^{\infty} e^{-0.11k} = 1,000 \cdot \frac{e^{-0.11}}{1 - e^{-0.11}} = \frac{1,000}{e^{0.11} - 1} \approx 8,600.$$

21. The Taylor coefficients are given by the following table:

k	$f^{(k)}(x)$	$f^{(k)}(0)$	$k!$	$f^{(k)}(0)/k!$
0	$(2+x)^{-1}$	$1/2$	1	$1/2$
1	$-(2+x)^{-2}$	$-1/2^2$	1	$-1/2^2$
2	$2(2+x)^{-3}$	$2/2^3$	2	$1/2^3$
3	$-3!(2+x)^{-4}$	$-3!/2^4$	3!	$-1/2^4$
4	$4!(2+x)^{-5}$	$4!/2^5$	4!	$1/2^5$

From this pattern we infer that $f^{(k)}(x) = (-1)^k k!(2+x)^{-k-1}$ and $\frac{f^{(k)}(0)}{k!} = \frac{(-1)^k}{2^{k+1}}$. To verify that this pattern continues, we check that if the kth derivative has this form, then

$$f^{(k+1)}(x) = \frac{d}{dx}(-1)^k k!(2+x)^{-k-1} = (-1)^{k+1}(k+1)!(2+x)^{-k-2}.$$

Thus, the pattern continues, and $\frac{f^{(k+1)}(0)}{(k+1)!} = \frac{(-1)^{k+1}}{2^{k+2}}$. Therefore, the Taylor series is $\sum_{k=0}^{\infty}(-1)^k \frac{1}{2^{k+1}}x^k$.

23. The Taylor coefficients are given by the following table:

k	$f^{(k)}(x)$	$f^{(k)}(1)$	$k!$	$f^{(k)}(1)/k!$
0	$\ln(2-x)$	0	1	0
1	$-\frac{1}{2-x}$	-1	1	-1
2	$-\frac{1}{(2-x)^2}$	-1	2	$-\frac{1}{2}$
3	$-\frac{2}{(2-x)^3}$	-2	3!	$-\frac{1}{3}$
4	$-\frac{3!}{(2-x)^4}$	$-3!$	4!	$-\frac{1}{4}$
5	$-\frac{4!}{(2-x)^5}$	$-4!$	5!	$-\frac{1}{5}$

From this table we see the pattern $f^{(k)}(x) = -\frac{(k-1)!}{(2-x)^k}$ and $\frac{f^{(k)}(1)}{k!} = -\frac{1}{k}$ for $k \geq 1$. To verify that this pattern continues, we check that if the kth derivative has this form, then

$$f^{(k+1)}(x) = -(k-1)!\frac{d}{dx}\frac{1}{(2-x)^k} = -\frac{(-k)(k-1)!}{(2-x)^{k+1}}\cdot(-1) = -\frac{k!}{(2-x)^{k+1}},$$

so that $\frac{f^{(k+1)}(1)}{(k+1)!} = -\frac{1}{k+1}$. Therefore, the Taylor series is $-\sum_{k=1}^{\infty}\frac{(x-1)^k}{k}$.

25. The Taylor coefficients are given by the following table:

k	$f^{(k)}(x)$	$f^{(k)}(0)$	$k!$	$f^{(k)}(1)/k!$
0	$e^{x\ln 2}$	1	1	1
1	$\ln 2 e^{x\ln 2}$	$\ln 2$	1	$\ln 2$
2	$(\ln 2)^2 e^{x\ln 2}$	$(\ln 2)^2$	2!	$\frac{(\ln 2)^2}{2!}$
3	$(\ln 2)^3 e^{x\ln 2}$	$(\ln 2)^3$	3!	$\frac{(\ln 2)^3}{3!}$
4	$(\ln 2)^4 e^{x\ln 2}$	$(\ln 2)^4$	4!	$\frac{(\ln 2)^4}{4!}$

From this table we infer that $\frac{f^{(k)}(0)}{k!} = \frac{(\ln 2)^k}{k!}$, and the Taylor series is $\sum_{k=0}^{\infty}\frac{(\ln 2)^k}{k!}x^k$.

27. The Taylor coefficients are given by the following table:

k	$f^{(k)}(x)$	$f^{(k)}(0)$	$k!$	$f^{(k)}(0)/k!$
0	$(1-3x)^{-2}$	1	1	1
1	$3\cdot 2(1-3x)^{-3}$	$3\cdot 2$	1	$3\cdot 2$
2	$3^2\cdot 6(1-3x)^{-4}$	$3^2\cdot 6$	2	$3^2\cdot 3$
3	$3^3\cdot 24(1-3x)^{-5}$	$3^3\cdot 24$	6	$3^3\cdot 4$
4	$3^4\cdot 120(1-3x)^{-6}$	$3^4\cdot 120$	24	$3^4\cdot 5$
5	$3^5\cdot 720(1-3x)^{-7}$	$3^5\cdot 720$	120	$3^5\cdot 6$

From this table we see the pattern $f^{(k)}(x) = 3^k(k+1)!(1-3x)^{-(k+2)}$ and $\frac{f^{(k)}(0)}{k!} = 3^k(k+1)$. To verify that this pattern continues, we check that if the kth derivative has this form, then

$$f^{(k+1)}(x) = \frac{d}{dx}3^k(k+1)!(1-3x)^{-(k+2)}$$

$$= 3^k(k+1)!(-1)(k+2)(1-3x)^{-(k+3)}\cdot(-3)$$

$$= 3^{k+1}(k+2)!(1-3x)^{-(k+3)},$$

so that $\frac{f^{(k+1)}(0)}{(k+1)!} = 3^{k+1}(k+2)$. Therefore, the Taylor series is $\sum_{k=0}^{\infty}(k+1)3^k x^k$.

29. The Taylor coefficients are given by the following table:

k	$f^{(k)}(t)$	$f^{(k)}(0)$	$k!$	$f^{(k)}(0)/k!$
0	$\ln(1-t)$	0	1	0
1	$-\frac{1}{1-t}$	-1	1	-1
2	$-\frac{1}{(1-t)^2}$	-1	2	$-\frac{1}{2}$
3	$-\frac{2}{(1-t)^3}$	-2	6	$-\frac{1}{3}$
4	$-\frac{3!}{(1-t)^4}$	$-3!$	$4!$	$-\frac{1}{4}$
5	$-\frac{4!}{(1-t)^5}$	$-4!$	$5!$	$-\frac{1}{5}$

From this table, we infer that for $k > 0$, $f^{(k)}(t) = -\frac{(k-1)!}{(1-t)^k}$ and $\frac{f^{(k)}(0)}{k!} = -\frac{1}{k}$. We can use a method like that of the previous exercises to verify that the pattern continues. Therefore, the Taylor series is $-\sum_{k=1}^{\infty}\frac{t^k}{k}$.

Formula (41) gives $\ln(1+t) = \sum_{k=1}^{\infty}(-1)^{k-1}\frac{t^k}{k}$. Replacing t by $-t$, we get

$$\ln(1-t) = \sum_{k=1}^{\infty}(-1)^{k-1}\frac{(-t)^k}{k} = \sum_{k=1}^{\infty}(-1)^{2k-1}\frac{t^k}{k} = -\sum_{k=1}^{\infty}\frac{t^k}{k},$$

which is the same as the Taylor series.

31. (a) The Taylor coefficients are given by the following table:

k	$f^{(k)}(x)$	$f^{(k)}(0)$	$k!$	$f^{(k)}(0)/k!$
0	$(1+x)^r$	1	1	1
1	$r(1+x)^{r-1}$	r	$1!$	$\frac{r}{1!}$
2	$r(r-1)(1+x)^{r-2}$	$r(r-1)$	$2!$	$\frac{r(r-1)}{2!}$
3	$r(r-1)(r-2)(1+x)^{r-3}$	$r(r-1)(r-2)$	$3!$	$\frac{r(r-1)(r-2)}{3!}$
4	$r(r-1)(r-2)(r-3)(1+x)^{r-4}$	$r(r-1)(r-2)(r-3)$	$4!$	$\frac{r(r-1)(r-2)(r-2)}{4!}$

From this table, we infer that $f^{(k)}(x) = r(r-1)\cdots(r-k+1)(1+x)^{r-k}$ and $\frac{f^{(k)}(0)}{k!} = \frac{r(r-1)(r-2)\cdots(r-k+1)}{k!}$. We can verify that this pattern continues by the same method we used in previous examples. Thus, the Taylor series is $1 + \sum_{k=1}^{\infty} \frac{r(r-1)\cdots(r-k+1)}{k!} x^k$.

(b) If r is a postive integer, then for any $k > r$ the Taylor coefficient is

$$\frac{f^{(k)}(0)}{k!} = \frac{r(r-1)\cdots(r-r)\cdots(r-k+1)}{k!} = 0.$$

Therefore, the Taylor series reduces to a polynomial of degree r.
If $r = 2$, the Taylor series is $1 + 2x + x^2$.
If $r = 3$, the Taylor series is $1 + 3x + 3x^2 + x^3$.
If $r = 4$, the Taylor series is $1 + 4x + 6x^2 + 4x^3 + x^4$.
If $r = 5$, the Taylor series is $1 + 5x + 10x^2 + 10x^3 + 5x^4 + x^5$.

(c) If $r = -1$, then the series takes the form

$$1 + \sum_{k=1}^{\infty} \frac{(-1)(-2)\cdots(-k)}{k!} x^k = 1 + \sum_{k=1}^{\infty} (-1)^k \frac{k!}{k!} x^k = 1 + \sum_{k=1}^{\infty} (-1)^k x^k = \sum_{k=0}^{\infty} (-x)^k.$$

It is a geometric series whose sum is $\frac{1}{1+x}$, provided $|x| < 1$.

(d) If $r = -2$, then the series takes the form

$$1 + \sum_{k=1}^{\infty} \frac{(-2)(-3)\cdots(-k-1)}{k!} x^k = 1 + \sum_{k=1}^{\infty} (-1)^k \frac{(k+1)!}{k!} x^k$$

$$= 1 + \sum_{k=1}^{\infty} (-1)^k (k+1) x^k$$

$$= \sum_{k=0}^{\infty} (-1)^k (k+1) x^k.$$

We can also write this series in the form $\sum_{k=1}^{\infty} (-1)^{k-1} k x^{k-1}$. (Observe that both ways of writing it leads to the same expansion: $1 - 2x + 3x^2 - 4x^3 + \cdots.$) In this form, we see that this series can be obtained from that of (c) by differentiating the previous series term by term and then taking the negative. That may seem reasonable, because if $r = -1$, then $f(x) = (1+x)^{-1}$; whereas, if $r = -2$, $f(x) = (1+x)^{-2} = -\frac{d}{dx}(1+x)^{-1}$.

33. For simplicity, we take n to be even, say $n = 2m$, and write Taylor's formula with remainder:

$$\cos x = 1 - \frac{1}{2!}x^2 + \cdots + (-1)^m \frac{1}{(2m)!}x^{2m} + R_{2m}(x).$$

The remainder satisfies $|R_{2m}(x)| \le \frac{M|x|^{2m+1}}{(2m+1)!}$, where M is the maximum of $|f^{2m+1}(t)|$ for t between 0 and x. But $|f^{2m+1}(t)| = |\sin t| \le 1$. Therefore, $M \le 1$ and $|R_{2m}(x)| \le \frac{|x|^{2m+1}}{(2m+1)!}$. Since $\lim_{m \to \infty} \frac{|x|^{2m+1}}{(2m+1)!} = 0$ (see formula (39)), it follows that $\lim_{m \to \infty} |R_{2m}(x)| = 0$, and the series converges to $\cos x$.

35. The Taylor coefficients are given by the following table:

k	$f^{(k)}(x)$	$f^{(k)}(0)$	$k!$	$f^{(k)}(0)/k!$
0	$\sin x$	0	1	0
1	$\cos x$	1	1!	1
2	$-\sin x$	0	2!	0
3	$-\cos x$	-1	3!	$-1/3!$
4	$\sin x$	0	4!	0
5	$\cos x$	1	5!	$1/5!$

At this point, we observe that $f^{(4)}(x) = f(x) = \sin x$, and so the pattern of derivatives will repeat in cycles of four: $\sin x$, $\cos x$, $-\sin x$, $-\cos x$, ... The numbers $f^{(k)}(0)$ will therefore have the repeating pattern: 0, 1, 0, -1, 0, 1, 0, -1, ..., and the Taylor series is

$$x - \frac{1}{3!}x^3 + \frac{1}{5!}x^5 - \frac{1}{7!}x^7 + \cdots = \sum_{k=0}^{\infty}(-1)^k \frac{x^{2k+1}}{(2k+1)!}.$$

37. In the $[-\frac{\pi}{2}, \frac{\pi}{2}] \times [-1.4, 1.2]$ window shown, all the graphs have been drawn, but the fit is so close it is virtually impossible to distinguish them.

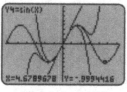

$[-2\pi, 2\pi] \times [-2, 2]$

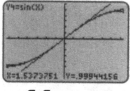

$[-\frac{\pi}{2}, \frac{\pi}{2}] \times [-1.4, 1.2]$

39. Using the **seq** and **sum** operations in the LIST editor:

41. In this and the next two exercises, we use the approximation

$$\ln(1 + t) = \sum_{k=1}^{n}(-1)^{k+1}\frac{t^k}{k}$$

for $-1 < k \le 1$.

43.

45. In this and the next exercise, we use the approximation

$$\sin t \approx \sum_{k=0}^{n}(-1)^{2k+1}\frac{t^{2k+1}}{(2k+1)!}.$$

EXERCISES 10.3

1. We have $c_k = \frac{1}{k}$ and $\lim_{k\to\infty}\left|\frac{c_{k+1}}{c_k}\right| = \lim_{k\to\infty}\frac{k}{k+1} = 1$. Therefore, the radius of convergence is $R = 1$.

3. We have $c_k = k^2$ and $\lim_{k\to\infty}\left|\frac{c_{k+1}}{c_k}\right| = \lim_{k\to\infty}\left|\frac{(k+1)^2}{k^2}\right| = 1$. The radius of convergence is $R = 1$.

5. We have $c_k = \frac{2k+1}{2^k}$ and

$$\lim_{k\to\infty}\left|\frac{c_{k+1}}{c_k}\right| = \lim_{k\to\infty}\frac{2(k+1)+1}{2^{k+1}}\cdot\frac{2^k}{2k+1} = \lim_{k\to\infty}\frac{1}{2}\cdot\frac{2k+3}{2k+1} = \frac{1}{2}.$$

Therefore, the radius of convergence is $R = 2$.

7. We have $c_k = \frac{k!}{10^k}$ and

$$\lim_{k\to\infty}\left|\frac{c_{k+1}}{c_k}\right| = \lim_{k\to\infty}\frac{(k+1)!}{10^{k+1}}\cdot\frac{10^k}{k!} = \lim_{k\to\infty}\frac{k+1}{10} = \infty.$$

Therefore, the radius of convergence is $R = 0$.

9. We have $c_k = \frac{k}{2k+1}$ and

$$\lim_{k\to\infty}\left|\frac{c_{k+1}}{c_k}\right| = \lim_{k\to\infty}\frac{k+1}{2(k+1)+1}\cdot\frac{2k+1}{k} = \lim_{k\to\infty}\frac{2k^2+3k+1}{2k^2+3k} = 1.$$

Therefore, the radius of convergence is $R = 1$.

11. The power series expansion $e^x = \sum_{k=0}^{\infty}\frac{x^k}{k!}$ is valid for every x. Therefore, $x^2 e^x = x^2\sum_{k=0}^{\infty}\frac{x^k}{k!} = \sum_{k=0}^{\infty}\frac{x^{k+2}}{k!}$ is valid for every x. This series expansion can also be written as

$$x^2 e^x = \sum_{k=2}^{\infty}\frac{x^k}{(k-2)!}.$$

13. The power series expansion $\frac{1}{1-t} = \sum_{k=0}^{\infty} t^k$ is valid on the interval $(-1,1)$. Substitituting $t = x^2$, we obtain $\frac{1}{1-x^2} = \sum_{k=0}^{\infty} x^{2k}$, which is valid for $(-1,1)$. Then $\frac{x}{1-x^2} = x \cdot \sum_{k=0}^{\infty} x^{2k} = \sum_{k=0}^{\infty} x^{2k+1}$ is valid on $(-1,1)$.

15. The power series expansions $\ln(1+x) = \sum_{k=1}^{\infty} (-1)^{k-1} \frac{x^k}{k}$ and

$$\ln(1-x) = \sum_{k=1}^{\infty} (-1)^{k-1} \frac{(-1)^k x^k}{k} = -\sum_{k=1}^{\infty} \frac{x^k}{k}$$

are both valid on the interval $(-1,1)$. Then

$$\ln\left(\frac{1+x}{1-x}\right) = \ln(1+x) - \ln(1-x)$$

$$= \sum_{k=1}^{\infty} (-1)^{k-1} \frac{x^k}{k} + \sum_{k=1}^{\infty} \frac{x^k}{k}$$

$$= \sum_{k=1}^{\infty} [(-1)^{k-1} + 1] \frac{x^k}{k}$$

is valid on $(-1,1)$. Observing that the coefficient is 0 if k is even and equals $2/k$ if k is odd, say $k = 2m + 1$, we can rewrite the power series expansion as

$$\ln\left(\frac{1+x}{1-x}\right) = \sum_{m=0}^{\infty} \frac{2x^{2m+1}}{2m+1}.$$

17. The power series expansion $\cos x = \sum_{k=0}^{\infty} (-1)^k \frac{1}{(2k)!} x^{2k}$ is valid for all x. Then

$$x \cos x = x \sum_{k=0}^{\infty} (-1)^k \frac{1}{(2k)!} x^{2k} = \sum_{k=0}^{\infty} (-1)^k \frac{1}{(2k)!} x^{2k+1}$$

is valid for all x.

19. The power series expansion $\sin x = \sum_{k=0}^{\infty} (-1)^k \frac{1}{(2k+1)!} x^{2k+1}$ is valid for all x. Then

$$\frac{\sin x}{x} = \frac{1}{x} \sum_{k=0}^{\infty} (-1)^k \frac{1}{(2k+1)!} x^{2k+1} = \sum_{k=0}^{\infty} (-1)^k \frac{1}{(2k+1)!} x^{2k}$$

is valid for all $x \neq 0$.

21. Recall that $\sin x = \sum_{k=0}^{\infty} (-1)^k \frac{x^{2k+1}}{(2k+1)!} = x - \frac{x^3}{3!} + \frac{x^5}{5!} - \frac{x^7}{7!} + \cdots$. Term-by-term differentiation gives

$$\sum_{k=0}^{\infty} \frac{d}{dx} (-1)^k \frac{x^{2k+1}}{(2k+1)!} = 1 - \frac{3x^2}{3!} + \frac{5x^4}{5!} - \frac{7x^6}{7!} + \cdots$$

$$= 1 - \frac{x^2}{2!} + \frac{x^4}{4!} - \frac{x^6}{6!} + \cdots = \sum_{k=0}^{\infty} (-1)^k \frac{x^{2k}}{(2k)!},$$

which is the Taylor series for $\cos x$.

23. By the geometric series formula, $\frac{1}{1-u} = \sum_{k=0}^{\infty} u^k$ for $-1 < u < 1$. Then

$$\frac{1}{(1-u)^2} = \frac{d}{du}\frac{1}{1-u} = \frac{d}{du}\sum_{k=0}^{\infty} u^k = \sum_{k=0}^{\infty} ku^{k-1} = \sum_{k=0}^{\infty}(k+1)u^k$$

Substituting $u = x^2$, we obtain $\frac{1}{(1-x^2)^2} = \sum_{k=0}^{\infty}(k+1)x^{2k}$, valid on the interval $(-1, 1)$.

25. By Exercise 12, $\ln(1+x^2) = \sum_{k=1}^{\infty}(-1)^{k-1}\frac{x^{2k}}{k}$ for $-1 \le x \le 1$. Applying Theorem 10.3.6, we take the antiderivative of each term to get

$$F(x) = \sum_{k=1}^{\infty}(-1)^{k-1}\frac{x^{2k+1}}{k(2k+1)},$$

with $F'(x) = \ln(1+x^2)$. Then

$$\int_0^{1/2} \ln(1+x^2)\,dx = F(1/2) - F(0) = \sum_{k=1}^{\infty}(-1)^{k-1}\frac{1}{k(2k+1)2^{2k+1}}.$$

To get a numerical approximation, we take a partial sum. For instance,

$$\int_0^{1/2} \ln(1+x^2)\,dx \approx \sum_{k=1}^{10}(-1)^{k-1}\frac{1}{k(2k+1)2^{2k+1}} \approx 0.038867$$

27. Let $y(t) = a_0 + a_1 t + a_2 t^2 + a_3 t^3 + a_4 t^4 + a_5 t^5 + \cdots$. Then

$$y'(t) = a_1 + 2a_2 t + 3a_3 t^2 + 4a_4 t^3 + 5a_5 t^4 + \cdots$$

$$y''(t) = 2a_2 + 6a_3 t + 12a_4 t^2 + 20a_5 t^3 + \cdots.$$

Substituting these power series into the differential equation and collecting terms, we obtain

$$(2a_2 + a_0) + (6a_3 + a_1)t + (12a_4 + a_2)t^2 + (20a_5 + a_3)t^3 + \cdots = 0.$$

Setting the coefficient of each power of t equal to zero leads to the equations

$$2a_2 = -a_0, \quad 6a_3 = -a_1, \quad 12a_4 = -a_2, \quad 20a_5 = -a_3.$$

From the initial conditions, we get $1 = y(0) = a_0$ and $1 = y'(0) = a_1$. Then

$$a_2 = -\frac{a_0}{2} = -\frac{1}{2}, \quad a_3 = -\frac{a_1}{6} = -\frac{1}{6}, \quad a_4 = -\frac{a_2}{12} = \frac{1/2}{12} = \frac{1}{24},$$

and $a_5 = -\frac{a_3}{20} = \frac{1/6}{20} = \frac{1}{120}$. Thus,

$$y(t) = 1 + t - \frac{1}{2}t^2 - \frac{1}{6}t^3 + \frac{1}{24}t^4 + \frac{1}{120}t^5 + \cdots.$$

29. Let $y(t)$, $y'(t)$, and $y''(t)$ be given by power series as in the exercise 27. Substituting these power series into the differential equation and collecting terms, we obtain

$$(2a_2 + a_0) + 6a_3 t + (12a_4 - a_2)t^2 + (20a_5 - 2a_3)t^3 + \cdots = 1 + t.$$

Equating coefficients of like powers of t on both sides, we get

$$2a_2 + a_0 = 1, \quad 6a_3 = 1, \quad 12a_4 - a_2 = 0, \quad 20a_5 - 2a_3 = 0.$$

From the initial conditions, we get $2 = y(0) = a_0$ and $0 = y'(0) = a_1$. Then

$$a_2 = \frac{1 - a_0}{2} = -\frac{1}{2}, \quad a_3 = \frac{1}{6}, \quad a_4 = \frac{a_2}{12} = -\frac{1}{24}, \quad a_5 = \frac{a_3}{10} = \frac{1}{60}.$$

Thus, $y(t) = 2 - \frac{1}{2}t^2 + \frac{1}{6}t^3 - \frac{1}{24}t^4 + \frac{1}{60}t^5 + \cdots$.

31. (a) The initial conditions are $h(0) = 0$ (no height at the start) and $h'(0) = 0$ (not growing at the start).

(b) Let $h(t) = a_0 + a_1 t + a_2 t^2 + a_3 t^3 + a_4 t^4 + a_5 t^5 + \cdots$. Then

$$h'(t) = a_1 + 2a_2 t + 3a_3 t^2 + 4a_4 t^3 + 5a_5 t^4 + \cdots$$

$$h''(t) = 2a_2 + 6a_3 t + 12a_4 t^2 + 20a_5 t^3 + \cdots.$$

Substituting these power series into the differential equation and collecting terms, we obtain

$$(2a_2 - 4a_0) + (6a_3 - 4a_1)t + (12a_4 - 4a_2)t^2 + (20a_5 - 4a_3)t^3 + \cdots = 6t - 5t^3.$$

Equating coefficients of like powers of t on both sides, we get

$$2a_2 - 4a_0 = 0, \quad 6a_3 - 4a_1 = 6, \quad 12a_4 - 4a_2 = 0, \quad 20a_5 - 4a_3 = -5.$$

From the initial conditions, we get $0 = y(0) = a_0$ and $0 = y'(0) = a_1$. Then

$$a_2 = 2a_0 = 0, \quad a_3 = \frac{1}{6}(6 + 4a_1) = 1, \quad a_4 = \frac{1}{3}a_2 = 0, \quad a_5 = \frac{1}{20}(4a_3 - 5) = -\frac{1}{20}.$$

Thus, $h(t) = t^3 - \frac{1}{20}t^5 + \cdots$.

(c) Since t is in years, the mean height after 3 months is given (in centimeters) by

$$h(1/4) = \left(\frac{1}{4}\right)^3 - \frac{1}{20} \cdot \left(\frac{1}{4}\right)^5 = \frac{319}{20480} \approx 0.0156.$$

CHAPTER 10: REVIEW EXERCISES

1. The Taylor coefficients are given by the following table:

k	$f^{(k)}(x)$	$f^{(k)}(1)$	$k!$	$f^{(k)}(1)/k!$
0	$x^3 + 2x^2 - 4x + 3$	2	1	2
1	$3x^2 + 4x - 4$	3	1	3
2	$6x + 4$	10	2	5
3	6	6	6	1

Therefore, $P_3(x) = 2 + 3(x - 1) + 5(x - 1)^2 + (x - 1)^3$.
To check, we expand $P_3(x)$ and collect terms, as follows:

$$P_3(x) = 2 + 3x - 3 + 5(x^2 - 2x + 1) + x^3 - 3x^2 + 3x - 1 = x^3 + 2x^2 - 4x + 3.$$

3. The Taylor coefficients are given by the following table:

k	$f^{(k)}(x)$	$f^{(k)}(0)$	$k!$	$f^{(k)}(0)/k!$
0	$e^{x/2}$	1	1	1
1	$\frac{1}{2}e^{x/2}$	$\frac{1}{2}$	1	$\frac{1}{2}$
2	$\frac{1}{4}e^{x/2}$	$\frac{1}{4}$	2	$\frac{1}{8}$
3	$\frac{1}{8}e^{x/2}$	$\frac{1}{8}$	6	$\frac{1}{48}$

$$P_3(x) = 1 + \frac{1}{2}x + \frac{1}{8}x^2 + \frac{1}{48}x^3.$$

5. The Taylor coefficients are given by the following table:

k	$f^{(k)}(x)$	$f^{(k)}(0)$	$k!$	$f^{(k)}(0)/k!$
0	$\sqrt{1+x}$	1	1	1
1	$\frac{1}{2}(1+x)^{-1/2}$	$\frac{1}{2}$	1	$\frac{1}{2}$
2	$-\frac{1}{4}(1+x)^{-3/2}$	$-\frac{1}{4}$	2	$-\frac{1}{8}$
3	$\frac{3}{8}(1+x)^{-5/2}$	$\frac{3}{8}$	6	$\frac{1}{16}$
4	$-\frac{15}{16}(1+x)^{-7/2}$	$-\frac{15}{16}$	24	$-\frac{5}{128}$

$$P_4(x) = 1 + \frac{1}{2}x - \frac{1}{8}x^2 + \frac{1}{16}x^3 - \frac{5}{128}x^4.$$

7. The Taylor coefficients are given by the following table:

k	$f^{(k)}(x)$	$f^{(k)}(\pi)$	$k!$	$f^{(k)}(\pi)/k!$
0	$\cos x$	-1	1	-1
1	$-\sin x$	0	1	0
2	$-\cos x$	1	2	$\frac{1}{2}$
3	$\sin x$	0	6	0
4	$\cos x$	-1	24	$-\frac{1}{24}$

$$P_4(x) = -1 + \frac{1}{2}(x-\pi)^2 - \frac{1}{24}(x-\pi)^4.$$

9. The Taylor coefficients through degree four are given by:

k	$f^{(k)}(x)$	$f^{(k)}(0)$	$k!$	$f^{(k)}(0)/k!$
0	$\frac{1}{1-x}$	1	1	1
1	$\frac{1}{(1-x)^2}$	1	1	1
2	$\frac{2}{(1-x)^3}$	2	2	1
3	$\frac{3!}{(1-x)^4}$	$3!$	$3!$	1
4	$\frac{4!}{(1-x)^5}$	$4!$	$4!$	1

The general pattern: $c_k = 1$ for $k = 0, 1, 2, \ldots$. To check it:

$$f^{(5)}(x) = \frac{5!}{(1-x)^6} \quad \text{and} \quad c_5 = \frac{f^{(5)}(0)}{5!} = \frac{5!}{5!} = 1$$

$$f^{(6)}(x) = \frac{6!}{(1-x)^7} \quad \text{and} \quad c_6 = \frac{f^{(6)}(0)}{6!} = \frac{6!}{6!} = 1.$$

11. The Taylor coefficients through degree four are given by:

k	$f^{(k)}(x)$	$f^{(k)}(\pi/2)$	$k!$	$f^{(k)}(\pi/2)/k!$
0	$\sin x$	1	1	1
1	$\cos x$	0	1	0
2	$-\sin x$	-1	2	$-\frac{1}{2!}$
3	$-\cos x$	0	3!	0
4	$\sin x$	1	4!	$\frac{1}{4!}$

The general pattern: $c_{2k} = (-1)^k \frac{1}{(2k)!}$ and $c_{2k+1} = 0$ for $k = 0, 1, 2, \ldots$. To check it:

$$f^{(5)}(x) = \cos x \quad \text{and} \quad c_5 = \frac{f^{(5)}(\pi/2)}{5!} = 0$$

$$f^{(6)}(x) = -\sin x \quad \text{and} \quad c_6 = \frac{f^{(6)}(\pi/2)}{6!} = -\frac{1}{6!}.$$

13. The Taylor coefficients through degree five are given by:

k	$f^{(k)}(x)$	$f^{(k)}(0)$	$k!$	$f^{(k)}(0)/k!$
0	e^{-x}	1	1	1
1	$-e^{-x}$	-1	1	-1
2	e^{-x}	1	2	$\frac{1}{2}$
3	$-e^{-x}$	-1	6	$-\frac{1}{6}$
4	e^{-x}	1	24	$\frac{1}{24}$
5	$-e^{-x}$	-1	120	$-\frac{1}{120}$

Thus, $P_5(x) = 1 - x + \frac{1}{2}x^2 - \frac{1}{6}x^3 + \frac{1}{24}x^4 - \frac{1}{120}x^5$, and

$$e^{-0.5} \approx 1 - (0.5) + \frac{1}{2}(0.5)^2 - \frac{1}{6}(0.5)^3 + \frac{1}{24}(0.5)^4 - \frac{1}{120}(0.5)^5 \approx 0.60651.$$

The error satisfies $|R_5(0.5)| \leq \frac{M|0.5-0|^6}{6!}$, where M is the maximum of $|f^{(6)}(x)| = e^{-x}$ on the interval $[0, 0.5]$. Since e^{-x} is a decreasing function, its maximum over the interval is 1, achieved at $x = 0$. Therefore, $M = 1$, and $|R_5(0.5)| \leq \frac{(0.5)^6}{6!} < 0.000022$

15. The Taylor coefficients through degree two are given by:

k	$f^{(k)}(x)$	$f^{(k)}(0)$	$k!$	$f^{(k)}(0)/k!$
0	$(1-x^2)^{-1/2}$	1	1	1
1	$x(1-x^2)^{-3/2}$	0	1	0
2	$(1-x^2)^{-3/2} + 3x^2(1-x^2)^{-5/2}$	1	2	$\frac{1}{2}$

Therefore, $P_2(x) = 1 + \frac{1}{2}x^2$, and

$$\int_0^{0.2} (1-x^2)^{-1/2}\, dx \approx \int_0^{0.2} \left(1 + \frac{1}{2}x^2\right) dx = \left(x + \frac{1}{6}x^3\right)\Big|_0^{0.2}$$

$$= 0.2 + \frac{1}{6}0.2^3 \approx 0.20133.$$

17. We are given $f(0) = 0$ and $f'(x) = (1+x^2)^{-1}$. By taking successive derivatives, we get the Taylor coefficients, as follow:

k	$f^{(k)}(x)$	$f^{(k)}(0)$	$k!$	$f^{(k)}(0)/k!$
0	—	0	1	0
1	$(1+x^2)^{-1}$	1	1	1
2	$-2x(1+x^2)^{-2}$	0	2	0
3	$-2(1+x^2)^{-2} + 8x^2(1+x^2)^{-3}$	-2	6	$-\frac{1}{3}$

Therefore, $P_3(x) = x - \frac{1}{3}x^3$.

19. We are given $y(0) = 0$ and $y' = e^y$. By taking successive derivatives, we get the Taylor coefficients, as follows:

k	$y^{(k)}(t)$	$y^{(k)}(0)$	$k!$	$y^{(k)}(0)/k!$
0	—	0	1	0
1	e^y	1	1	1
2	$y'e^y$	1	2	$\frac{1}{2}$
3	$[y'' + (y')^2]e^y$	2	6	$\frac{1}{3}$
4	$[y''' + 3y''y' + (y')^3]e^y$	6	24	$\frac{1}{4}$
5	$[y^{(4)} + 4y'''y' + 3(y'')^2 + 6y''(y')^2 + (y')^4]e^y$	24	120	$\frac{1}{5}$

Note: In evaluating $y^{(k)}(0)$, we start with the initial condition $y(0) = 0$. Then $y'(0) = e^{y(0)} = e^0 = 1$, $y''(0) = y'(0)e^{y(0)} = 1 \cdot e^0 = 1$, and so forth. Thus, $P_5(t) = t + \frac{1}{2}t^2 + \frac{1}{3}t^3 + \frac{1}{4}t^4 + \frac{1}{5}t^5$, and

$$y(0.5) \approx P_5(0.5) = 0.5 + \frac{1}{2}(0.5)^2 + \frac{1}{3}(0.5)^3 + \frac{1}{4}(0.5)^4 + \frac{1}{5}(0.5)^5 \approx 0.68854.$$

21. This series is a geometric series of the form $\sum_{k=0}^{\infty}(-2)^k$. It diverges, since $|-2| = 2 > 1$.

23. Rewrite the series in the form $\sum_{k=0}^{\infty} \frac{4}{3}(-1/3)^k = \frac{4}{3}\sum_{k=0}^{\infty}(-1/3)^k$. Since $|-1/3| = \frac{1}{3} < 1$, the series converges to $\frac{4}{3} \cdot \frac{1}{1-(-1/3)} = 1$.

25. $2.222\ldots = 10 \cdot (0.2222\ldots) = 10 \cdot \frac{2}{10}\sum_{k=0}^{\infty}(1/10)^k = 2 \cdot \frac{1}{1-1/10} = \frac{20}{9}$

27. $1.0\overline{321} = 1 + 0.1 \cdot (0.\overline{321}) = 1 + 0.1 \cdot \frac{321}{1,000}\sum_{k=0}^{\infty}\left(\frac{1}{1,000}\right)^k = 1 + 0.1 \cdot \frac{321}{1,000} \cdot \frac{1}{1-\frac{1}{1,000}} = 1 +$

$0.1 \cdot \frac{321}{999} = 1 + \frac{107}{3,330} = \frac{3,437}{3,330}$

29. In this case, the amount removed in the first year is $a = 1.3 \times 10^6$, the amount in the second year is $a(0.99)$, the amount in the third year is $a(0.99)^2$, and so forth. (Each year the amount is 99% of the previous year.) Adding these over the first k years, we get

$$a(1 + 0.99 + 0.99^2 + 0.99^3 + \cdots + 0.99^{k-1}) = a\frac{1 - 0.99^k}{1 - 0.99}.$$

Letting $k \to \infty$, we get $a \cdot \sum_{k=0}^{\infty} 0.99^k = \frac{1.3 \times 10^6}{1 - 0.99} = 100 \cdot (1.3 \times 10^6) = 1.3 \times 10^8$ Since this is less than the 5.2×10^8 cubic meters of mineral in the parcel, the mining can continue forever without exhausting the source.

31. The Taylor coefficients are given by the following table:

k	$f^{(k)}(x)$	$f^{(k)}(0)$	$k!$	$f^{(k)}(0)/k!$
0	$\frac{2}{(2+x)^2}$	$\frac{2}{2^2}$	1	$\frac{1}{2}$
1	$-\frac{2 \cdot 2}{(2+x)^3}$	$-\frac{2}{2^2}$	1	$-\frac{2}{2^2}$
2	$\frac{2 \cdot 2 \cdot 3}{(2+x)^4}$	$\frac{3!}{2^3}$	2	$\frac{3}{2^3}$
3	$-\frac{2 \cdot 4!}{(2+x)^5}$	$-\frac{4!}{2^4}$	3!	$-\frac{4}{2^4}$
4	$\frac{2 \cdot 5!}{(2+x)^6}$	$\frac{5!}{2^5}$	4!	$\frac{5}{2^5}$
5	$-\frac{2 \cdot 6!}{(2+x)^7}$	$-\frac{6!}{2^6}$	120	$-\frac{6}{2^6}$

From this table, we see the pattern $c_k = \frac{f^{(k)}(0)}{k!} = (-1)^k \frac{k+1}{2^{k+1}}$. Therefore, the Taylor series is $\sum_{k=0}^{\infty} (-1)^k \frac{k+1}{2^{k+1}} x^k$.

33. Since $e^x = \sum_{k=0}^{\infty} \frac{x^k}{k!}$ we have

$$(x + 1)e^x = (x + 1)\sum_{k=0}^{\infty} \frac{x^k}{k!} = x\sum_{k=0}^{\infty} \frac{x^k}{k!} + \sum_{k=0}^{\infty} \frac{x^k}{k!} = \sum_{k=0}^{\infty} \frac{x^{k+1}}{k!} + \sum_{k=0}^{\infty} \frac{x^k}{k!}.$$

The two series on the right-hand side are

$$\sum_{k=0}^{\infty} \frac{x^{k+1}}{k!} = \frac{x}{0!} + \frac{x^2}{1!} + \frac{x^3}{2!} + \frac{x^4}{3!} + \cdots \quad \text{and} \quad \sum_{k=0}^{\infty} \frac{x^k}{k!} = 1 + \frac{x}{1!} + \frac{x^2}{2!} + \frac{x^3}{3!} + \frac{x^4}{4!} + \cdots.$$

Adding them we get

$$1 + \left(\frac{1}{0!} + \frac{1}{1!}\right)x + \left(\frac{1}{1!} + \frac{1}{2!}\right)x^2 + \left(\frac{1}{2!} + \frac{1}{3!}\right)x^3 + \cdots = 1 + \sum_{k=1}^{\infty}\left[\frac{1}{(k-1)!} + \frac{1}{k!}\right]x^k,$$

which is the Taylor series.

35. Since $\cos x = \sum_{k=0}^{\infty} (-1)^k \frac{1}{(2k)!} x^{2k}$ and $\sin x = \sum_{k=0}^{\infty} (-1)^k \frac{1}{(2k+1)!} x^{2k+1}$, we have

$$\cos x + x \sin x = \sum_{k=0}^{\infty} (-1)^k \frac{1}{(2k)!} x^{2k} + \sum_{k=0}^{\infty} (-1)^k \frac{1}{(2k+1)!} x^{2k+2}.$$

These two series can be written out as follows:

$$\sum_{k=0}^{\infty}(-1)^k\frac{1}{(2k)!}x^{2k} = 1 - \frac{x^2}{2!} + \frac{x^4}{4!} - \frac{x^6}{6!} + \frac{x^6}{6!} - \frac{x^8}{8!} + \cdots$$

$$\sum_{k=0}^{\infty}(-1)^k\frac{1}{(2k+1)!}x^{2k+2} = \frac{x^2}{1!} - \frac{x^4}{3!} + \frac{x^6}{5!} - \frac{x^8}{8!} + \cdots.$$

Adding them gives

$$1 - \left(\frac{1}{2!} - \frac{1}{1!}\right)x^2 + \left(\frac{1}{4!} - \frac{1}{3!}\right)x^4 + \left(\frac{1}{6!} - \frac{1}{5!}\right)x^6 - \left(\frac{1}{8!} - \frac{1}{7!}\right)x^8 + \cdots$$

$$= 1 + \sum_{k=1}^{\infty}(-1)^k\left[\frac{1}{(2k)!} - \frac{1}{(2k-1)!}\right]x^{2k},$$

which is the Taylor series.

37. (a) The Taylor coefficients through degree five are given by the following table:

k	$f^{(k)}(x)$	$f^{(k)}(0)$	$k!$	$f^{(k)}(0)/k!$
0	$(1+x)^{1/2}$	1	1	1
1	$\frac{1}{2}(1+x)^{-1/2}$	$\frac{1}{2}$	1	$\frac{1}{2}$
2	$-\frac{1}{2^2}(1+x)^{-3/2}$	$-\frac{1}{2^2}$	2	$-\frac{1}{2!2^2}$
3	$\frac{1\cdot3}{2^3}(1+x)^{-5/2}$	$\frac{1\cdot3}{2^3}$	6	$\frac{1\cdot3}{3!2^3}$
4	$-\frac{1\cdot3\cdot5}{2^4}(1+x)^{-7/2}$	$-\frac{1\cdot3\cdot5}{2^4}$	24	$-\frac{1\cdot3\cdot5}{4!2^4}$
5	$\frac{1\cdot3\cdot5\cdot7}{2^5}(1+x)^{-9/2}$	$\frac{1\cdot3\cdot5\cdot7}{2^5}$	120	$\frac{1\cdot3\cdot5\cdot7}{5!2^5}$

From this table, we infer that

$$\frac{f^{(k)}(0)}{k!} = (-1)^{k+1}\frac{1\cdot3\cdot5\cdots(2k-3)}{k!2^k} \quad\text{for}\quad k = 2, 3, \ldots.$$

Therefore, the Taylor series is $1 + \frac{1}{2}x + \sum_{k=2}^{\infty}(-1)^{k+1}\frac{1\cdot3\cdot5\cdots(2k-3)}{k!2^k}x^k$.

(b) $\sqrt{0.9} = (1 - 0.1)^{1/2} \approx 1 + \frac{1}{2}(-0.1) - \frac{1}{2!2^2}(-0.1)^2 + \frac{1\cdot3}{3!2^3}(-0.1)^3 - \frac{1\cdot3\cdot5}{4!2^4}(-0.1)^4$

$$= 0.948683594.$$

39. We have $c_k = \frac{1}{k(k+1)}$ and

$$\lim_{k\to\infty}\left|\frac{c_{k+1}}{c_k}\right| = \lim_{k\to\infty}\frac{1}{(k+1)(k+2)}\cdot\frac{k(k+1)}{1} = \lim_{k\to\infty}\frac{k}{k+2} = 1.$$

Therefore, the radius of convergence is $R = 1$.

41. We have $c_k = (-1)^k\frac{1}{2k+1}$ and

$$\lim_{k\to\infty}\left|\frac{c_{k+1}}{c_k}\right| = \lim_{k\to\infty}\frac{1}{2(k+1)+1}\cdot\frac{2k+1}{1} = \lim_{k\to\infty}\frac{2k+1}{2k+3} = 1.$$

Therefore, the radius of convergence is $R = 1$.

43. We have $c_k = \frac{2^{k-1}}{3^{k+1}}$ and

$$\lim_{k\to\infty}\left|\frac{c_{k+1}}{c_k}\right| = \lim_{k\to\infty}\frac{2^k}{3^{k+2}}\cdot\frac{3^{k+1}}{2^{k-1}} = \lim_{k\to\infty}\frac{2}{3} = \frac{2}{3}.$$

Therefore, the radius of convergence is $R = \frac{3}{2}$.

45. We start with the formula $\ln(x+1) = \sum_{k=1}^{\infty}(-1)^{k-1}\frac{x^k}{k}$ for $-1 < x < 1$. Then $x^2\ln(x+1) = x^2\sum_{k=1}^{\infty}(-1)^{k-1}\frac{x^k}{k} = \sum_{k=1}^{\infty}(-1)^{k-1}\frac{x^{k+2}}{k}$ for $-1 < x < 1$. This series can also be written as $\sum_{k=3}^{\infty}(-1)^{k-1}\frac{x^k}{k-2}$

47. We start with $\cos x = \sum_{k=0}^{\infty}(-1)^k\frac{x^{2k}}{(2k)!} = 1 - \frac{x^2}{2!} + \frac{x^4}{4!} - \frac{x^6}{6!} + \cdots$, valid for $-\infty < x < \infty$. Then $1 - \cos x = \frac{x^2}{2!} - \frac{x^4}{4!} + \frac{x^6}{6!} - \cdots$, and

$$\frac{1-\cos x}{x} = \frac{x}{2!} - \frac{x^3}{4!} + \frac{x^5}{6!} - \cdots = \sum_{k=0}^{\infty}(-1)^k\frac{x^{2k+1}}{(2k+2)!}.$$

49. Using the geometric series, we obtain

$$\frac{x^2}{1-x^2} = x^2\frac{1}{1-x^2} = x^2\sum_{k=0}^{\infty}(x^2)^k = \sum_{k=0}^{\infty}x^{2k+2},$$

valid for $-1 < x < 1$. Then

$$\int_0^{1/2}\frac{x^2}{1-x^2}\,dx = \int_0^{1/2}\sum_{k=0}^{\infty}x^{2k+2}\,dx = \sum_{k=0}^{\infty}\int_0^{1/2}x^{2k+2}\,dx$$

$$= \sum_{k=0}^{\infty}\frac{1}{(2k+3)}x^{2k+3}\bigg|_0^{1/2} = \sum_{k=0}^{\infty}\frac{1}{(2k+3)2^{2k+3}}.$$

For a numerical approximation, we compute a partial sum. For instance, taking $n = 10$ gives

$$\int_0^{1/2}\frac{x^2}{1-x^2}\,dx \approx \sum_{k=0}^{10}\frac{1}{(2k+3)2^{2k+3}} \approx 0.04930614.$$

51. Let $y(t) = a_0 + a_1 t + a_2 t^2 + a_3 t^3 + a_4 t^4 + a_5 t^5 + \cdots$. Then

$$y'(t) = a_1 + 2a_2 t + 3a_3 t^2 + 4a_4 t^3 + 5a_5 t^4 + \cdots$$

$$y''(t) = 2a_2 + 6a_3 t + 12a_4 t^2 + 20a_5 t^3 + \cdots.$$

Substituting these power series into the differential equation and collecting terms, we obtain

$$2a_2 + (6a_3 - a_0)t + (12a_4 - a_1)t^2 + (20a_5 - a_2)t^3 + \cdots = 1.$$

Equating coefficients of like powers of t on both sides, we get

$$2a_2 = 1, \quad 6a_3 - a_0 = 0, \quad 12a_4 - a_1 = 0, \quad 20a_5 - a_2 = 0.$$

From the initial conditions, we get $1 = y(0) = a_0$ and $0 = y'(0) = a_1$. Then

$$a_2 = \frac{1}{2}, \quad a_3 = \frac{a_0}{6} = \frac{1}{6}, \quad a_4 = \frac{a_1}{12} = 0, \quad a_5 = \frac{a_2}{20} = \frac{1/2}{20} = \frac{1}{40}.$$

Therefore, $y(t) = 1 + \frac{1}{2}t^2 + \frac{1}{6}t^3 + \frac{1}{40}t^5 + \cdots$.

CHAPTER 10: PRACTICE EXAM

1. The Taylor coefficients through degree three are as follows:

k	$f^{(k)}(x)$	$f^{(k)}(0)$	$k!$	$f^{(k)}(0)/k!$
0	xe^{2x}	0	1	0
1	$(2x+1)e^{2x}$	1	1	1
2	$(4x+4)e^{2x}$	4	2	2
3	$(8x+12)e^{2x}$	12	6	2

Thus, $P_1(x) = x$, $P_2(x) = x + 2x^2$, and $P_3(x) = x + 2x^2 + 2x^3$.

3. The Taylor coefficients are given by the following table:

k	$f^{(k)}(x)$	$f^{(k)}(1)$	$k!$	$f^{(k)}(1)/k!$
0	$1 + x + x^2 + x^3$	4	1	4
1	$1 + 2x + 3x^2$	6	1	6
2	$2 + 6x$	8	2	4
3	6	6	6	1

$P_3(x) = 4 + 6(x-1) + 4(x-1)^2 + (x-1)^3$.

5. The Taylor coefficients are given by the following table:

k	$f^{(k)}(r)$	$f^{(k)}(1)$	$k!$	$f^{(k)}(1)/k!$
0	$r \ln r$	0	1	0
1	$\ln r + 1$	1	1	1
2	$\frac{1}{r}$	1	2	$\frac{1}{2}$

$P_2(r) = (r-1) + \frac{1}{2}(r-1)^2$.

7. Starting with the formula $e^x = 1 + x + \frac{x^2}{2!} + \cdots$, we substitute $x = -t^4$ to get $e^{-t^4} = 1 - t^4 + \frac{t^8}{2!} \cdots$. Therefore, the fourth-degree Taylor polynomial of e^{-t^4} is $P_4(t) = 1 - t^4$. Therefore,

$$\int_0^{0.2} e^{-t^4}\, dt \approx \int_0^{0.2} (1 - t^4)\, dt = \left(t - \frac{1}{5}t^5 \right) \Big|_0^{0.2} = 0.2 - \frac{1}{5}(0.2)^5 = 0.199936.$$

9. Since $e^t = \sum_{k=0}^{\infty} \frac{t^k}{k!}$ for $-\infty < t < \infty$, substituting $t = -\frac{x}{2}$ gives

$$e^{-x/2} = \sum_{k=0}^{\infty} \frac{1}{k!} \left(-\frac{x}{2} \right)^k = \sum_{k=0}^{\infty} (-1)^k \frac{1}{k!2^k} x^k,$$

valid for $-\infty < x < \infty$.

11. At the kth step, the amount spent will be $(0.8)^k \cdot 50$ billion dollars. Adding these for $k = 1, 2, 3, \ldots$, we get

$$\sum_{k=1}^{\infty} (0.8)^k 50 = 50 \sum_{k=1}^{\infty} (0.8)^k = 50 \cdot \frac{0.8}{1 - 0.8} = 200$$

billion dollars.

13. We are given $B(0) = 3$ and $B' = 0.3B(1 - \frac{B}{9}) - \frac{0.4B^2}{1+B^2}$. Then, substituting $t = 0$, we get

$$B'(0) = 0.3B(0)\left(1 - \frac{B(0)}{9}\right) - \frac{0.4B(0)^2}{1 + B(0)^2} = 0.9\left(1 - \frac{3}{9}\right) - \frac{(0.4) \cdot 9}{1 + 9} = 0.24.$$

Next, by taking the derivative of B' (with respect to t), using the product and chain rules, and then evaluating at $t = 0$ (with $B(0) = 3$ and $B'(0) = 0.24$), we obtain the following

k	$B^{(k)}(t)$	$B^{(k)}(0)$	$k!$	$B^{(k)}(0)/k!$
0	—	3	1	3
1	$0.3B(1 - B/9) - \frac{0.4B^2}{1+B^2}$	0.24	1	0.24
2	$0.3B' - \frac{1}{15}BB' - \frac{0.8BB'}{(1+B^2)^2}$	0.01824	2	0.00912

Therefore, $P_2(t) = 3 + 0.24t + 0.00912t^2$, and

$$B(10) \approx P_2(10) = 3 + 0.24 \cdot 10 + 0.00912 \cdot (10^2) = 6.312.$$

CHAPTER 11
Probability and Statistics

EXERCISES 11.1

Write S for the sample space of an experiment.

1. The sample space S consists of 12 outcomes. It can be represented as follows:

$$S = \left\{ \begin{array}{l} (H,1),(H,2),(H,3),(H,4),(H,5),(H,6), \\ (T,1),(T,2),(T,3),(T,4),(T,5),(T,6) \end{array} \right\}.$$

3. In this case there are no repeats. If we assume the order in which the chips are drawn does not matter, then there are 6 outcomes, and

$$S = \{(1,2),(1,3),(1,4),(2,3),(2,4),(3,4)\}.$$

5. In this case there are no repeats, and the order in which the letters are written matters. There are 24 outcomes, listed below in alphabetical order.

$$S = \left\{ \begin{array}{l} abc, abd, acb, acd, adb, adc, bac, bad, \\ bca, bcd, bda, bdc, cab, cad, cba, cbd, \\ cda, cdb, dab, dac, dba, dbc, dca, dcb \end{array} \right\}$$

7. Letting H represent a head and T a tail, every outcome is a string of letters ending in H and with all previous entries (if any) being Ts. There are infinitely many outcomes, and $S = \{H, TH, TTH, TTTH, TTTTH, \ldots\}$.

9. The time it takes to be served is a positive real number. With no advance knowledge, we assume that it can be any positive real number. Thus, there are infinitely many outcomes, and $S = (0, \infty)$, the set of all positive real numbers.

11. (a) We can take the sample space to be $(0, \infty)$ the set of all positive real numbers (similar to exercise 9).

(b) $E \cap F$: more than 5 and less than 8 minutes
F' : at least 8 minutes
$E' \cap G$: more than 4.5 but not more than 5 minutes
$F \cap G'$: not more than 4.5 minutes
$E \cup G'$: either less than or equal to 4.5 minutes or else more than 5 minutes

(c) $G' \cap E = \emptyset$, $E \cup G = G$, $F \cup G = S$, $E \cap G = E$

13. $E = \{HHH, HHT, HTH, THH\}$,
$F = \{HHT, HTH, HTT, THH, THT, TTH, TTT\}$, and
$G = \{HTT, THT, TTH, TTT\}$.
Thus, E and G is the only mutually exclusive pair.

15. First, $P(\{s_1, s_2, s_3, s_4\}) = P(\{s_1, s_3, s_4\}) + P(\{s_2\}) = 0.4 + 0.1 = 0.5$.
Then $P(\{s_5, s_6\}) = P(\{s_1, s_2, s_3, s_4\}') = 1 - P(\{s_1, s_2, s_3, s_4\}) = 1 - 0.5 = 0.5$.

17. Jane has invited 9 quests.

    (a) Eight of Jane's guests are her relatives. Thus, the probability is $\frac{8}{9}$.

    (b) Three guests are her siblings. Thus, the probability is $\frac{3}{9} = \frac{1}{3}$.

    (c) The blood relatives of Jane's father are her sisters and her son. Thus, the probability is $\frac{4}{9}$.

19. As shown in Example 11.1.6, the sample space for the experiment of rolling two fair dice has 36 outcomes.

    (a) $A = \{(1,6),(2,5),(3,4),(4,3),(5,2),(6,1)\}$. Thus, $P(A) = \frac{6}{36} = \frac{1}{6}$.

    (b) $B = \{(5,6),(6,5)\}$. Thus, $P(B) = \frac{2}{36} = \frac{1}{18}$.

    (c) C is the event that the numbers on the top faces have the same parity—that is, either both are odd or both are even. Looking at the list of ordered pairs in Example 11.1.6, we see that half the outcomes in each row have that property, for a total of 18. Thus, $P(C) = \frac{18}{36} = \frac{1}{2}$.

    (d) Again referring to the list in Example 11.1.6, we see that there are exactly 9 outcomes with both entries odd—that is, half the outcomes in each of the first, third, and fifth rows. Thus, $P(D) = \frac{9}{36} = \frac{1}{4}$.

    (e) Again referring to the list in Example 11.1.6, we see that there are exactly 27 outcomes with at least one entry odd—that is, all entries in the first, third, and fifth rows, and half the entries in the second, fourth, and sixth rows. Thus, $P(E) = \frac{27}{36} = \frac{3}{4}$.

21. The sample space consists of all ordered pairs of the form (i,j), where i and j are integers from one to four. Here is an explicit list of the 16 outcomes:

$$\begin{array}{cccc}
(1,1) & (1,2) & (1,3) & (1,4) \\
(2,1) & (2,2) & (2,3) & (2,4) \\
(3,1) & (3,2) & (3,3) & (3,4) \\
(4,1) & (4,2) & (4,3) & (4,4)
\end{array}$$

    (a) $\{(2,2),(2,4),(4,2),(4,4)\}$. The probability is $\frac{4}{16} = \frac{1}{4}$.

    (b) $\{(1,2),(1,4),(2,1),(2,2),(2,3),(2,4),(3,2),(3,4),(4,1),(4,2),(4,3),(4,4)\}$. The probability is $\frac{12}{16} = \frac{3}{4}$.

    (c) $\{(2,4),(3,3),(4,2)\}$. The probability is $\frac{3}{16}$.

    (d) $\{(1,1),(2,2),(3,3),(4,4)\}$. The probability is $\frac{4}{16} = \frac{1}{4}$.

    (e) $\{(1,2),(1,3),(2,1),(2,2),(2,4),(3,1),(3,3),(3,4),(4,2),(4,3)\}$. The probability is $\frac{10}{16} = \frac{5}{8}$.

23.

Father/Mother	N	t
N	NN	Nt
N	NN	Nt

    (a) $\dfrac{2}{4} = \dfrac{1}{2}$ **(b)** $\dfrac{0}{4} = 0$

25. $P(\text{at least one head}) = 1 - P(TT) = 1 - (0.45)^2 = 0.7975$

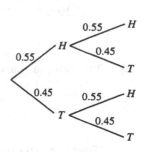

27. (a) $P(E \cup F) = P(E) + P(F) - P(E \cap F) = 0.6 + 0.5 - 0.4 = 0.7.$

(b) $P(E) = P((E \cap F) \cup (E \cap F')) = P(E \cap F) + P(E \cap F'),$ since $(E \cap F) \cap (E \cap F') = \emptyset.$ Therefore,

$$P(E \cap F') = P(E) - P(E \cap F) = 0.6 - 0.4 = 0.2.$$

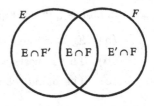

29. Let E be the event "enjoyed the math class" and M the event "math major." Then we are given

$$P(E) = 0.38, \quad P(M) = 0.26, \quad P(M \cap E) = 0.20.$$

(a) $P(M \cap E') = P(M) - P(M \cap E) = 0.26 - 0.20 = 0.06$

(b) $P(M \cup E) = P(M) + P(E) - P(M \cap E) = 0.38 + 0.26 - 0.20 = 0.44$

(c) $P(M') = 1 - P(M) = 1 - 0.26 = 0.74$

(d) $P(M' \cap E) = P(E) - P(M \cap E) = 0.38 - 0.20 = 0.18.$

31. There are $2^6 = 64$ outcome.

(a) There is only 1 outcome in this event. Thus, the probability is $\frac{1}{64}$

(b) There is one outcome with all heads, and there are 6 outcomes with one tail and five heads (tail on the first and the rest heads, tail on the second and the rest heads, etc.). Thus, the probability is $\frac{7}{64}$.

(c) "At most one head" is the same as "at least 5 tails." By symmetry, that has the same probability as "at least 5 heads," and in part (b) we saw that the probability is $\frac{7}{64}$.

33. There are $2^4 = 16$ outcomes in the sample space.

(a) This event has 6 outcomes. Listing them in alphabetical order: $\{BBGG, BGBG, BGGB, GBBG, GBGB, GGBB\}$. Thus, the probability is $\frac{6}{16} = \frac{3}{8}$.

(b) There are 4 outcomes with three girls and one boy (first is a boy and the rest girls, second is a boy and the rest girls, etc.) By symmetry, there are also 4 outcomes with three boys and one girl. Therefore, this event has 8 outcomes, and its probability is $\frac{8}{16} = \frac{1}{2}$.

Thus, we see that (b) has greater probability.

EXERCISES 11.2

1. (a) The sample space of this experiment is $S = \{HT, TH, HH, TT\}$. The event that at least one toss is a head can be described as $\{HT, TH, HH\}$. Thus, the probability of that event is $\frac{3}{4}$.

(b) Let E be the event that at least one toss is a head and G the event that the second toss is a tail. Then $G = \{HT, TT\}$ and $E \cap G = \{HT\}$, so that $P(E \cap G) = \frac{1}{4}$ and $P(G) = \frac{2}{4} = \frac{1}{2}$. Hence

$$P(E|G) = \frac{P(E \cap G)}{P(G)} = \frac{1/4}{1/2} = \frac{1}{2}.$$

3. The sample space, with each outcome in descending order of age, is

$$S = \{bbbb, bbbg, bbgb, bbgg, bgbb, bgbg, bggb, bggg,$$

$$gbbb, gbbg, gbgb, gbgg, ggbb, ggbg, gggb, gggg\}.$$

There are 16 outcomes.

(a) The event that at least three children are girls is

$$E = \{bggg, gbgg, ggbg, gggb, gggg\}.$$

Thus, $P(E) = \frac{5}{16}$.

(b) The event that the youngest is a girl is

$$G = \{bbbg, bbgg, bgbg, bggg, gbbg, gbgg, ggbg, gggg\}.$$

and $E \cap G = \{bggg, gbgg, ggbg, gggg\}$. Thus,

$$P(E|G) = \frac{P(E \cap G)}{P(G)} = \frac{4/16}{8/16} = \frac{4}{8} = \frac{1}{2}.$$

(c) The event that exactly three children are girls is

$$F = \{bggg, gbgg, ggbg, gggb\},$$

and $F \cap G = \{bggg, gbgg, ggbg\}$. Thus,

$$P(F|G) = \frac{P(F \cap G)}{P(G)} = \frac{3/16}{8/16} = \frac{3}{8}.$$

5. The sample space has 36 outcomes (see example 11.1.6). Let E be the event "both dice are even" and G be the event "the dice add up to six." Then $G = \{(1,5), (2,4), (3,3), (4,2), (5,1)\}$,

$$E = \{(2,2), (2,4), (2,6), (4,2), (4,4), (4,6), (6,2), (6,4), (6,6)\},$$

and $E \cap G = \{(2,4), (4,2)\}$

(a) $P(E|G) = \dfrac{P(E \cap G)}{P(G)} = \dfrac{2/36}{5/36} = \dfrac{2}{5}$ (b) $P(G|E) = \dfrac{P(E \cap G)}{P(E)} = \dfrac{2/36}{9/36} = \dfrac{2}{9}$

7. There are 28 students (16 boys, 12 girls) in the English class. 10 students are seniors (6 boys, 4 girls) and 18 are not. Let E be the event "the selected student is a senior" and B the event "the selected student is a boy."

(a) $P(E') = \dfrac{18}{28} = \dfrac{9}{14}$

(b) $P(B) = \frac{16}{28} = \frac{4}{7}$ and $P(E' \cap B) = \frac{10}{28} = \frac{5}{14}$ (the number of boys who are not seniors is $16 - 6 = 10$). Thus,

$$P(E'|B) = \frac{P(E' \cap B)}{P(B)} = \frac{5/14}{4/7} = \frac{5}{8}.$$

(c) $P(E|B) = \dfrac{P(E \cap B)}{P(B)} = \dfrac{6/28}{4/7} = \dfrac{3}{8}$

(d) The event "the selected student is a girl" is the same as B' (the student is not a boy). Then $P(B') = \frac{12}{28} = \frac{3}{7}$ and $P(E \cap B') = \frac{4}{28}$. Therefore,

$$P(E|B') = \frac{P(E \cap B')}{P(B')} = \frac{4/28}{3/7} = \frac{1}{3}.$$

9. The multiplication principle states that

$$P(E \cap G) = P(G) \cdot P(E|G)$$

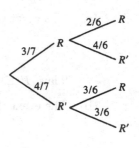

for any events E and G. Let R_i be the event the ith slip is red. Then

$$P(R_1) = \frac{3}{7}, \quad P(R_2|R_1) = \frac{2}{6} = \frac{1}{3}$$

$$P(R_1') = \frac{4}{7}, \quad P(R_2'|R_1') = \frac{3}{6} = \frac{1}{2}.$$

(a) $P(R_1 \cap R_2) = P(R_1) \cdot P(R_2|R_1) = \frac{3}{7} \cdot \frac{1}{3} = \frac{1}{7}$

(b) $P(R_1' \cap R_2') = P(R_1') \cdot P(R_2'|R_1') = \frac{4}{7} \cdot \frac{1}{2} = \frac{2}{7}$

(c) The event that at least one of the slips is red is the complement of the event that neither of the slips is red. That is, $R_1 \cup R_2 = (R_1' \cap R_2')'$. Thus,

$$P(R_1 \cup R_2) = 1 - P(R_1' \cap R_2') = 1 - \frac{2}{7} = \frac{5}{7}.$$

11. In this case, the events are not independent, and

$$P(C_1) = \frac{1}{4}, \quad P(C_1') = \frac{3}{4}$$

$$P(C_2|C_1) = \frac{12}{51}, \quad P(C_2'|C_1) = \frac{39}{51}$$

$$P(C_2|C_1') = \frac{13}{51}, \quad P(C_2'|C_1') = \frac{38}{51}.$$

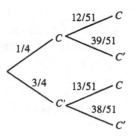

(a) $P(C_1 \cap C_2) = P(C_1) \cdot P(C_2|C_1) = \frac{1}{4} \cdot \frac{12}{51} = \frac{3}{51} = \frac{1}{17}$

(b) $P(C_1' \cap C_2') = P(C_1') \cdot P(C_2'|C_1') = \frac{3}{4} \cdot \frac{38}{51} = \frac{19}{34}$

(c) $P((C_1 \cap C_2') \cup (C_1' \cap C_2)) = P(C_1 \cap C_2') + P(C_1' \cap C_2)$

$$= P(C_1) \cdot P(C_2'|C_1) + P(C_1') \cdot P(C_2|C_1') = \frac{1}{4} \cdot \frac{39}{51} + \frac{3}{4} \cdot \frac{13}{51} = \frac{13}{34}$$

13. (a) 0.138

(b) 0.056

(c) $P(\text{Oriental}) \cdot P(\text{Connecticut}|\text{Oriental}) = (0.138) \cdot (0.056) = 0.007728$

15. Let D be the event "a component is defective" and G the event "a component passes the quality control test." Then $P(D) = 0.03$ (3% of the components are defective, and $P(G|D) = 0.1$ (10% of the defectives pass inspection). The event "you will buy a defective component" is $D \cap G$, (the component is defective *and* it passes inspection.) Then

$$P(D \cap G) = P(G|D) \cdot P(D) = 0.1 \cdot 0.03 = 0.003.$$

17. The sample space has 8 outcomes (see Example 11.1.5), and the events in question are:

$$A = \{HHH, HHT, HTH, HTT\} \quad \text{with} \quad P(A) = \frac{1}{2}$$

$$B = \{HHH, HHT, HTH, THH\} \quad \text{with} \quad P(B) = \frac{1}{2}$$

$$C = \{HHH, TTT\} \quad \text{with} \quad P(C) = \frac{1}{4}$$

$$A \cap B = \{HHH, HHT, HTH\} \quad \text{with} \quad P(A \cap B) = \frac{3}{8}$$

$$A \cap C = \{HHH\} \quad \text{with} \quad P(A \cap C) = \frac{1}{8}$$

$$B \cap C = \{HHH\} \quad \text{with} \quad P(B \cap C) = \frac{1}{8}.$$

We see that

$$P(A \cap B) \neq P(A)P(B), \quad P(A \cap C) = P(A)P(C), \quad P(B \cap C) = P(B)P(C).$$

Thus, A and C are independent; B and C are independent; but A and B are not independent.

19. If A and B are independent, then $P(A \cap B) = P(A)P(B)$. If they are mutually exclusive, then $P(A \cap B) = 0$. Thus $0 = P(A \cap B) = P(A)P(B)$. But then either $P(A) = 0$ or $P(B) = 0$. So the answer is no.

21. Let A be the event that the computer will fail to work and B the event that the printer will fail to work. Then
$$P(A \cap B) = P(A) \cdot P(B) = (0.15) \cdot (0.20) = 0.03.$$

23. There are 36 outcomes (see Example 11.1.6 for a list of them). Let E_1 be the event that the first die is even and E_2 the event that the second die is even. Then $P(E_1) = P(E_2) = \frac{1}{2}$.

(a) $E_1 \cap E_2 = \{(2,2), (2,4), (2,6), (4,2), (4,4), (4,6), (6,2), (6,4), (6,6)\}$. Therefore, $P(E_1 \cap E_2) = \frac{9}{36} = \frac{1}{4}$.

(b) Since E_1 and E_2 are independent,
$$P(E_1 \cap E_2) = P(E_1)P(E_2) = \frac{1}{2} \cdot \frac{1}{2} = \frac{1}{4}.$$

25. (a) Let E_1 be the event that the gene inherited from father will be negative and E_2 the event that the gene inherited from mother will be negative. Then $P(E_1) = P(E_2) = 1 - 0.61 = 0.39$. For a person to have Rh negative blood, both genes have to be negative, which is the event $E_1 \cap E_2$. Since E_1 and E_2 are independent, $P(E_1 \cap E_2) = P(E_1)P(E_2) = (0.39) \cdot (0.39) = 0.1521$.

(b) The probability of type O blood is 0.46, and the probablity of Rh negative is 0.1521. Since they are independent, the probability that a randomly chosen person has O negative blood is equal to $(0.46) \cdot (0.1521) = 0.069966$.

27. Let G be the event that a randomly chosen student will pass the first test and E the event that a randomly chosen student will pass the second test. Then $P(G) = 0.72$, $P(E|G) = 0.91$, $P(E|G') = 0.12$. In addition, we have $P(G') = 1 - P(G) = 1 - 0.72 = 0.28$.

(a) The probability that a randomly chosen student will pass both tests is $P(G \cap E) = P(G) \cdot P(E|G) = (0.72) \cdot (0.91) = 0.6552$.

(b) By Bayes formula,

$$P(G|E) = \frac{P(E|G)P(G)}{P(E|G)P(G) + P(E|G')P(G')}$$

$$= \frac{(0.91) \cdot (0.72)}{(0.91) \cdot (0.72) + (0.12) \cdot (0.28)} \approx 0.9512.$$

29. Let E be the event that a selected car will be defective, A the event that a selected car comes from plant A, and B the event that a selected car comes from plant B. Then we have $P(E|A) = 0.03$, $P(E|B) = 0.04$, $P(A) = 0.6$, and $P(B) = 0.4$.

(a) The probability that the defective car comes from plant A is

$$P(A|E) = \frac{P(E|A)P(A)}{P(E|A)P(A) + P(E|B)P(B)}$$

$$= \frac{(0.03)(0.6)}{(0.03)(0.6) + (0.04)(0.4)} = \frac{9}{17} \approx 0.53.$$

(b) The probability that the defective car comes from plant B is

$$P(B|E) = \frac{P(E|B)P(B)}{P(E|B)P(B) + P(E|A)P(A)}$$

$$= \frac{(0.04)(0.4)}{(0.04)(0.4) + (0.03)(0.6)} = \frac{8}{17} \approx 0.47.$$

31. Let I be the event that a person is innocent and M the event that the person's DNA matches the sample. Taking the sample space to be the population of 5,000,000, the given data tells us that

$$P(I) = \frac{4,999,999}{5,000,000}, \quad P(M) = \frac{6}{5,000,000}, \quad P(M \cap I) = \frac{5}{5,000,000}.$$

(a) $P(M|I) = \dfrac{P(M \cap I)}{P(I)} = \dfrac{5}{4,999,999}$ (b) $P(I|M) = \dfrac{P(M \cap I)}{P(M)} = \dfrac{5}{6}$

33. First, by Theorem 11.1.2, $P(E \cap G) \leq P(G)$. Therefore

$$P(E|G) = \frac{P(E \cap G)}{P(G)} \leq 1.$$

Since $P(E \cap G) \geq 0$ and $P(G) > 0$, we also have $P(E|G) \geq 0$.

Second, $P(S|G) = \frac{P(S \cap G)}{P(G)} = \frac{P(G)}{P(G)} = 1$, and

$$P(\emptyset|G) = \frac{P(\emptyset \cap G)}{P(G)} = \frac{P(\emptyset)}{P(G)} = \frac{0}{P(G)} = 0.$$

Third, if E and F are mutually exclusive, then $E \cap G$ and $F \cap G$ are mutually exclusive. Also, $(E \cup F) \cap G) = (E \cap G) \cup (F \cap G)$. Therefore,

$$P(E \cup F|G) = \frac{P((E \cup F) \cap G)}{P(G)} = \frac{P((E \cap G) \cup (F \cap G))}{P(G)}$$

$$= \frac{P(E \cap G) + P(F \cap G)}{P(G)} = \frac{P(E \cap G)}{P(G)} + \frac{P(F \cap G)}{P(G)}$$

$$= P(E|G) + P(F|G).$$

EXERCISES 11.3

1. (a)

x	0	1	2	3
$P(X = x)$	$\frac{1}{8}$	$\frac{3}{8}$	$\frac{3}{8}$	$\frac{1}{8}$

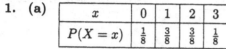

(b) $E(X) = 0(\frac{1}{8}) + 1(\frac{3}{8}) + 2(\frac{3}{8}) + 3(\frac{1}{8}) = 1.5$

$$\text{Var}(X) = (0 - 1.5)^2 \cdot \frac{1}{8} + (1 - 1.5)^2 \cdot \frac{3}{8} + (2 - 1.5)^2 \cdot \frac{3}{8} + (3 - 1.5)^2 \frac{1}{8} = 0.75 = \frac{3}{4}$$

$$\sigma(X) = \sqrt{\text{Var}(X)} = \sqrt{\frac{3}{4}} = \frac{\sqrt{3}}{2}$$

3. See Example 11.1.6 for a description of the sample space of this experiment. There are 36 outcomes

There are 6 outcomes with $X = 0$:

$$\{(1,1), (2,2), (3,3), (4,4), (5,5), (6,6)\}$$

There are 10 outcomes with $X = 1$:

$$\{(6,5), (5,4), (4,3), (3,2), (2,1), (5,6), (4,5), (3,4), (2,3), (1,2)\}$$

Continuing in this way, we obtain:

x	0	1	2	3	4	5
$P(X = x)$	$\frac{6}{36} = \frac{1}{6}$	$\frac{10}{36} = \frac{5}{18}$	$\frac{8}{36} = \frac{2}{9}$	$\frac{6}{36} = \frac{1}{6}$	$\frac{4}{36} = \frac{1}{9}$	$\frac{2}{36} = \frac{1}{18}$

$$E(X) = 0 \cdot \frac{1}{6} + 1 \cdot \frac{5}{18} + 2 \cdot \frac{2}{9} + 3 \cdot \frac{1}{6} + 4 \cdot \frac{1}{9} + 5 \cdot \frac{1}{18} = \frac{35}{18}$$

$$\text{Var}(X) = 0^2 \cdot \frac{1}{6} + 1^2 \cdot \frac{5}{18} + 2^2 \cdot \frac{2}{9} + 3^2 \cdot \frac{1}{6} + 4^2 \cdot \frac{1}{9} + 5^2 \cdot \frac{1}{18} - \left(\frac{35}{18}\right)^2 = \frac{665}{324}.$$

5. X is the number of bikes sold in a given day. The range of X is $\{6, 7, 8, 9, 10\}$.

x	6	7	8	9	10
$P(X = x)$	$\frac{40}{400} = \frac{1}{10}$	$\frac{60}{400} = \frac{3}{20}$	$\frac{160}{400} = \frac{2}{5}$	$\frac{80}{400} = \frac{1}{5}$	$\frac{60}{400} = \frac{3}{20}$

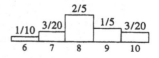

$$E(X) = 6 \cdot \frac{1}{10} + 7 \cdot \frac{3}{20} + 8 \cdot \frac{2}{5} + 9 \cdot \frac{1}{5} + 10 \cdot \frac{3}{20} = 8.15$$

$$\text{Var}(X) = 6^2 \cdot \frac{1}{10} + 7^2 \cdot \frac{3}{20} + 8^2 \cdot \frac{2}{5} + 9^2 \cdot \frac{1}{5} + 10^2 \cdot \frac{3}{20} - (8.15)^2 = 1.3275$$

$$\sigma(X) = \sqrt{1.3275} \approx 1.1522$$

7. Let X be the distance from the center of the gathering place to the center of an individual home. There were 53 homes whose distances from the center were measured. On the basis of those measurements, we construct the following probability distribution table and histogram:

x	$10 - 19$	$20 - 29$	$30 - 39$	$40 - 49$	$50 - 59$
$P(X = x)$	$\frac{15}{53}$	$\frac{12}{53}$	$\frac{14}{53}$	$\frac{7}{53}$	$\frac{5}{53}$

The average distance (in meters) is estimated to be

$$14.5 \frac{15}{53} + 24.5 \frac{12}{53} + 34.5 \frac{14}{53} + 44.5 \frac{7}{53} + 54.5 \frac{5}{53} \approx 29.783.$$

9. Let X be the net gain on a single play. The range of X is $\{-1, 1, 2, 3\}$. There are 4 aces in the deck, 4 tens, and the total number of jacks, queens and kings is 12. That leaves 32 other cards. Thus, the probability distribution is as follows:

x	-1	1	2	3
$P(X = x)$	$\frac{32}{52}$	$\frac{4}{52}$	$\frac{12}{52}$	$\frac{4}{52}$

(a) $E(X) = (-1)\frac{32}{52} + 1 \cdot \frac{4}{52} + 2 \cdot \frac{12}{52} + 3 \cdot \frac{4}{52} = \frac{2}{13}$

(b) If you play the game 100 times, then you would expect to win $\frac{2}{13} \cdot 100 \approx 15.38$ dollars.

11. Let $p_1 = P(X = 1)$, $p_2 = P(X = 2)$, and $p_3 = P(X = 4)$. Then

$$E(X) = 1 \cdot p_1 + 2 \cdot p_2 + 4 \cdot p_3 = 3$$

and

$$\text{Var}(X) = (1 - 3)^2 \cdot p_1 + (2 - 3)^2 \cdot p_2 + (4 - 3)^2 \cdot p_3 = \frac{3}{2},$$

or

$$4 \cdot p_1 + p_2 + p_3 = \frac{3}{2}.$$

In addition, $p_1 + p_2 + p_3 = 1$. We thus have a system of three equations in the unknowns p_1, p_2, and p_3, whose unique solution is $p_1 = \frac{1}{6}$, $p_2 = \frac{1}{4}$, $p_3 = \frac{7}{12}$. The probability distribution is

x	1	2	4
$P(X = x)$	$\frac{1}{6}$	$\frac{1}{4}$	$\frac{7}{12}$

13. We first observe that $X = x_i \iff Y = ax_i + b$. Therefore, the range of Y is the set of numbers $\{ax_1 + b, ax_2 + b, \ldots, ax_n + b\}$, and

$$P(Y = ax_i + b) = P(X = x_i) = p_i.$$

It follows that

$$E(Y) = (ax_1 + b)p_1 + (ax_2 + b)p_2 + \cdots + (ax_n + b)p_n$$

$$= a(x_1p_1 + x_2p_2 + \cdots + x_np_n) + b(p_1 + p_2 + \cdots + p_n)$$

$$= aE(X) + b.$$

(Recall that $E(X) = x_1p_1 + x_2p_2 + \cdots + x_np_n$ and $p_1 + p_2 + \cdots + p_n = 1$.)

If we write μ_X and μ_Y for $E(X)$ and $E(Y)$ respectively, then we have just seen that $\mu_Y = a\mu_X + b$. The variance of Y is given by

$$\mathrm{Var}(Y) = (ax_1 + b - \mu_Y)^2 p_1 + \cdots + (ax_n + b - \mu_Y)^2 p_n$$

$$= (ax_1 + b - [a\mu_X + b])^2 p_1 + \cdots + (ax_n + b - [a\mu_X + b])^2 p_n$$

$$= (ax_1 - a\mu_X)^2 p_1 + \cdots + (ax_n - a\mu_X)^2 p_n$$

$$= a^2[(x_1 - \mu_X)^2 p_1 + \cdots + (x_n - \mu_X)^2 p_n] = a^2 \mathrm{Var}(X).$$

EXERCISES 11.4

1. Let X be the number of heads in three tosses. Then X is a binomial random variable with parameters $n = 3$ and $p = 0.2$, and

$$P(X = k) = \binom{3}{k}(0.2)^k(1 - 0.2)^{3-k} \text{ for } k = 0, 1, 2, 3.$$

x	0	1	2	3
$P(X = k)$	0.512	0.384	0.096	0.008

3. Let X be the number of fives in four rolls of a fair die. Then X is a binomial random variable with parameters $n = 4$ and $p = \frac{1}{6}$, and

$$P(X = k) = \binom{4}{k}\left(\frac{1}{6}\right)^k\left(\frac{5}{6}\right)^{4-k} \text{ for } k = 0, 1, 2, 3, 4.$$

(a) $P(X = 1) = \binom{4}{1}\left(\frac{1}{6}\right)\left(\frac{5}{6}\right)^3 = \dfrac{500}{1,296} = \dfrac{125}{324} \approx 0.3858$

(b) $P(X \le 1) = P(X = 0) + P(X = 1) = \binom{4}{0}\left(\frac{5}{6}\right)^4 + \binom{4}{1}\left(\frac{1}{6}\right)\left(\frac{5}{6}\right)^3 = \dfrac{1,125}{1,296} \approx 0.868$

5. Let X be the number of dogs choosing the new flavor. Then X is a binomial random variable with parameters $n = 10$ and $p = \frac{1}{2}$, and

$$P(X = 9) = \binom{10}{9}\left(\frac{1}{2}\right)^9\left(\frac{1}{2}\right) = \dfrac{10}{2^{10}} = \dfrac{5}{512} \approx 0.01.$$

7. **(a)** $\binom{5}{3} = 10.$ **(b)** $\binom{6}{4} = 15.$ **(c)** $\binom{9}{5} = 126.$

9. Let X be the number of boys in a family of five children. Then X is a binomial random variable with parameters $n = 5$ and $p = \frac{1}{2}$ (assuming that a new-born is equally likely to be a boy or girl). Thus, the probability of exactly four boys is

$$P(X = 4) = \binom{5}{4}\left(\frac{1}{2}\right)^4\left(\frac{1}{2}\right) = \frac{5}{32} = 0.15625,$$

and the probability of at least four boys is

$$P(X \geq 4) = P(X = 4) + P(X = 5) = \binom{5}{4}\left(\frac{1}{2}\right)^4\left(\frac{1}{2}\right) + \binom{5}{0}\left(\frac{1}{2}\right)^5$$

$$= \frac{\binom{5}{4} + \binom{5}{0}}{2^5} = \frac{6}{32} = 0.1875.$$

11. **(a)** With $n = 50$ and $p = \frac{1}{9}$, **(b)** $P(X > 10) = 1 - P(X \leq 10)$
$P(X \leq 4) \approx 0.3348$ ≈ 0.0196

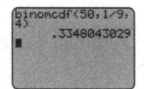

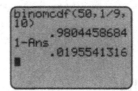

13. Let X be the number of votes favoring the Republican. Then X is a binomial random variable with $p = 0.58$.

(a) In this case, $n = 12$ and

$$P(X = 7) = \binom{12}{7}(0.58)^7(1 - 0.58)^5 = 792 \cdot (0.58)^7(0.42)^5 \approx 0.2285.$$

(b) In this case, $n = 50$ and $E(X) = np = 50 \cdot (0.58) = 29.$

15. Let X be the number of games the Notre Dame football team will lose next season. Then X is a binomial random variable with parameters $n = 10$ and $p = 0.2$.

(a) $P(X = 0) = \binom{10}{0}(1 - 0.2)^{10} = (0.8)^{10} \approx 0.1074.$

(b) $P(X = 1) = \binom{10}{1}(0.2)(1 - 0.2)^9 = 10 \cdot (0.2) \cdot (0.8)^9 \approx 0.2684.$

(c) $P(X < 3) = P(X = 0) + P(X = 1) + P(X = 2)$

$$= \binom{10}{0}(0.8)^{10} + \binom{10}{1}(0.2)(0.8)^9 + \binom{10}{2}(0.2)^2(0.8)^8 \approx 0.6778.$$

17. **(a)** $\binom{14}{2} = \frac{14 \cdot 13}{2} = 91$ **(b)** $\binom{15}{3} = \frac{15 \cdot 14 \cdot 13}{3 \cdot 2 \cdot 1} = 455$

(c) $\binom{20}{4} = \frac{20 \cdot 19 \cdot 18 \cdot 17}{4 \cdot 3 \cdot 2 \cdot 1} = 4,845$ **(d)** $\binom{100}{3} = \frac{100 \cdot 99 \cdot 98}{3 \cdot 2 \cdot 1} = 161,700.$

19. With $n = 100$ and $p = 0.52$, $E(X) = np = 100 \cdot 0.52 = 52$, and

$$\text{Var}(X) = np(1 - p) = 100 \cdot (0.52)(1 - 0.52) = 24.96.$$

21. Let X be the number of T-shirts sold to adults. Then X is a binomial random variable with parameters $n = 6,000$ and $p = 0.69$.

(a) $\mu = E(X) = 6,000 \cdot 0.69 = 4,140$

(b) Since the profit on each shirt is \$8, the expected profit (in dollars) is

$$8E(X) = 8 \cdot 4,140 = 33,120.$$

(c) Since $\text{Var}(X) = np(1 - p) = 6,000 \cdot (0.69) \cdot (0.31) = 1,283.4$, the standard deviation is $\sigma(X) = \sqrt{\text{Var}(X)} = \sqrt{1,283.4} \approx 35.8$. Thus, $\mu - 2\sigma \approx 4,140 - 2 \cdot 35.8 \approx 4,068.4$, which means they will sell at least 4,069; and $\mu + 2\sigma \approx 4,140 + 2 \cdot 35.8 \approx 4,211.6$, which means they will sell at most 4,211. The lower bound on the profit is $4,069 \cdot 8 = 32,552$ dollars, and the upper bound on the profit is $4,211 \cdot 8 = 33,688$ dollars.

23. $P(X < 48) = P(X \leq 47) \approx 3.5 \times 10^{-60}$

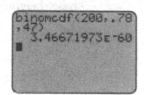

25. There are exactly n strings with 1 H and $n - 1$ T's. To see why, think of where to place the H in the string. We can place it first to get $HTT \cdots T$, second to get $THTT \cdots T$, third to get $TTHTT \cdots T$, and so forth. There are n such arrangements, and they exhaust all the possibilities for a string with 1 H and $n - 1$ T's. Therefore, $\binom{n}{1} = n$.

EXERCISES 11.5

In exercises 1, 3, and 5, we need to show that

$$f(x) \geq 0 \text{ for } a \leq x \leq b \quad \text{and} \quad \int_a^b f(x)\, dx = 1.$$

1. $f(x) = \frac{x^2}{9} \geq 0$ for all x, and

$$\int_0^3 \frac{x^2}{9}\, dx = \frac{x^3}{27}\bigg|_0^3 = \frac{1}{27}(3^3 - 0) = 1.$$

3. $f(x) = \frac{1}{x^2} > 0$ for all $x \neq 0$, and

$$\int_1^\infty f(x)\, dx = \int_1^\infty \frac{1}{x^2}\, dx = \lim_{b \to \infty} \int_1^b \frac{1}{x^2}\, dx$$

$$= \lim_{b \to \infty} \left(-\frac{1}{x}\right)\bigg|_1^b = \lim_{b \to \infty} \left(-\frac{1}{b} + 1\right) = 1.$$

5. $f(x) = xe^{-x} \geq 0$ for all $x \geq 0$, and

$$\int_0^\infty f(x)\, dx = \lim_{b\to\infty} \int_0^b xe^{-x}\, dx \quad \text{(use integration by parts)}$$

$$= \lim_{b\to\infty} \left. (-xe^{-x} - e^{-x}) \right|_0^b \quad \text{(with } u = x \text{ and } dv = e^{-x}\, dx)$$

$$= \lim_{b\to\infty} (-be^{-b} - e^{-b} + 1) = 1.$$

7. **(a)** If $x \leq 1$, then $2x \leq 2$, and

$$2 - 2x \geq 0.$$

(b) $\int_0^2 (2 - 2x)\, dx$ is the area of the right triangle formed by the points $(0,0)$, $(0,2)$, and $(1,0)$, which equals

$$\frac{1}{2} \cdot \text{base} \cdot \text{height} = \frac{1}{2} \cdot 1 \cdot 2 = 1.$$

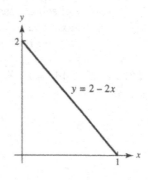

(c) The probability that at least 60% of the students finish is the area under the graph of $f(x)$ for $0.6 \leq x \leq 1$ (shaded in the figure). That is the area of a right triangle with base length $1 - 0.6 = 0.4$ and height $f(0.6) = 0.8$, which equals $\frac{1}{2} \cdot (0.4) \cdot (0.8) = 0.16$.

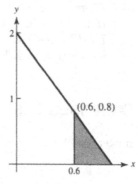

9. First, if $0 \leq x \leq 2$, then $f(x) = -\frac{1}{2}(x - 2) \geq 0$. Second,

$$\int_0^2 f(x)\, dx = \int_0^2 \left(-\frac{1}{2}x + 1 \right) dx = \left. \left(-\frac{1}{4}x^2 + x \right) \right|_0^2 = -1 + 2 = 1.$$

Therefore, $f(x)$ is a probability density function.

11. First, $f(x) = \frac{3}{2}x^2 \geq 0$ for all x. Second,

$$\int_{-1}^1 f(x)\, dx = \int_{-1}^1 \frac{3}{2}x^2\, dx = \left. \frac{1}{2}x^3 \right|_{-1}^1 = \frac{1}{2} + \frac{1}{2} = 1.$$

13. $f(x) = x(1 - x)$, $0 < x < 1$ is not a probability density function, since

$$\int_0^1 x(1 - x)\, dx = \int_0^1 (x - x^2)\, dx = \left. \left(\frac{1}{2}x^2 - \frac{1}{3}x^3 \right) \right|_0^1 = \frac{1}{2} - \frac{1}{3} = \frac{1}{6} \neq 1.$$

15. $f(x) = e^{-0.1x} \, dx$, $x > 0$, is not a probability density function, since

$$\int_0^\infty e^{-0.1x} \, dx = \lim_{b \to \infty} \int_0^b e^{-0.1x} \, dx = \lim_{b \to \infty} -10 e^{-0.1x} \Big|_0^b$$

$$= \lim_{b \to \infty} \left(-10 e^{-0.1b} + 10 \right) = 10 \neq 1.$$

17. We must have $1 = \int_0^1 cx^2(1 - x) \, dx = c(\frac{1}{3}x^3 - \frac{1}{4}x^4)\big|_0^1 = \frac{c}{12}$. Therefore, $c = 12$. We also observe that if $c = 12$, then $f(x) \geq 0$ for $x \leq 1$.

19. **(a)** We must have $\int_{1,000}^{4,000} f(x) \, dx = \int_{1,000}^{4,000} cx^{-2} \, dx = 1$. Note that

$$1 = \int_{1,000}^{4,000} cx^{-2} \, dx = -cx^{-1} \Big|_{1,000}^{4,000} = c \left(-\frac{1}{4,000} + \frac{1}{1,000} \right) = \frac{3c}{4,000}.$$

Therefore, $c = \frac{4,000}{3}$.

(b) $P(3,000 \leq X \leq 4,000) = \displaystyle\int_{3,000}^{4,000} \frac{4,000}{3} x^{-2} \, dx = -\frac{4,000}{3} x^{-1} \Big|_{3,000}^{4,000}$

$$= -\frac{4,000}{3} \left(\frac{1}{4,000} - \frac{1}{3,000} \right) = \frac{1}{9}$$

21. Using integration by parts with $u = x$ and $dv = e^{-x} \, dx$, we have

$$\int xe^{-x} \, dx = -xe^{-x} + \int e^{-x} \, dx = -(x + 1)e^{-x} + c.$$

(a) $P(X \geq 1) = \displaystyle\int_1^\infty xe^{-x} \, dx = \lim_{b \to \infty} \left[-(x + 1)e^{-x} \right] \Big|_1^b$

$$= \lim_{b \to \infty} \left[-(b + 1)e^{-b} + 2e^{-1} \right] = \frac{2}{e}$$

(b) $P(0 \leq X < 5) = \displaystyle\int_0^5 xe^{-x} \, dx = -(x + 1)e^{-x} \Big|_0^5$

$$= 1 - 6e^{-5} = \frac{e^5 - 6}{e^5}$$

23. First, $f(x) = \frac{1}{2\sqrt{x}} > 0$ for all $x > 0$. Second,

$$\int_1^4 f(x) \, dx = \int_1^4 \frac{1}{2\sqrt{x}} \, dx = \sqrt{x} \Big|_1^4 = \sqrt{4} - \sqrt{1} = 1.$$

Therefore, $f(x)$ is a probability density function.

If $1 < x < 4$, then $F(x) = \int_1^x \frac{1}{2\sqrt{t}} \, dt = \sqrt{t} \big|_1^x = \sqrt{x} - 1$. Thus,

$$F(x) = \begin{cases} 0 & x \leq 1 \\ \sqrt{x} - 1 & 1 < x \leq 4 \\ 1 & x > 4. \end{cases}$$

25. First, if $-1 \le x \le 0$, then $f(x) = 1 + x \ge 0$; and if $0 < x \le 1$, then $f(x) = 1 - x \ge 0$. Second,

$$\int_{-1}^{1} f(x)\, dx = \int_{-1}^{0} (1+x)\, dx + \int_{0}^{1} (1-x)\, dx$$

$$= \left(x + \frac{1}{2}x^2 \right)\Big|_{-1}^{0} + \left(x - \frac{1}{2}x^2 \right)\Big|_{0}^{1}$$

$$= \frac{1}{2} + \frac{1}{2} = 1.$$

Therefore, $f(x)$ is a probability density function.
  If $-1 < x < 0$, then

$$F(x) = \int_{-1}^{x} (1+t)\, dt = \left(t + \frac{1}{2}t^2 \right)\Big|_{-1}^{x} = x + \frac{1}{2}x^2 + \frac{1}{2};$$

and if $0 < x < 1$, then

$$F(x) = \int_{-1}^{0} (1+t)\, dt + \int_{0}^{x} (1-t)\, dt = \frac{1}{2} + \left(t - \frac{t^2}{2} \right)\Big|_{0}^{x} = \frac{1}{2} + x - \frac{1}{2}x^2.$$

Thus,

$$F(x) = \begin{cases} 0 & x \le -1 \\[2mm] x + \dfrac{1}{2}x^2 + \dfrac{1}{2} & -1 < x \le 0 \\[2mm] \dfrac{1}{2} + x - \dfrac{1}{2}x^2 & 0 < x \le 1 \\[2mm] 1 & x > 1. \end{cases}$$

27. (a) $P(X \le 3) = F(3) = \dfrac{1}{2}\sqrt{3-2} = \dfrac{1}{2}$

(b) $P(4 \le X \le 5) = P(4 < X \le 5) = P(X \le 5) - P(X \le 4)$

$$= F(5) - F(4) = \frac{1}{2}\sqrt{5-2} - \frac{1}{2}\sqrt{4-2} = \frac{\sqrt{3} - \sqrt{2}}{2}$$

(c) $f(x) = F'(x) = \dfrac{1}{4\sqrt{x-2}},\ 2 < x < 6$

29. $\mu = E(X) = \displaystyle\int_{0}^{1} x(2 - 2x)\, dx = \left(x^2 - \frac{2}{3}x^3 \right)\Big|_{0}^{1} = 1 - \frac{2}{3} = \frac{1}{3}$

$$\mathrm{Var}(X) = \int_{0}^{1} x^2 f(x)\, dx - \mu^2 = \int_{0}^{1} x^2(2 - 2x)\, dx - \left(\frac{1}{3} \right)^2$$

$$= \left(\frac{2}{3}x^3 - \frac{1}{2}x^4 \right)\Big|_{0}^{1} - \frac{1}{9} = \frac{2}{3} - \frac{1}{2} - \frac{1}{9} = \frac{1}{18}$$

31. $\mu = E(X) = \dfrac{3}{4} \displaystyle\int_0^2 x(2x - x^2)\, dx = \dfrac{3}{4}\left(\dfrac{2}{3}x^3 - \dfrac{x^4}{4}\right)\Big|_0^2$

$\qquad = \dfrac{3}{4}\left(\dfrac{16}{3} - \dfrac{16}{4}\right) = \dfrac{3}{4}\cdot\dfrac{16}{12} = 1$

$\qquad\qquad \mathrm{Var}(X) = \displaystyle\int_0^2 x^2 f(x)\, dx - \mu^2 = \dfrac{3}{4}\displaystyle\int_0^2 x^2(2x - x^2)\, dx - 1$

$\qquad\qquad\qquad = \dfrac{3}{4}\left(\dfrac{1}{2}x^4 - \dfrac{x^5}{5}\right)\Big|_0^2 - 1 = \dfrac{3}{4}\left(8 - \dfrac{32}{5}\right) - 1 = \dfrac{1}{5}$

33. By integration by parts, we have

$$\int 2xe^{-2x}\, dx = -\left(x + \dfrac{1}{2}\right)e^{-2x} + c \quad \text{and}$$

$$\int 2x^2 e^{-2x}\, dx = -x^2 e^{-2x} + \int 2xe^{-2x}\, dx = -\left(x^2 + x + \dfrac{1}{2}\right)e^{-2x} + c.$$

Then

$$\mu = E(X) = \int_0^\infty 2xe^{-2x}\, dx = \lim_{b\to\infty}\int_0^b 2xe^{-2x}\, dx$$

$$= \lim_{b\to\infty}\left[-\left(x + \dfrac{1}{2}\right)e^{-2x}\right]_0^b = \lim_{b\to\infty}\left[-\left(b + \dfrac{1}{2}\right)e^{-2b} + \dfrac{1}{2}\right] = \dfrac{1}{2}$$

$$\mathrm{Var}(X) = \int_0^\infty x^2 f(x)\, dx - \mu^2 = \lim_{b\to\infty}\int_0^b 2x^2 e^{-2x}\, dx - \dfrac{1}{4}$$

$$= \lim_{b\to\infty}\left[-\left(x^2 + x + \dfrac{1}{2}\right)e^{-2x}\right]_0^b - \dfrac{1}{4}$$

$$= \lim_{b\to\infty}\left[-\left(b^2 + b + \dfrac{1}{2}\right)e^{-2b} + \dfrac{1}{2}\right] - \dfrac{1}{4} = \dfrac{1}{2} - \dfrac{1}{4} = \dfrac{1}{4}.$$

35. By integration by parts, $\int x^2 e^{-x}\, dx = -(x^2 + 2x + 2)e^{-x} + c$. Then, since $f(x) = xe^{-x}$, the mean time (in minutes) is the expected value

$$E(X) = \int_0^\infty xf(x)\, dx = \lim_{b\to\infty}\int_0^b x^2 e^{-x}\, dx = \lim_{b\to\infty} -(x^2 + 2x + 2)e^{-x}\Big|_0^b$$

$$= \lim_{b\to\infty}[-(b^2 + 2b + 2)e^{-b} + 2] = 2.$$

37. By integration by parts, we have $\int \dfrac{1}{10}xe^{-x/10}\, dx = -(x + 10)e^{-x/10} + c$ and

$$\int \dfrac{1}{10}x^2 e^{-x/10}\, dx = -(x^2 + 20x + 200)e^{-x/10} + c.$$

Then

$$\mu = E(X) = \int_0^\infty xf(x)\, dx = \lim_{b\to\infty}\int_0^b \dfrac{1}{10}xe^{-x/10}\, dx$$

$$= \lim_{b\to\infty} -(x + 10)e^{-x/10}\Big|_0^b = \lim_{b\to\infty}[-(b + 10)e^{-\frac{1}{10}b} + 10] = 10,$$

$$\text{Var}(X) = \int_0^\infty x^2 f(x)\, dx - \mu^2 = \lim_{b \to \infty} \int_0^b \frac{1}{10} x^2 e^{-x/10}\, dx - 100$$

$$= \lim_{b \to \infty} \left. -(x^2 + 20x + 200)e^{-x/10} \right|_0^b - 100$$

$$= \lim_{b \to \infty} [-(b^2 + 20b + 200)e^{-b/10} + 200] - 100 = 100,$$

and $\sigma(X) = \sqrt{\text{Var}(X)} = \sqrt{100} = 10$.

39. $E(X) = \int_0^3 x f(x)\, dx = \int_0^3 \frac{1}{3} x\, dx = \left. \frac{1}{6} x^2 \right|_0^3 = \frac{3}{2}$

If m is the median of X, then $\frac{1}{2} = \int_0^m \frac{1}{3}\, dx = \frac{1}{3} m$. Therefore, $m = \frac{3}{2}$.

41. $E(X) = \int_0^1 x f(x)\, dx = \int_0^1 3x^3\, dx = \left. \frac{3}{4} x^4 \right|_0^1 = \frac{3}{4}$

If m is the median of X, then $\frac{1}{2} = \int_0^m 3x^2\, dx = m^3$. Thus, $m = \sqrt[3]{\frac{1}{2}}$

43. By integration by parts, $\int x e^{-x}\, dx = -(x+1)e^{-x} + c$. Then

$$E(X) = \int_0^\infty x f(x)\, dx = \lim_{b \to \infty} \int_0^b x e^{-x}\, dx = \lim_{b \to \infty} \left. [-(x+1)e^{-x}] \right|_0^b$$

$$= \lim_{b \to \infty} [-(b+1)e^{-b} + 1] = 1.$$

If m is the median of X, then $\frac{1}{2} = \int_0^m e^{-x}\, dx = 1 - e^{-m}$. Thus, $m = \ln 2$.

45. If the graph of $f(x)$ is symmetric about the line $x = \mu$, then $m = \mu$; i.e., the mean of X is equal to the median.

47. To estimate the mean, we would take the average of the scores, that is

$$\frac{100 + 94 + 91 + \cdots + 42 + 36 + 20}{21} \approx 71.143.$$

To estimate the median, we would take the score in the middle—that is the one with the same number of scores above and below it—which in this case is 73.

EXERCISES 11.6

1. The waiting time (in minutes) is a random variable X with a uniform distribution over the interval $[0, 2]$. Thus, $f(x) = \frac{1}{2}$ for $0 < x < 2$. Since 30 seconds equal $\frac{1}{2}$ minute, the probability of waiting 30 seconds or less is given by

$$P\left(X \le \frac{1}{2}\right) = \int_0^{1/2} \frac{1}{2}\, dx = \left. \frac{1}{2} x \right|_0^{1/2} = \frac{1}{4}.$$

3. Let X be the distance from an accident point to the eastern end of the highway. Then X is a random variable with a uniform distribution over $(0, 30)$. The probability density function is $f(x) = \frac{1}{30}$, $0 < x < 30$.

(a) $P(0 \leq X \leq 3) = \int_0^3 \frac{1}{30}\, dx = \frac{1}{30} x \Big|_0^3 = \frac{1}{10}$

(b) Since the accidents occur independently, the probability that all four occur in that same 3-mile stretch is $\left(\frac{1}{10}\right)^4 = 0.0001$.

5. Let X be the waiting time between the arrivals of two successive aircraft. The probability density function is $f(x) = 20e^{-20x}$, $x \geq 0$.

(a) $E(X) = \frac{1}{\lambda} = \frac{1}{20}$ hour, or 3 minutes.

(b) $P\left(0 \leq X < \frac{1}{20}\right) = \int_0^{1/20} 20e^{-20x}\, dx = -e^{-20x} \Big|_0^{1/20} = 1 - e^{-1} \approx 0.63$

7. He would estimate the mean waiting time by averaging the ten given waiting times—that is,

$$E(X) \approx \frac{12 + 20 + 14 + 10 + 23 + 18 + 5 + 9 + 15 + 18}{10} = 14.4.$$

Then $\lambda \approx \frac{1}{14.4} \approx 0.0694$.

9. (a) Since the event $(X \geq a + b) \cap (X \geq b)$ is the same as $X \geq a + b$, we have

$$P(X \geq a + b | X \geq b) = \frac{P(X \geq a + b)}{P(X \geq b)} = \frac{\int_{a+b}^{\infty} \lambda e^{-\lambda x}\, dx}{\int_b^{\infty} \lambda e^{-\lambda x}\, dx}$$

$$= \frac{e^{-\lambda(a+b)}}{e^{-\lambda b}} = e^{-\lambda a}.$$

(b) $P(X \geq a) = \int_a^{\infty} \lambda e^{-\lambda x}\, dx = e^{-\lambda a}$. Thus,

$$P(X \geq a + b | X \geq b) = P(X \geq a).$$

(c) We can interpret the result of part (b) as saying that the probability of waiting an additional a time units given that you have already waited b units is the same as the unconditional probability of waiting a time units. In other words, the probability of waiting a time units does not change no matter how long you have already waited.

11. Using integration by parts with $u = x$ and $dv = \lambda e^{-\lambda x}$, we obtain

$$E(X) = \int_0^{\infty} x f(x)\, dx = \lim_{b \to \infty} \int_0^b \lambda x e^{-\lambda x}\, dx = \lim_{b \to \infty} -\left(xe^{-\lambda x} + \frac{1}{\lambda} e^{-\lambda x}\right) \Big|_0^b$$

$$= \lim_{b \to \infty} \left[-\left(be^{-\lambda b} + \frac{1}{\lambda} e^{-\lambda b}\right) + \frac{1}{\lambda}\right] = \frac{1}{\lambda}.$$

Next, using integration by parts twice, first with $u = x^2$ and $dv = \lambda e^{-\lambda x}$ and second with $u = x$ and $dv = e^{-\lambda x}$, we obtain

$$\text{Var}(X) = \int_0^\infty x^2 f(x)\, dx - E(X)^2 = \lim_{b \to \infty} \int_0^b \lambda x^2 e^{-\lambda x}\, dx - \frac{1}{\lambda^2}$$

$$= \lim_{b \to \infty} -\left(x^2 + \frac{2}{\lambda}x + \frac{2}{\lambda^2}\right) e^{-\lambda x}\Big|_0^b - \frac{1}{\lambda}$$

$$= \lim_{b \to \infty} \left[-\left(b^2 + \frac{2}{\lambda}b + \frac{2}{\lambda^2}\right) e^{-\lambda b} + \frac{2}{\lambda^2}\right] - \frac{1}{\lambda^2} = \frac{2}{\lambda^2} - \frac{1}{\lambda^2} = \frac{1}{\lambda^2}.$$

13. $P(1.2 \le Z \le 1.6) = \Phi(1.6) - \Phi(1.2) \approx 0.9452 - 0.8849 = 0.0603$

15. $P(Z < -0.48) = P(Z > 0.48) = 1 - \Phi(0.48) \approx 1 - 0.6844 = 0.3156$

17. $P(-1.64 \le Z \le 0.85) = P(Z \le 0.85) - P(Z < -1.64)$

$$= P(Z \le 0.85) - (1 - P(Z \le 1.64))$$

$$\approx 0.8023 - (1 - 0.9495) = 0.7518$$

19. Given $\mu = 5$ and $\sigma = 0.2$ let $Z = \frac{X - \mu}{\sigma} = \frac{X - 5}{0.2}$. Then Z has a standard normal distribution.

(a) $P(X \ge 5.3) = P\left(\frac{X - 5}{0.2} \ge \frac{5.3 - 5}{0.2}\right) = P(Z \ge 1.5)$

$$= 1 - P(Z < 1.5) = 1 - \Phi(1.5) \approx 1 - 0.9332 = 0.0668$$

(b) $P(X < 4.56) = P\left(\frac{X - 5}{0.2} < \frac{4.56 - 5}{0.2}\right) = P(Z < -2.2)$

$$= P(Z > 2.2) = 1 - P(Z \le 2.2)$$

$$= 1 - \Phi(2.2) \approx 1 - 0.9861 = 0.0139$$

(c) $P(4.88 \le X \le 5.26) = P\left(\frac{4.88 - 5}{0.2} \le \frac{X - 5}{0.2} \le \frac{5.26 - 5}{0.2}\right)$

$$= P(-0.6 \le Z \le 1.3) = P(Z \le 1.3) - P(Z < -0.6)$$

$$= P(Z \le 1.3) - [1 - P(Z < 0.6)]$$

$$= \Phi(1.3) - [1 - \Phi(0.6)] \approx 0.9032 - (1 - 0.7257)$$

$$= 0.6289$$

21. Let X be the number of ounces of the orange juice in a container, and set $Z = \frac{X - 64}{0.1}$. Then Z has a standard normal distribution.

$$P(0 \le X < 63.9) = P\left(\frac{0 - 64}{0.1} \le Z < \frac{63.9 - 64}{0.1}\right)$$

$$= P(-640 \le Z < -1)$$

$$= P(1 \le Z \le 640) = \Phi(640) - \Phi(1)$$

$$\approx 1 - 0.8413 = 0.1587$$

23. Let X be the blood clotting time, and set $Z = \frac{X-8.1}{\sqrt{4.8}}$. Then Z has a standard normal distribution.

(a) $P(0 \leq X < 2) = P\left(\frac{0-8.1}{\sqrt{4.8}} \leq Z < \frac{2-8.1}{\sqrt{4.8}}\right)$

$\approx P(-3.70 \leq Z < -2.78)$

$= \Phi(3.70) - \Phi(2.78) \approx 1 - 0.9973 = 0.0027$

(b) $P(X > 10) = P\left(Z > \frac{10-8.1}{\sqrt{4.8}}\right) = P(Z > 0.87)$

$= 1 - \Phi(0.87) \approx 1 - 0.8078 = 0.1922$

25. Let X be the exam score, and set $Z = \frac{X-\mu}{\sigma} = \frac{X-75}{7}$. Then Z has a standard normal distribution.

$$P(X \geq 89) = P\left(\frac{X-75}{7} \geq \frac{89-75}{7}\right)$$

$$= P(Z > 2) = 1 - \Phi(2) \approx 1 - 0.9772 = 0.0228$$

Thus, 2.28% of the students may get an A on that exam.

27. Let X be the number of seeds in a sunflower head, and set $Z = \frac{X-500}{40}$. Then Z has a standard normal distribution.

(a) $P(X > 600) = P\left(Z > \frac{600-500}{40}\right) = P(Z > 2.5)$

$= 1 - \Phi(2.5) \approx 1 - 0.9938 = 0.0062$

(b) We assume that the number of seeds in the two sunflower heads are independent. Then the probability that each one will have more than 600 seeds is equal to $[1 - \Phi(2.5)]^2$, which is approximately equal to $(0.0062)^2 = 0.00003844$.

29. (a) $F(Z) = P(Z \leq z) = P\left(\frac{X-\mu}{\sigma} \leq z\right)$

$= P(X \leq \mu + \sigma z) = \frac{1}{\sigma\sqrt{2\pi}} \int_{-\infty}^{\mu+\sigma z} e^{-(x-\mu)^2/2\sigma^2}\, dx$

(b) $F'(Z) = \frac{d}{dz} \frac{1}{\sigma\sqrt{2\pi}} \int_{-\infty}^{\mu+\sigma z} e^{-(x-\mu)^2/2\sigma^2}\, dx$

$= \frac{1}{\sigma\sqrt{2\pi}} e^{-1/2(\frac{\mu+\sigma z - \mu}{\sigma})^2} \frac{d}{dz}(\mu + \sigma z) = \frac{1}{\sigma\sqrt{2\pi}} e^{-z^2/2} \cdot \sigma$

$= \frac{1}{\sqrt{2\pi}} e^{-z^2/2}$

CHAPTER 11: REVIEW EXERCISES

1. There are 12 outcomes. The sample space (listed in alphabetical order) is

$$S = \{ab, ac, ad, ba, bc, bd, ca, cb, cd, da, db, dc\}.$$

3. An outcome is a string of H's and T's that ends at the second H or fourth letter, whichever comes first. Thus,

$$S = \{HH, HTH, THH, HTTH, HTTT, THTH, THTT, TTHH, TTHT, TTTH, TTTT\}.$$

There are 11 outcomes.

5. $A \cup B = \{HHHH, HHHT, HHTH, HHTT, HTHH, HTHT, HTTH, HTTT, THHT,$
$\qquad THTT, TTHT, TTTT\}.$

$A \cap B = \{HHHT, HHTT, HTHT, HTTT\}.$

$A' = \{THHT, THTH, TTHH, THTT, TTHT, TTTH, TTTT, THHH\}.$

Since $HHHH$ is an outcome in both A and B', we conclude that A and B' are not mutually exclusive.

7. We have

$$P(s_2) = P(s_1)$$

$$P(s_3) = \frac{1}{3}P(s_2) = \frac{1}{3}P(s_1)$$

$$P(s_4) = 2P(s_3) = \frac{2}{3}P(s_1)$$

Since $1 = P(s_1) + P(s_2) + P(s_3) + P(s_4)$, these substitutions give

$$1 = P(s_1) + P(s_1) + \frac{1}{3}P(s_1) + \frac{2}{3}P(s_1) = 3P(s_1).$$

Therefore, $P(s_1) = \frac{1}{3}$, and it follows that $P(s_2) = \frac{1}{3}$, $P(s_3) = \frac{1}{9}$, and $P(s_4) = \frac{2}{9}$.

9. The sample space is

$$S = \{HHH, THH, HTH, HHT, TTH, THT, HTT, TTT\}.$$

(a) $P(\{HHH\}) = \dfrac{1}{8}$

(b) $P(\{HTT, THT, TTH, TTT\}) = \dfrac{4}{8} = \dfrac{1}{2}$

(c) $P(\{HHH, HHT, HTH, HTT, THH, THT, TTH\}) = \dfrac{7}{8}$

Another way to answer (c) is to observe that the complement of "at least one head" is "all tails" or $\{TTT\}$, whose probability is $\frac{1}{8}$. Therefore, the probability of all tails is $1 - \frac{1}{8} = \frac{7}{8}$.

11. Let W_i be the event "the ith is white." Then $P(W_1) = \frac{5}{8}$ and $P(W_2|W_1) = \frac{4}{7}$. Therefore, by the multiplication principle

$$P(W_1 \cap W_2) = P(W_1) \cdot P(W_2|W_1)$$

$$= \frac{5}{8} \cdot \frac{4}{7} = \frac{5}{14}.$$

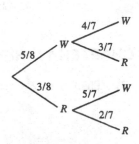

13. Since $P(E) = P(E \cap F) + P(E \cap F')$, we have

$$P(E \cap F) = P(E) - P(E \cap F') = 0.3 - 0.1 = 0.2.$$

Therefore, $P(E \cup F) = P(E) + P(F) - P(E \cap F) = 0.3 + 0.5 - 0.2 = 0.6.$

15. The sample space consists of $2^6 = 64$ outcomes (all 6-letter strings of B's and G's).

   (a) There is only one outcome, $GGGGGG$ in this event. Therefore, the probability is $\frac{1}{64}$

   (b) There are six outcomes: $BGGGGG, GBGGGG, \ldots, GGGGGB$. (There are six ways to insert a single B in a six-letter string with all the rest being G's.) Therefore, the probability is $\frac{6}{64} = \frac{3}{32}$.

17. There are 36 outcomes. (See Example 11.1.6 for a list of them.)

   (a) Let G be the event "both faces odd" and E the event "the sum is eight." Then

$$G = \{(1,1), (1,3), (1,5), (3,1), (3,3), (3,5), (5,1), (5,3), (5,5)\},$$
$$E = \{(2,6), (3,5), (4,4), (5,3), (6,2)\}, \text{ and } E \cap G = \{(3,5), (5,3)\}.$$

     Therefore, $P(E|G) = \frac{P(E \cap G)}{P(G)} = \frac{2/36}{9/36} = \frac{2}{9}$.

   (b) The event " at least one even" is G', the complement of "both odd," and $E \cap G' = \{(2,6), (4,4), (6,2)\}$. Therefore, the probability that the sum is eight given that at least one is even is

$$P(E|G') = \frac{P(E \cap G')}{P(G')} = \frac{3/36}{27/36} = \frac{1}{9}.$$

   (c) $P(G|E) = \dfrac{P(G \cap E)}{P(E)} = \dfrac{2/36}{5/36} = \dfrac{2}{5}$

19. **(a)** If the first is an ace, then there are 3 aces among 51 cards left in the deck. Therefore, the probability of the second being an ace given that the first is an ace is 3/51, or 1/17.

   (b) If the first is not an ace, then there are 4 aces among 51 cards left in the deck. Therefore, the probability of the second being an ace given that the first is not an ace is 4/51.

21. No, the equality does not hold in general. Exercise 19 provides a counter example. In drawing two cards from a deck without replacement, let A be the event "the first is an ace" and B the event "the second is an ace." Then, as explained in-exercise 19, $P(B|A) = \frac{3}{51}$ and $P(B|A') = \frac{4}{51}$.

23. Let A be the event that the chef will prepare veal piccante and B the event that the chef will prepare sweet corn tamales. We are given that $P(A) = 0.9$ and $P(B|A) = 0.85$. Then

$$P(A \cap B) = P(A) \cdot P(B|A) = (0.9) \cdot (0.85) = 0.765 > 0.75.$$

Therefore, the assistants should start preparing these dishes.

25. Let S_i be the event "spade on the ith draw" for $i = 1, 2$. Similarly, D_i for "diamond on the ith draw," H_i for heart, and C_i for club.

   (a) If the first is a heart, then there remain 13 spades in 51 cards. Therefore, $P(S_2|H_1) = \frac{13}{51}$. Similarly, $P(S_2|C_1) = P(S_2|D_1) = \frac{13}{51}$.

     On the other hand, if the first is a spade, then there remain 12 spades in 51 cards. Therefore $P(S_2|S_1) = \frac{12}{51}$.

(b) $P(H_1 \cap S_2) = P(H_1) \cdot P(S_2|H_1) = \dfrac{1}{4} \cdot \dfrac{13}{51} = \dfrac{13}{204}$

$P(C_1 \cap S_2) = P(C_1) \cdot P(S_2|C_1) = \dfrac{1}{4} \cdot \dfrac{13}{51} = \dfrac{13}{204}$

$P(D_1 \cap S_2) = P(D_1) \cdot P(S_2|D_1) = \dfrac{1}{4} \cdot \dfrac{13}{51} = \dfrac{13}{204}$

$P(S_1 \cap S_2) = P(S_1) \cdot P(S_2|S_1) = \dfrac{1}{4} \cdot \dfrac{12}{51} = \dfrac{12}{204} = \dfrac{3}{51}$

(c) $S_2 = (H_1 \cap S_2) \cup (C_1 \cap S_2) \cup (D_1 \cap S_2) \cup (S_1 \cap S_2)$, the union of four events, each pair of which is mutually exclusive. Therefore,

$$P(S_2) = P(H_1 \cap S_2) + P(C_1 \cap S_2) + P(D_1 \cap S_2) + P(S_1 \cap S_2)$$

$$= \frac{13}{204} + \frac{13}{204} + \frac{13}{204} + \frac{12}{204} = \frac{51}{204} = \frac{1}{4}.$$

27. (a) Let A be the event that the first is a head and B the event that the third is a head. Then $P(A) = \frac{1}{2}$, $P(B) = \frac{1}{2}$, and $P(A \cap B) = \frac{2}{8} = \frac{1}{4}$. Since $P(A \cap B) = P(A)P(B)$, A and B are independent.

(b) Let A be the event that the third is a head and B the event that there is at least one head. Then $P(A) = \frac{1}{2}$, $P(B) = \frac{7}{8}$, and $P(A \cap B) = P(A) = \frac{1}{2}$. Since $P(A \cap B) \neq P(A)P(B)$, A and B are not independent.

(c) Let A be the event that the third is a head and B the event that there is at least one head and one tail. Then $P(A) = \frac{1}{2}$, $P(B) = \frac{6}{8} = \frac{3}{4}$, and $P(A \cap B) = \frac{3}{8}$. Since $P(A \cap B) = P(A)P(B)$, A and B are independent.

(d) Let A be the event that all three are the same and B the event that the first is a head. $P(A) = \frac{2}{8} = \frac{1}{4}$, $P(B) = \frac{1}{2}$, $P(A \cap B) = \frac{1}{8}$. Since $P(A \cap B) = P(A)P(B)$, A and B are independent.

(e) Let A be the event that all three are the same and B the event that there is at least one head. $P(A) = \frac{2}{8} = \frac{1}{4}$, $P(B) = \frac{7}{8}$, $P(A \cap B) = \frac{1}{8}$. Since $P(A \cap B) \neq P(A)P(B)$, A and B are not independent.

29. Each pair of calls is independent. Therefore,

- the probability that both will be watching is $(0.64)^2 = 0.4096$;
- the probability that neither will be watching is $(1 - 0.64)^2 = 0.1296$;
- the probability that none of the three households called will be watching the game is $(1 - 0.64)^3 = 0.046656$.

31. $E(X) = (-3)(0.08) + (-2)(0.1) + (-1)(0.12) + 0(0.15) + 1 \cdot (0.25) + 2(0.2) + 3(0.1) = 0.39$
$\mathrm{Var}(X) = (-3)^2(0.08) + (-2)^2(0.1) + (-1)^2(0.12) + 0^2(0.15) + 1^2(0.25) + 2^2(0.2) + 3^2(0.1) - 0.39^2 = 3.0379$

33. (a)

x	0	1	2	3	4	5	6	7	8
$P(X = x)$	$\frac{1}{20}$	$\frac{3}{40}$	$\frac{3}{20}$	$\frac{9}{40}$	$\frac{7}{40}$	$\frac{3}{20}$	$\frac{1}{10}$	$\frac{1}{20}$	$\frac{1}{40}$

(b) $E(X) = 0 \cdot \dfrac{1}{20} + 1 \cdot \dfrac{3}{40} + 2 \cdot \dfrac{3}{20} + 3 \cdot \dfrac{9}{40} + 4 \cdot \dfrac{7}{40} + 5 \cdot \dfrac{3}{20} + 6 \cdot \dfrac{1}{10} + 7 \cdot \dfrac{1}{20} + 8 \cdot \dfrac{1}{40} = 3.65$

$\text{Var}(X) = 0 \cdot \dfrac{1}{20} + 1 \cdot \dfrac{3}{40} + 4 \cdot \dfrac{3}{20} + 9 \cdot \dfrac{9}{40} + 16 \cdot \dfrac{7}{40} + 25 \cdot \dfrac{3}{20} + 36 \cdot \dfrac{1}{10}$

$\qquad\qquad + 49 \cdot \dfrac{1}{20} + 64 \cdot \dfrac{1}{40} - (3.65)^2 = 3.5775$

$\sigma(X) = \sqrt{3.5775} \approx 1.8914$

(c) $P(X > 6) = \dfrac{1}{20} + \dfrac{1}{40} = \dfrac{3}{40}$

35. Let X be the number of a's appearing in the word. Then X is a binomial random variable satisfying the binomial distribution with parameters $n = 6$ and $p = \frac{1}{26}$. For $k = 0, 1, 2, 3, 4, 5, 6$,

$$P(X = k) = \binom{6}{k} \left(\frac{1}{26}\right)^k \left(\frac{25}{26}\right)^{6-k}.$$

37. (a) $\dbinom{9}{4} = \dfrac{9 \cdot 8 \cdot 7 \cdot 6}{4 \cdot 3 \cdot 2 \cdot 1} = 126$ **(b)** $\dbinom{14}{6} = \dfrac{14 \cdot 13 \cdot 12 \cdot 11 \cdot 10 \cdot 9}{6 \cdot 5 \cdot 4 \cdot 3 \cdot 2 \cdot 1} = 3{,}003$

(c) $\dbinom{3{,}124}{0} = 1$ **(d)** $\dbinom{2{,}714}{1} = 2{,}714$

39. Let X be the number of 5's in an 18-digit number whose digits are all from 1 to 9. Then X is a binomial random variable with parameters $n = 18$ and $p = \frac{1}{9}$.

(a) $P(X = 0) = \dbinom{18}{0} \left(\dfrac{8}{9}\right)^{18} = \left(\dfrac{8}{9}\right)^{18}$

(b) $E(X) = np = 18 \cdot \dfrac{1}{9} = 2$

(c) $\text{Var}(X) = np(1 - p) = 18 \cdot \frac{1}{9} \cdot \left(\frac{8}{9}\right) = \frac{16}{9}$ and $\sigma(X) = \sqrt{\frac{16}{9}} = \frac{4}{3}$. Then $E(X) + \sigma(X) = 2 + \frac{4}{3} = \frac{10}{3}$ and $E(X) - \sigma(X) = 2 - \frac{4}{3} = \frac{2}{3}$. Thus, the upper and lower bounds for the number of 5's are 3 and 1, respectively.

41. First, $2x^{-3} > 0$ for all $x > 0$ and, in particular, for $x \geq 1$. Second,

$$\int_1^\infty 2x^{-3} \, dx = \lim_{b \to \infty} \int_1^b 2x^{-3} \, dx = \lim_{b \to \infty} (-x^{-2}) \Big|_1^b = \lim_{b \to \infty} (1 - b^{-2}) = 1.$$

Therefore, $f(x)$ is a probability density function. If $x > 1$, then

$$F(x) = \int_{-\infty}^x f(t) \, dt = \int_1^x 2t^{-3} \, dt = -t^{-2} \Big|_1^x = 1 - x^{-2}.$$

Thus,

$$F(x) = \begin{cases} 1 - x^{-2} & \text{if } x \geq 1 \\ 0 & \text{if } x < 1. \end{cases}$$

If $x > 1$, then $F'(x) = \frac{d}{dx}(1 - x^{-2}) = 2x^{-3}$. If $x < 0$, then $F'(x) = 0$. Thus, $F'(x) = f(x)$ for $x \neq 1$.

43. (a) $P\left(\frac{1}{4} \leq x \leq \frac{3}{4}\right) = F\left(\frac{3}{4}\right) - F\left(\frac{1}{4}\right) = \left[1 - \sqrt{1 - \frac{3}{4}}\right] - \left[1 - \sqrt{1 - \frac{1}{4}}\right] = \frac{\sqrt{3} - 1}{2}$

(b) $f(x) = F'(x) = \frac{d}{dx}[1 - \sqrt{1 - x}] = \frac{1}{2\sqrt{1 - x}}$ for $0 < x < 1$

45. $\mu = E(X) = \int_1^\infty x \cdot \frac{4}{x^5} \, dx = \lim_{b \to \infty} \int_1^b \frac{4}{x^4} \, dx = \lim_{b \to \infty} \left.-\frac{4}{3x^3}\right|_1^b = \lim_{b \to \infty} \left(\frac{4}{3} - \frac{4}{3b^3}\right) = \frac{4}{3}$

$\mathrm{Var}(X) = \int_1^\infty x^2 \cdot \frac{4}{x^5} \, dx - \mu^2 = \lim_{b \to \infty} \int_1^b \frac{4}{x^3} \, dx - \frac{16}{9} = \lim_{b \to \infty} \left(2 - \frac{2}{b^2}\right) - \frac{16}{9} = 2 - \frac{16}{9} = \frac{2}{9}$

47. $\mu = E(X) = \int_0^1 x\left(x + \frac{1}{2}\right) dx = \frac{1}{3}x^3 + \frac{1}{4}x^2 \Big|_0^1 = \frac{1}{3} + \frac{1}{4} = \frac{7}{12}$

$\mathrm{Var}(X) = \int_0^1 x^2 \left(x + \frac{1}{2}\right) dx - \mu^2 = \left(\frac{1}{4}x^4 + \frac{1}{6}x^3\right)\Big|_0^1 - \frac{49}{144}$

$= \frac{5}{12} - \frac{49}{144} = \frac{11}{144}$

49. $E(X) = \int_0^4 \frac{x}{4\sqrt{x}} \, dx = \frac{1}{6}x^{3/2} \Big|_0^4 = \frac{1}{6} \cdot 4^{3/2} = \frac{4}{3}$

If m is the median of X, then $\frac{1}{2} = \int_0^m \frac{1}{4\sqrt{x}} \, dx = \frac{1}{2}\sqrt{m}$. Therefore, $m = 1$.

51. Using integration by parts twice, we obtain

$$E(X) = \int_0^\infty x^2 e^{-x} \, dx = \lim_{b \to \infty} \left.-(x^2 + 2x + 2)e^{-x}\right|_0^b$$

$$= \lim_{b \to \infty} \left(2 - (b^2 + 2b + 2)e^{-b}\right) = 2.$$

To find the median m, we must solve the equation

$$\frac{1}{2} = \int_0^m xe^{-x} \, dx = -(x + 1)e^{-x}\Big|_0^m = 1 - (m + 1)e^{-m}.$$

Using the equation solver on a graphing calculator, we get $m \approx 1.68$.

53. $E(X) = \frac{a+b}{2}$ and $\sigma = \sqrt{\frac{(b-a)^2}{12}} = \frac{b-a}{2\sqrt{3}}$. Then we are looking for the probability that X is in the interval $\left(\frac{a+b}{2} - \frac{b-a}{2\sqrt{3}}, \frac{a+b}{2} + \frac{b-a}{2\sqrt{3}}\right)$. The length of that interval is $(b - a)/\sqrt{3}$. Therefore, the probability is

$$\frac{(b - a)/\sqrt{3}}{b - a} = \frac{1}{\sqrt{3}}$$

(the ratio of the length of the interval $\left(\frac{a+b}{2} - \frac{b-a}{2\sqrt{3}}, \frac{a+b}{2} + \frac{b-a}{2\sqrt{3}}\right)$ to that of the entire domain (a, b)).

55. (a) $P(X < 1) = \int_0^1 2e^{-2x} \, dx = (-e^{-2x})\big|_0^1 = 1 - e^{-2}$

(b) $E(X) = \frac{1}{\lambda} = \frac{1}{2}$ minute

57. Let $Z = \frac{X-10}{2.8}$. Then Z has a standard normal distribution.

(a) $P(X \leq 12) = P\left(Z \leq \frac{12-10}{2.8}\right) \approx P(Z \leq 0.71) = \Phi(0.71) \approx 0.7611$

(b) $P(8 < X < 11) = P\left(\frac{8-10}{2.8} < Z < \frac{11-10}{2.8}\right)$

$\approx P(-0.71 < Z < 0.36) = \Phi(0.36) - [1 - \Phi(0.71)]$

$\approx 0.6406 - (1 - 0.7611) = 0.4017$

(c) $P(X > 9) = P\left(Z > \frac{9-10}{2.8}\right) \approx P(Z > -0.36) = P(Z \leq 0.36) = \Phi(0.36) \approx 0.6406$

59. Let X_A and X_B be the purchase amounts of a customer in the grocery stores A and B, respectively. Then $\mu_A = 41.25$, $\sigma_A = \sqrt{101}$, $\mu_B = 27.93$, $\sigma_B = \sqrt{52.70}$.

$$P(X_A > 30) = P\left(\frac{X_A - 41.25}{\sqrt{101}} > \frac{30 - 41.25}{\sqrt{101}}\right)$$

$$\approx P(Z > -1.12) = \Phi(1.12) \approx 0.8686$$

$$P(X_B > 30) = P\left(\frac{X_B - 27.93}{\sqrt{52.70}} > \frac{30 - 27.93}{\sqrt{52.70}}\right)$$

$$\approx P(Z > 0.29) = 1 - \Phi(0.29) \approx 0.3859$$

CHAPTER 11: EXAM

1. (a) Since $P(A \cup B) = P(A) + P(B) - P(A \cap B)$, we have

$$P(A \cap B) = P(A) + P(B) - P(A \cup B) = 0.5 + 0.7 - 0.8 = 0.4.$$

(b) Since $P(B') = 1 - P(B) = 0.3$ and

$$P(A \cap B') = P(A) - P(A \cap B) = 0.5 - 0.4 = 0.1$$

we have $P(A \cup B') = P(A) + P(B') - P(A \cap B') = 0.5 + 0.3 - 0.1 = 0.7.$

3. Let A be the event that a student has seen Toy Story and B the event that a student has seen Toy Story II. Then $P(A) = 0.78$, $P(B) = 0.64$, and $P(A \cap B) = 0.56$. The probability that a student did not see either is $A' \cap B'$, which is the complement of $A \cup B$. Therefore, its probability is $1 - P(A \cup B) = 1 - [P(A) + P(B) - P(A \cap B)] = 1 - (0.78 + 0.64 - 0.56) = 0.14.$

5. See Example 11.6.1 for a list of the sample space, which contains 36 outcomes. Using the list, we find that

$$P(A) = \frac{6}{36} = \frac{1}{6}, \quad P(B) = \frac{9}{36} = \frac{1}{4}, \quad P(A \cap B) = \frac{3}{36} = \frac{1}{12}.$$

(a) $P(A|B) = \frac{P(A \cap B)}{P(B)} = \frac{1/12}{1/4} = \frac{1}{3}$ **(b)** $P(B|A) = \frac{P(A \cap B)}{P(A)} = \frac{1/12}{1/6} = \frac{1}{2}$

7. $P(A) = \frac{6}{36} = \frac{1}{6}$, $P(B) = \frac{5}{36}$, $P(C) = \frac{6}{36} = \frac{1}{6}$

(a) Since $P(A \cap B) = \frac{1}{36} \neq P(A)P(B)$, A and B are not independent.

(b) Since $P(A \cap C) = \frac{1}{36} = P(A)P(C)$, A and C are independent.

9. The probability distribution table is as follows:

x	0	1	2
$P(X = x)$	0.6	0.3	0.1

(a) $E(X) = 0 \cdot (0.6) + 1 \cdot (0.3) + 2 \cdot (0.1) = 0.5$

(b) $\text{Var}(X) = 0 \cdot (0.6) + 1 \cdot (0.3) + 4 \cdot (0.1) - (0.5)^2 = 0.45$

(c) $\sigma(X) = \sqrt{0.45} \approx 0.67$

11. Let X be the number of white chips in 20 draws. Then X is a binomial random variable with parameters $n = 20$ and $p = \frac{2}{5}$.

(a) For $k = 0, 1, 2, \dots, 20$,

$$P(X = k) = \binom{20}{k} \left(\frac{2}{5}\right)^k \left(1 - \frac{2}{5}\right)^{20-k} = \binom{20}{k} \left(\frac{2}{5}\right)^k \left(\frac{3}{5}\right)^{20-k}.$$

(b) $E(X) = np = 20 \cdot \dfrac{2}{5} = 8$

13. (a) If $f(x) = c(x + 2)^{-2}$, $x > 0$, is a probability density function, then

$$1 = \int_0^\infty c(x+2)^{-2}\, dx = c \cdot \lim_{b \to \infty} \left[-(x+2)^{-1}\right]\Big|_0^b = c \cdot \lim_{b \to \infty} \left[\frac{1}{2} - \frac{1}{b+2}\right] = \frac{c}{2}.$$

Therefore, $c = 2$. We also observe that if $c = 2$, then $f(x) > 0$.

(b) $P(1 \le X \le 3) = \displaystyle\int_1^3 2(x+2)^{-2}\, dx = -2(x+2)^{-1}\Big|_1^3 = \dfrac{4}{15}$

15. The probability density function is $f(x) = F'(x) = \frac{x}{2}$, $0 < x < 2$. Then $E(X) = \int_0^2 \frac{x^2}{2}\, dx = \frac{x^3}{6}\big|_0^2 = \frac{4}{3}$. If m is the median of X, then $\frac{1}{2} = F(m) = \frac{m^2}{4}$. Therefore, $m = \sqrt{2}$.

17. X is a uniformly distributed random variable over $(1, 3)$. Therefore,

$$E(X) = \frac{1+3}{2} = 2 \quad \text{and} \quad \text{Var}(X) = \frac{(3-1)^2}{12} = \frac{1}{3}.$$

19. Set $Z = \frac{X-80}{5}$. Then Z has a standard normal distribution.

(a) $P(X > 88) = P\left(\dfrac{X - 80}{5} > \dfrac{88 - 80}{5}\right) = P(Z > 1.6)$

$$= 1 - P(Z \le 1.6) = 1 - \Phi(1.6) \approx 1 - 0.9452 = 0.0548$$

(b) $P(X < 78) = P\left(\dfrac{X - 80}{5} < \dfrac{78 - 80}{5}\right) = P(Z < -0.4)$

$$= 1 - \Phi(0.4) \approx 1 - 0.6554 = 0.3446$$

(c) $P(X = 80) = 0$

NOTES

NOTES

NOTES

NOTES

NOTES

NOTES

NOTES